On the Backs of Tortoises

Darwin, the Galápagos, and the
Fate of an Evolutionary Eden

ELIZABETH HENNESSY

Yale UNIVERSITY PRESS   NEW HAVEN AND LONDON

Published with assistance from the Louis Stern Memorial Fund.

Yale University Press books may be purchased in quantity
for educational, business, or promotional use. For
information, please e-mail sales.press@yale.edu (U.S. office)
or sales@yaleup.co.uk (U.K. office).

Frontispiece: A giant tortoise wades in a pond at Campo
Duro in the highlands of Isabela Island, July 2018.
(Photo by the author)

Set in Scala type by IDS Infotech Ltd., Chandigarh, India.
Printed in the United States of America.

Library of Congress Control Number: 2019936401
ISBN 978-0-300-23274-5 (hardcover : alk. paper)

A catalogue record for this book is available from the
British Library.

This paper meets the requirements of ANSI/NISO
Z39.48–1992 (Permanence of Paper).

10   9   8   7   6   5   4   3   2   1

for JPH

Contents

On the Backs of Tortoises

Preface and Acknowledgments

When I tell people in the United States, where I was born and live, that I do research in the Galápagos archipelago, their faces usually light up. I imagine they are picturing sun-drenched tropical islands inhabited by gigantic tortoises and playful sea lions. They usually ask whether I have been to this world-renowned vacation spot—indeed, I am fortunate to have visited several times. Then they often ask me something that for a long time surprised me: Where do you stay when you visit the islands? It's a practical question—one that I think comes from a desire to picture me there. But it took me a while to understand why they were interested in what I saw as a rather banal detail. What I realized, though, after many versions of this conversation, is that it is common among North Americans to believe that the archipelago is uninhabited. That would make my answer much more exciting: Might I camp on a remote island like biologists or spend my time on a ship cruising the archipelago, as many tourists do? But my answer never lives up to the adventure that I think people want to hear about. They often seem a bit deflated, and surprised too, when I tell them that I stay in hotels or rent apartments for longer stays. When I explain that the islands have several towns that are home to some thirty thousand people, I feel like I am throwing a wet blanket on their imagination of a pristine wilderness. Sometimes a chagrined look crosses their faces, and they say something about how it is a shame that so many people now live in this place that really ought to be protected.

This is not the impression of the Galápagos that I hope to convey. It is striking that people who live so far from the islands have an emotional stake in them—it speaks to the power of our imaginations of this storied place. Yet I am often dismayed by the assumption that the presence of towns must be detrimental and that people do not belong in this place that is often called Darwin's natural laboratory of evolution. I understand why people think this way. It is a view influenced by news reports about a crisis of development in a place that most North Americans know from the many nature documentaries that take people on armchair vacations. But news reports and nature documentaries offer very partial views of life in the archipelago. They portray a world of extremes—either pristine nature or a crisis. Neither is the reality that I have seen.

The idea that people do not really belong in the Galápagos not only is held by outsiders, but also has seeped into the discourses of daily life on the islands, where it is not uncommon to hear people talk about themselves as invasive species or for longtime residents to lament the presence of newer migrants. An older Ecuadorian farmer once told me that he was concerned that so many people were moving to the island where he lived, lest the weight of the population cause the island to flip over. The interview was in Spanish, and I had to confirm with a friend later that I had heard the man correctly. I had. I think he meant it literally, as if the island was floating. But metaphorically, his fear was exactly in line with that of conservationists and global publics who fret that the islands have reached a tipping point of overpopulation and ecological destruction from which there can be no return.

In many ways, this book is my response to such ways of understanding the Galápagos. It is also my response to would-be tourists who have asked me whether it's alright to visit the islands, or whether they would be better left alone. I write with North Americans in mind not because I believe the gringo perspective is universal, but rather because I am one. I write with them in mind also because North Americans' perspectives are powerful—they make up the largest market of Galápagos tourists, driving an industry that strives to meet visitor expectations. But this tour I offer is not one that highlights untouched wilderness. Nor do I offer lamentations about overpopulation on fragile island ecosystems. I am not unmoved by these concerns, but I think they are too quickly adopted and too comfortably rest on the false imagination that the Galápagos Islands are, or should be, pristine. In the pages that follow, I ask readers to slow down and to consider the long-standing, dense entanglements among people and nature that have shaped the Galápagos over the past five centuries, if not longer. Doing so, I believe, is crucial for safeguarding the future of this special place.

I am able to do this because of the many people in the archipelago who have opened their doors and shared their stories with me over the past twelve years.

This book is based on research I conducted in the Galápagos during the summers of 2007, 2008, and 2009; for about six months during 2011–2012; and during short trips in 2015 and 2018. I owe them all a profound debt. I do not list them all here, in part to protect the privacy of sources. I owe particular thanks to Washington Tapia, without whose support and knowledge the project never would have gotten off the ground. My utmost thanks go as well to Don Fausto Llerena, Moises Villafuerte, José "Pepe" Villa, Wilfrido Michuy, Steve Divine, Fredy Cabrera, Paola Pozo, Swen Lorenz, Guido, Fernando Ortiz, Galo Quezada, Cruz Marquez, Sixto Naranjo, Arturo Izurieta, Washington Ramos, Graham Watkins, Danny Rueda, Oscar Carvajal, Felipe Cruz, Henri Moreno, Angel Arias, Johnson Arias, Rolando Loyola, Galo Torres, Washington Llerena, Domingo Navarrette, Carlos Zapata, and Ivone Torres. I wish I could have told in this book all the stories that you shared with me.

To the many Galápagos scientists who took time to speak with me, opening their labs and their living rooms, I thank you for your willingness to teach and talk with me and your patience for this book, which has been a long time coming. In particular, I thank Linda Cayot, Gisella Caccone, Steve Blake, Craig MacFarland, Howard and Heidi Snell, Tom Fritts, Peter Kramer, Gunther Reck, and Tjite de Vries. The first Galápagos researcher I talked to when I hatched the idea of a tortoise-focused study was James Gibbs, who turned out to be exactly the right person. This project would not have been possible without his support.

I also owe thanks to many archivists, historians, and scientists who work on Galápagos history. At the California Academy of Sciences I am grateful for the assistance of Alan Leviton, Barbara West, and Jens Vindum, who took me back into the storeroom to see taxidermied giant tortoises that have now sat on the institution's shelves for more than a century. In the archives, I would have been lost without the aid of Becky Morin, Christina Fidler, Rebekah Kim, and Seth Cotterell. I also thank Madeleine Thompson at the Wilderness Conservation Society; Alexandre Coutelle at UNESCO; Paul Martyn Cooper and Daisy Cunynghame at the British Natural History Museum; Erika Loor Orozco, the now late Elizabeth Knight, and Edgardo Civallero at the Charles Darwin Research Station; and archivists at the American Museum of Natural History and the Smithsonian Institution. At the San Diego Zoo and Safari Park, thank you to Amy Jankowski, Kim Lovich, Tommy Owens, Jenna Lyons, and Oliver Ryder.

The funding that made this research possible came from the American Council of Learned Societies/Mellon Foundation, the Social Science Research Council/Mellon Foundation, the National Science Foundation (Grant No. 608195277), the Association of American Geographers, the Institute for Study of the Americas at the University of North Carolina at Chapel Hill, the Institute for Research in the Humanities and the Graduate School at the University of Wisconsin, and the Rachel Carson Center, where I was fortunate to spend several months as I wrote this manuscript. I am deeply grateful to Christof Mauch for creating a space that showcases the best, most generous side of academic life. Munich would not have been nearly as wonderful without Lisa Brady, Paula Ungar, Monica Vasile, Ruth Morgan, Deb Klebesadel, Gregg Mitman, Felix Mauch, Annka Liepold, and Jenny Carlson, who has been by my side, in cyberspace at least, since they made us leave and go back to our real lives.

I owe a host of thanks to friends and colleagues in Chapel Hill who shaped my thinking and life in such fundamental ways. To Wendy Wolford, whom I once told I was uninterested in nature, I owe the deepest gratitude. I first went to the Galápagos with Wendy as well as Liza Guzman, Patricia Polo, Elena Steponaitis, and Flora Lu. Thank you too to Kim Engie, Amy McCleary, Laura Brewington, and Gaby Valdivia for further adventures in the field. Larry Grossberg, Altha Cravey, Steve Walsh, Scott Kirsch, John Pickles, Margaret Weiner, Arturo Escobar, and Barbara Herrnstein Smith both opened doors and helped me cross through them. Holly Worthen, Sara Safransky, Ashley Carse, and Brenda Baletti all shaped this project at formative stages. My thanks too to Maggie Carrel, Freya Thimsen, Elizabeth Robbins, Murat Es, Rolien Hoyng, Joseph Palis, Joe Wiltberger, Alice Brooke Wilson, Dennis Arnold, Autumn Thoyre, Conor Harrison, and Mike Dimpfl.

In addition to being the best interviewer I have ever seen in action, Patricia Polo Almeida has been the dearest of friends since our days in Chapel Hill and has made Quito feel like a second home for me. In Ecuador, I am most grateful to Diego Quiroga and Carlos Mena of the Universidad San Francisco de Quito as well as Ana Sevilla and Elisa Sevilla for support, friendship, and inspiring work. Thank you too to the Viejo Sabio of Galápagos history, Octavio Latorre, for welcoming us to his home and archive. I am also indebted to Fernando Torres Espinoza and Luis Esteban Vizuete Marcilla for research assistance.

For the past five years, I have been fortunate to call the University of Wisconsin–Madison my home. The Center for Culture, History, and Environment (CHE) has been an incredibly welcoming and enriching community since I arrived in Wisconsin, for which I am most grateful. I thank Bill Cronon for the many UClub lunches and editorial advice—my writing is stronger for it. My fellow public-school kid, Gregg Mitman, I thank you for paving paths through the academy and for opening them to others. Paul Robbins, you are always welcome in my office at the dark end of the hall. Thank you for making the Nelson Institute a place where I am proud to work. Caroline Gottschalk Druschke's arrival at Wisconsin brought with it a surge wave of her enthusiasm, and I thank her as well as Laura Perry, Kata Beilin, Sai Suryanarayanan, Zhe Yu Lee, Claudia Calderon, Mario Ortiz-Robles, Shari Wilcox, Ian Baird, and all the Multispecies Justice folks for providing an inspiring intellectual home—and for your thoughtful comments on part of the manuscript. I also thank the undergraduate and graduate students in my classes on Latin American environmental history, animal histories, and political ecology who read and commented on drafts. The book is stronger for your input. I have benefited greatly from the opportunity to share and discuss pieces of this project over the years with my colleagues in CHE, Geography, Forest and Wildlife Ecology, the Holtz Center, and the History of Science, Medicine, and Technology. I thank particularly Samer Alatout, Nicole Nelson, Lynn Nyhart, and Florence Hsia.

I deeply appreciate feedback on my work from audiences in several venues, including conferences of the Association of American Geographers, the American and European Societies for Environmental History, la Sociedad Latinoamericana y Caribeña de Historia Ambiental, the Society for Social Studies of Science, and the Latin American Studies Association. In Colombia, I am indebted to my hosts German Palacio and Claudia Leal and Shawn van Ausdal, as well as those at the History Colloquium at the Universidad de los Andes and at the History, Environment and Politics Colloquium at the Universidad Nacional Sede Amazonia. At the other Columbia, in New York, I benefited from discussions at the Biodiversity and Its Histories Workshop and particularly thank Megan Raby and Deborah Coen. At Ludwig Maximilian University of Munich, I thank Eveline Dürr and other anthropologists for thoughtful questions at the Amerikas Colloquium. In Oaxaca, I am most grateful to Lety Gomez Salinas as well as Holly Worthen, Joe Bryan, and Tad Muttersbaugh, who read the first, very long, draft of my first chapter. Joe Bryan, I will always see you as the leader of the Dark Side; you shaped my sense of what was possible with this project.

Lisa Campbell, Gregory Cushman, Bill Cronon, Gregg Mitman, Samer Alatout, Paul Robbins, Alberto Vargas, and Sara Guyer read a full draft of the manuscript at a crucial stage. So did Carlos Lozada, who has been a friend and mentor for going on twenty years now (how can that be?). I am deeply grateful to the University of Wisconsin Center for the Humanities First Book program for allowing me to bring these great minds together to stretch the limits of my thinking. Your close reading and comments are woven into the fabric of the book. James Gibbs and Diego Quiroga also read full drafts of the nearly final manuscript, for which I am most grateful.

At Yale University Press, Jean Thomson Black's unwavering enthusiasm has been a lifeline, as have been Michael Deneen's ready answers to my questions. I also thank Jeffrey Schier and Jessie Dolch for their efforts on the book. It takes quite an effort to pull together a book manuscript, and I owe considerable thanks to Laura Perry, Bailey Albretch, Holly Worthen, Meghan Kelly, Nathan Jandl, and my mom, who all have helped me compile the manuscript at various stages. I am sure, too, that I am forgetting many others—my appreciation is stronger than my memory.

Thanks to my family as well (I kept my nose to the grindstone, John!), especially Jean, to whom this book is dedicated. You opened the world's horizons for me and created a sanctuary where I always feel at home. My thanks as well to the many friends who are also like family, many of whom have been cautiously asking for years now how the book is going: Lisa, Lynne Marie, Jenny, Annie, Anne, Mary, and Beth. To Levi Van Sant, thank you for being a friend and collaborator since we met in Lexington so many years ago. Kim and Jolyon Thomas and Jessi Lehman, thank you for being both dear friends and inspiring thinkers. Many thanks from me, and Belly, to Rafi Arefin and Travis De Wolfe. Daegan Miller, I am wiser for our talks on writing and life. Ann, Colleen, and Rich have taken great care of me in Madison. Also in Madison, I must particularly thank Jen Gaddis: I would not have been able to pull this off over the past five years without you. And finally to my mom, Linda, for always being there, for your support and encouragement.

On the Backs of Tortoises

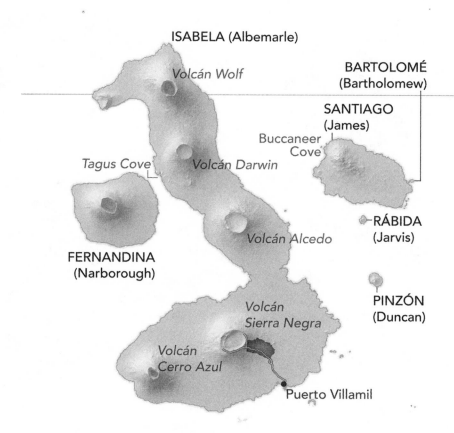

PINTA (Abingdon)

PACIFIC

OCEAN

ISABELA (Albemarle)

Volcán Wolf

BARTOLOMÉ
(Bartholomew)

SANTIAGO
(James)

Buccaneer
Cove

Tagus Cove   *Volcán Darwin*

Volcán Alcedo

RÁBIDA
(Jarvis)

FERNANDINA
(Narborough)

PINZÓN
(Duncan)

*Volcán
Sierra Negra*

*Volcán
Cerro Azul*

Puerto Villamil

■ Urban and Agricultural Areas

20 Miles

ISLAND NAME (Old English Name)

Galápagos Archipelago, national park and urban and agricultural areas. (Map designed by Meghan Kelly, University of Wisconsin Cartography Lab)

MARCHENA
(Bindloe)

GENOVESA
(Tower)

CENTRAL
AMERICA

ECUADOR

Quito

Guayaquil

Galápagos
(Marine Reserve)

SOUTH
AMERICA

Equator

PACIFIC

OCEAN

BALTRA (Seymour)

SANTA CRUZ
(Indefatigable)

SAN CRISTÓBAL
(Chatham)

El Chato

SANTA FÉ
(Barrington)

Puerto
Ayora

Puerto
Baquerizo
Moreno

Puerto
Velasco
Ibarra

Post Office Bay

FLOREANA (Charles)

Asilo de Paz

ESPAÑOLA
(Hood)

PACIFIC

OCEAN

PINTA
abingdonii

ISABELA
Volcán Wolf
becki ▲

SANTIAGO
darwini

Volcán Darwin
microphyes ▲

Volcán Alcedo
vandenburghi ▲

◇—RÁBIDA

FERNANDINA
phantasticus

Volcán Sierra Negra
guntheri ▲

PINZÓN
duncanensis

Volcán Cerro Azul
vicina ▲

•Puerto Villamil

■ Extant
■ Extinct
□ No Population
≡ Re-Tortoised

ISLAND NAME *species name*

20 Miles

Giant tortoise species populations and survival status, Galápagos Archipelago.
(Map designed by Meghan Kelly, University of Wisconsin Cartography Lab)

MARCHENA

GENOVESA

CENTRAL AMERICA

ECUADOR

Quito

Galápagos
(Marine Reserve)

Guayaquil

SOUTH AMERICA

———— Equator ————

PACIFIC

OCEAN

BALTRA

SANTA CRUZ
porteri (Western Island)
donfaustoi (Eastern Island)

Puerto Ayora

SANTA FÉ
undescribed

Puerto Baquerizo Moreno

SAN CRISTÓBAL
chathamensis

Puerto Velasco Ibarra

FLOREANA
niger

ESPAÑOLA
hoodensis

I · What We Stand On

In which the tortoises are an oft-told tale

There is an old story I have heard many times now, as likely legend as true. It supposedly took place in an auditorium or a university lecture hall in the early twentieth century, where the philosopher William James—or maybe it was philosopher and mathematician Bertrand Russell—stood before a crowded room, giving a talk on recent theories of cosmology. At the end of the lecture, as members of the audience started asking questions, an elderly woman in the back made her way to the front of the room. When her turn to speak came, she declared, "Your talk was very interesting, Mr. James, but it is wrong. We all know that the Earth really rests on the back of a turtle." James raised his eyebrows, but nodded respectfully and replied: "Ah . . . but then what does the turtle stand on?" The old lady replied, "That's easy: the turtle stands on the back of a second, far larger, turtle." Amused, James persisted, to see what might happen if he asked one more time: "But what does that second turtle stand on?" The old woman smiled and shook her head, "Very clever, Mr. James, but it's turtles all the way down!"[1]

Every time I have heard this story, the teller interprets it slightly differently. How are we supposed to understand the old woman's claim that the world rests on the backs of turtles? What is it about this curious story that makes it such an often-told tale? The old woman's insistence on the stack of turtles makes the story memorable. For some readers today, the story may be amusing partly because the stack of turtles seems like something out of a Dr. Seuss story. Yet we laugh a little uneasily at the punch line because it invites us to doubt our own certainties. Do we really know that the old woman is wrong? In a number of non-Western cultures, turtles play foundational roles in origin stories. In drawing a contrast between two ways of knowing the world, the story points to a deeper truth about the ground we stand on, the things we take for granted without really knowing them to be true. Take, for instance, how the physicist Stephen Hawking interpreted the story to introduce his *Brief History of Time.* In a book that explains research on some of the most challenging contemporary questions about the nature of the universe, including the Big Bang and black holes, Hawking opens with the old woman and her turtles to remind readers that the basis for our knowledge of the cosmos may turn out to be not much sturdier than her tower of turtles. We may find her claim to be ridiculous, he writes, "but why do we think we know better?"[2] Our knowledge is uncertain, Hawking reminds us, and is always a work in progress. The ground on which we stand is not as solid as we might like to imagine. Yet still we must proceed, keeping track of our uncertainty so that we remain open to new questions, new ways of understanding the world.

The turtles-all-the-way-down story is a tale, a metaphor, but this book is about a real place where the old woman's claim rings true: the Galápagos Islands, where the world as we know it *does* rest on the backs, not of turtles, but of giant tortoises. I do not mean that anyone who lives in the Galápagos believes that the Earth or the islands rest on an infinite pile of the enormous reptiles for which the archipelago is so well known. Rather I mean that these land-dwelling animals have fundamentally shaped the history of the islands.[3] The Galápagos archipelago would not be the same place without them. The giant tortoises are, today, charismatic icons of conservation. But they are much more than that. I have been visiting these islands for more than a decade now, and the more I return, the more deeply I am struck by the many different meanings and roles the tortoises carry. What it means to rest on their backs is far more complicated, ethically ambiguous, and fascinating than I could have imagined when I first stepped ashore. Much like Stephen Hawking's pondering of the implications of black holes for our understanding of time, I have learned that the turtles-all-the-way-down history of the Galápagos is far less certain and straightforward than it appears.

For centuries, explorers, buccaneers, and whalers sailing the Pacific made their way to the islands in search of tortoises to eat—indeed, that is how the archipelago got its name. In the sixteenth century, in the midst of the conquest of the Americas, Spanish sailors named the equatorial archipelago for its tastiest resources: *galápagos,* an old Spanish word for tortoise. The Spanish, like indigenous Americans before them, did not colonize the islands, but the name marked them for centuries of plunder.[4] The remote islands, six hundred miles off the coast of what is now Ecuador, were an ideal place for buccaneers to rest and restock their vessels between raids on Spanish ships and ports. As Spanish control of the Pacific ebbed during the eighteenth and nineteenth centuries, it became European and North American whalers and sealers who pillaged the islands, which were storehouses for early sailors where they searched for freshwater and collected timber, fresh fruit, vegetables—and tortoises. Sailors devoured the animals by the hundreds, stocking the hulls of their ships with these slow-moving, portable, and by all accounts delicious sources of fresh meat and fat.

One of these young sailors was Charles Darwin, whose visit aboard the *Beagle* in 1835 would change the fate of the islands and their namesake animals. During the *Beagle*'s five-week sojourn in the Galápagos, Darwin was fascinated by the tortoises, mockingbirds, finches, and other wildlife he encountered—so much so that he is commonly, though erroneously, thought to have come up with the idea of evolution in this place that seemed to reveal "that mystery of mysteries—the first appearance of new beings on this earth."[5] Although Darwin historians have refuted the notion that natural selection came to Darwin as he studied the beaks of finches or the backs of tortoises, biologists since have shown that these animals and several other Galápagos species are striking evidence of evolution. Fifteen different species of giant tortoises, for example, once populated the Galápagos, each adapting in form over hundreds of thousands of years to survive in the harsh conditions of the archipelago's volcanic islands. The animals also reshaped the nature of life on these volcanoes—they are what biologists now call keystone species and ecosystem engineers, megafaunal herbivores who structure the ecology of the islands as they plod up and down volcanic hillsides, munching on grasses and fruits and depositing seeds along the way.

Since Darwin's visit, however, the tortoises have also become evidence of human-caused extinction. Three of the fifteen species now exist only as historical records, casualties of sailors and of the colonists who settled the islands during the nineteenth century. By the early twentieth century, sailors and settlers had killed some two hundred thousand giant tortoises, hunting populations on three islands until they disappeared and devastating those on other islands. The remaining tortoises are "living fossils," remnants of a prehistoric world when giant reptiles and other megafauna roamed the Earth.[6] Giant tortoises could once be found on islands around the world and on every continent except Australia and Antarctica. They are believed to have been hunted to extinction by early humans—part of the Pleistocene extinction of megafauna around the world ten thousand to fifteen thousand years ago. By the time the *Beagle* circumnavigated the globe, however, they existed only on the Aldabra atoll in the Indian Ocean and on the Galápagos archipelago. Tucked away on isolated islands, these giants had managed to survive when other megafauna had not. The giant tortoises were remnants of a lost world.

Over the course of the twentieth century, scientific desire to preserve this lost world made the tortoises objects not of plunder, but of conservation. A turning point for the fate of the tortoises, as well as for the islands more broadly, came in 1959—the centenary of the publication of Darwin's *On the Origin of Species*. That year, an international coalition of scientists and the Ecuadorian government established the Galápagos National Park and the Charles Darwin Research Station. Giant tortoises were enshrined in each institution's logo, making them mascots of science and conservation.

But conserving the Galápagos was about far more than protecting endangered and charismatic animals. The scientists and politicians behind the park and Darwin station argued that the Galápagos was a "living laboratory of evolution" and "monument to Charles Darwin."[7] They fought to protect the place Darwin had called a "little world within itself" where so many biological oddities—gigantic tortoises, iguanas who feed in the sea, and cormorants with stubby wings that prevent them from flying—had evolved as they adapted to austere island environments. How these creatures managed to arrive on volcanic islands hundreds of miles removed from the South American mainland, and how their forms adapted to new environments, had fascinated Darwin as he reflected on his time in the islands. "Considering the small size of the islands," he wrote, "we feel the more astonished at the number of aboriginal beings, and at their confined range."[8] The remoteness of oceanic islands, unlike islands that were once part of a continental landmass, means they have relatively few total species. This biological scarcity shaped unusual evolutionary pathways—and made the Galápagos a place where scientists could retrace these pathways with unusual precision by examining slight changes in the form of plants and animals on each island. For Darwin, the coloring of mockingbirds and the shape of finches' beaks illustrated processes of geographical differentiation that resulted in slightly different species on different islands, what is now called adaptive radiation. Today biologists recognize that many Galápagos species fit this pattern—tortoises, snails, lava lizards, and trees in the daisy family that reach more than sixty-five feet into the sky. Then there are what have become known as "Darwin's finches," textbook examples of why, in science writer David Quammen's words, islands can act as a "flywheel of evolution" because of their geographical isolation, which propels species to differentiate not just over millennia, but also within a human timescale of decades.[9]

As flywheels, the Galápagos Islands are remarkable among oceanic archipelagos because of their unusual ecological conditions, shaped because the islands sit just so at a conflux of tectonic plates and ocean currents. The archipelago is among the most active areas of volcanism in the world, located at the intersection of three tectonic plates: the Pacific, the Nazca, and the Cocos. The islands are young, geologically speaking, having arisen from undersea volcanoes some 700,000 to 4.2 million years ago.[10] Yet geological change happened not only in the deep past. Volcanoes continue to erupt regularly on the youngest, western-most islands, and the ground itself shifts, sometimes dramatically—in 1968, the floor of the caldera on Fernandina collapsed, and in 1954 a third of a square mile of reef uplifted thirteen feet out of the sea on Isabela.[11] This is hardly the stable environment you would expect of a laboratory, but these changes make fascinating "experiments" for geologists and for ecologists who examine how wildlife responds to such changes.

The archipelago's climate also shapes the peculiar features of these equatorial islands. Three oceanic currents and associated winds keep them unusually cool and dry for much of the year. The Humboldt Current (named, like many places in Latin America, for the German explorer) brings cold, nutrient-rich upwelled waters north along the South American coast, making equatorial waters hospitable to the world's only equatorial penguins and nourishing a rich marine life; the subsurface Equatorial Current brings cold water and an upwelling of nutrients from the west; and the seasonally warm Panama Current from the north provides more typically tropical weather in the austral spring. In addition to these currents, the El Niño and La Niña phenomena also have a substantial impact on Galápagos climate, making the archipelago a laboratory for climate change as well as evolution. El Niño events reduce the strength of the trade winds and the Equatorial Subcurrent, bathing the islands in much warmer, and less nutrient-dense, water, which can have drastic effects on marine life. The results on land are no less dramatic: the islands are drenched in heavy downpours which create rivers that rush down beds of dry lava with the strength to wash boulders into the sea—and likely giant tortoises with them, perhaps accounting for the dispersal of the tortoises from island to island.[12] The currents are also responsible for populating the islands with their flora and fauna, much of which washed, or flew, or swam ashore during the millennia since the volcanoes first emerged from the sea.

Today, 97 percent of the terrestrial area of the Galápagos—an area a bit smaller than the big island of Hawai'i—is protected in the national park, surrounded by one of the largest marine reserves in the world. But protecting this living museum of evolution was no matter of letting nature take its course. Over the past six decades now, conservationists have done their utmost to go "back to Eden" by restoring the islands—or trying to—to their condition in 1534, the year before they entered the annals of Western history when a Spanish bishop's ships accidentally drifted into the archipelago.[13] This book uses the story of the giant tortoises to trace a history of these conservation goals, and their feasibility. Why and how have conservationists sought to protect the Galápagos by restoring Eden—and what have been the effects?

If the Galápagos finches are textbook examples of island evolution, it is the giant tortoises who are the most telling mascots of the archipelago's conservation history. Take, for example, the single most famous Galápagos giant tortoise, Lonesome George. His story encapsulates the great lengths conservationists have gone to over the past half century to save the endangered species. George was found on a small island called Pinta in the north of the archipelago in 1971 where the tortoise species was thought to be long-since extinct. Park guards brought the adult male tortoise into the captive refuge of the breeding center on Santa Cruz Island in 1972, where he lived alongside tortoises from other islands who conservationists were helping to breed to replenish lost populations. He became a favorite with tourists—famous not because he was the largest tortoise, nor the oldest of these animals who can live 150 to 200 years. Instead, George was famous because he was the last of his species. But conservationists did all they could to keep his species going. They searched Pinta repeatedly for a female tortoise but found only the remains of tortoises who had fallen into deep crevasses formed from collapsed lava tubes. They offered a $10,000 reward to zoos around the world for the return of a Pinta female, but none was ever found. Keepers were undeterred. They put two female tortoises from another island with George, in hopes of producing a half-blood heir. But George showed little interest. Veterinarians put him on a special diet, thinking that the extra chub he had gained in captivity might explain his lack of interest in mating. They even recruited a biology student to teach him *how* to mate. But *Solitario Jorge*'s nickname fit him—he was never a very social, let alone amorous, tortoise, especially compared with other breeding studs at the center who have each sired a thousand offspring.[14]

Although George was not keen to reproduce, the Galápagos tortoise-breeding program has been among the most successful conservation breeding projects in the world. Since efforts began in the mid-1960s, conservationists have bred more than eight thousand juvenile tortoises and "repatriated" them to their endangered home populations across the archipelago. Today, scientists estimate that at least twenty thousand tortoises live in the wild in the Galápagos. This is likely only a tenth of the population that once was, but many of the species are now thriving and reproducing independently. They are able to do so because conservationists have waged all-out war on invasive species—rats, goats, and boars, to name a few—who prey on juvenile tortoises or compete with them for limited vegetation. These foreign species were introduced, sometimes intentionally, by sailors and colonists and have few if any natural predators on the islands, so their populations ballooned.[15] In response, conservationists themselves have taken on the role of predators—blanketing islands with poison designed specifically to target rodents and killing hundreds of thousands of goats, sometimes from helicopters with semiautomatic precision hunting rifles. This killing is largely hidden from the view of tourists who visit the islands, unlike the breeding center where the tortoises are on proud display, but it is no less central to the work it takes to restore the islands to an ecological state that some consider to be more desirable.

Breeding and eradication work, however, can go only so far toward restoring the past. On a Sunday morning in June 2012 the limitations of "restoring the tortoise dynasty" came sharply into focus.[16] While making his usual rounds to check on the tortoises, Don Fausto Llerena, the national park guard who had cared for the captive animals for three decades, found that Lonesome George had died during the night.[17] His death was mourned around the world; *Nature*, the *New York Times*, and the BBC News all ran eulogies for this tortoise known by hundreds of thousands of former tourists. Mourning him was about mourning the history of human-caused extinction. But it should also be about mourning the shortcomings of the approach to conservation that made him famous.

The Galápagos have a tortoise problem. I do not mean a problem with saving the endangered animals themselves, which conservationists have been quite successful at addressing. Instead I mean a problem with what the giant tortoises represent. To explain, let me turn to another version of the turtles-all-the-way-down story, this one told by philosopher of science Isabelle Stengers. In a book that focuses on the power of scientific theories to modify society, and the power of social thought to shape science, she uses the story to question the foundation of modern knowledge. She recounts much the same story as Hawking, asking whether there is "much difference between the old lady's turtles and the fundamental laws of physics."[18] Both theories posit a world that could, in principle, be fully explained. What differs are the elements of that explanation, be they an infinite stack of turtles or the quarks that physicists currently understand to be the fundamental building blocks of matter. Both reflect what philosophers call the "turtle problem" of infinite regress, a never-ending quest for solid ground. In the story, both the old woman and Hawking have us searching endlessly for an ever-larger turtle or an ever-smaller sub-atomic particle as the basis for knowledge.

The tortoise problem of conservation is similar—the stubbornly enticing and yet impossible desire to restore a world of prehistoric nature, to go back to Eden. What is at stake in saving the Galápagos is not only saving giant tortoises and other unique wildlife from extinction; it is also saving the foundational site of a secular, scientific origin story—"a biological and geological Eden." It is about saving the place that Darwin once wrote was the "origin . . . of all my views."[19] As a BBC documentary series put it, these are "the islands that changed the world," the place that "changed our understanding of life on earth."[20] Since the mid-twentieth century, scientists have worked to conserve *evolution* in this "natural laboratory."[21] Darwin never called the islands an Eden, though he might well have liked the metaphor—he carried John Milton's story of Adam and Eve, *Paradise Lost*, with him whenever he went ashore from the *Beagle*. Yet the idea of an evolutionary Eden is a contradiction. Ironically, in trying to preserve the Galápagos as an Eden, conservation biologists have attempted to restore the very foundation of life that was shattered by Darwin's emphasis on evolutionary process, all the way down.

In the Galápagos, giant tortoises anchor conservationist desire to restore the islands to a prehuman world. Their plodding, creaky gait, leathery skin, and worn armor certainly make them look the part of animals from a land before time. By many measures, conservationists have been quite successful at saving this place. To conserve evolution in this place often considered one of the last bastions of pristine nature, biologists have sought to protect the islands' historical isolation. Conservationists estimate that the islands retain 95 percent of their "original," or pre-1535, endemic biodiversity. This is an impressive statistic—a much higher rate of surviving endemic species than in Hawai'i, for example. The Galápagos archipelago is among the world's most striking conservation success stories. Yet despite all the good that conservation has achieved, this way of trying to save life will never bring back Eden. Evolution, as conservation biologists well know, cannot be rewound. And as the tortoises' own histories make clear, the islands are far from pristine. But more significant than that, their stories call into question the way of thinking about nature which suggests that the islands even *could* be pristine.

They will help us understand how that vision of the Galápagos became so hegemonic, and how we might think about the archipelago, and conservation goals, differently.

The tortoise problem is not unique to the Galápagos. It pervades the field of conservation biology. That science evaluates the state of the ecological world in comparison to a priori baselines, such as the Columbian encounter with the "new world," a time and place seen as more pure before the degrading effects of European conquest. The tacit referent is Eden before the Fall, a stance for which the field, along with restoration ecology and invasion biology, has been criticized as an "Edenic science."[22] These fields are premised on a fundamental and idealized distinction between nature and culture as discrete realms. This distinction structures how many Westerners understand the world, suggesting that histories of evolutionary adaptation and social histories of development are distinct sets of processes and, thus, that natural history and cultural history are fundamentally different, each knowable through its own fields of study: biology or history, ecology or anthropology. Yet such division of the world is continuously challenged both at home in our own skins by the knowledge that humans are animals, and around the world by global warming, species extinctions, ocean acidification, and myriad other troubles that are at once both social and natural. The result of such breaches of the divide is, for conservation biology, a view of nature as perpetually in crisis.[23] Yet rather than question the premises of such a framework, the crisis orientation only escalates as the stable ground of pure nature slips farther and farther out of grasp.

The Galápagos archipelago is a paradigm case for this bifurcated way of thinking about nature and conservation. Despite emphasis on the dynamic forces of life, the islands are often cast as strangely timeless, "anachronistic spaces" of nature that exist nearly outside of history. As a conservationist put it in the mid-twentieth century, the islands are—or should be—the land "where time stood still."[24] The BBC series, for example, presents a typical narrative of the islands' human history, told in dramatic prose by the actor Tilda Swinton. The episodes show resplendent footage of wildlife where no human touch is visible and re-created scenes of Darwin's explorations. The series ends by contrasting the image of Darwin stroking a tortoise with scenes of contemporary life on the islands, suggesting that population growth is a new challenge for this "crucible of life."[25] The message is that, aside from Darwin, the presence of people in the archipelago is unnatural—the evolutionary Eden that holds the secrets of the origins of life is a place where the purity of nature should be protected.[26]

The idea that the Galápagos archipelago is an evolutionary Eden is a "geographical imagination"—a powerful way of understanding and engaging with particular places that reflects social, class, and political positions and has profound material effects.[27] The evolutionary Eden trope draws from a long history of romantic ideas about desert island Edens in Western culture. Islands are allegories for the world—concentrated spaces that put the contours of modernity into relief. As such, they seem to offer clairvoyance, acting as harbingers of things to come. Islands have played a central role in the construction of Western environmental knowledge for centuries. In the seventeenth century, naturalists began to recognize the devastating effects of colonization on island environments. In the nineteenth century, Alfred Russel Wallace famously conceived of evolution by natural selection at the same time as Darwin while working in the Malay archipelago. And in the twentieth century, systems ecologists' theories of island biogeography depended on the closed isolation of islands as ideal natural laboratories.[28] In the Galápagos, the modern dream of a clear divide between nature and culture has underwritten tropes of Edenic nature and imaginations of the islands as a space that had been, and could be, kept isolated from social influences.[29] Protecting the Galápagos, like other national parks, in this way is meant to compensate for a sense of wildness and untainted origins that is lacking in modern life.[30] This is the conceptual framework that has powered conservationist desire to restore Eden and that has divided the Galápagos into zones of 97 percent nature and 3 percent society, as if a boundary fence might be capable of holding strong this divide so that conservationists can save the tortoises—and tourists can photograph them too.

Yet even seemingly isolated islands are not outside the reach of modernity—particularly in the so-called Anthropocene, which is marked by human imprint not only on wildlife habitats and the global climate, but on the very geological bedrock of our world.[31] The idea that islands are isolated is an artifact of modern culture that belies the central role that they played in the development of Western modernity. Instead of emphasizing the supposed static insularity of islands, the stories I tell in this book instead speak to the connected, mobile geographies of archipelagos.[32] This approach is more amenable to understanding the dynamic nature of the Galápagos as well as the ways that life in the archipelago has been shaped by human histories. This dynamism has pushed some conservation biologists away from strict adherence to the protection of pristine ecosystems and toward "novel" understandings of nature shaped by human influence.[33] Yet debates about novel ecosystems have been controversial, and their implications for island management are far from straightforward.

Attention to the relationships among natures and cultures is what Isabelle Stengers believes can get us out of the problem of endlessly searching for ground to stand on. In response to the turtle problem of a search for origins, she argued that we need to ask not "what [the turtles] rested on, but with whom they lived."[34] I take her provocation literally. Our foundation for knowledge is shaped as much by where we stand as by what and who stands around us. Neither turtles, nor quarks, nor giant tortoises stand alone. To address the tortoise problem, we need to understand the relationships among Galápagos tortoises and the many others with whom they have lived over the past five centuries. And to do that, we need to turn to one more version of the turtle story.

This tale is a bit different. It was told by the anthropologist Clifford Geertz, who heard the turtle story not as a lecture on theories of the universe, but as an Indian story about an exchange between an Englishman and a local, a setting with overtones of British colonialism. The Indian man told the Englishman that the Earth rested on a giant platform that rested on the back of an elephant, which in turn rested on the back of a turtle. The Englishman asked the familiar question: "What did the turtle rest on?" Another turtle, came the reply. And what did that turtle rest on? "Ah, Sahib, after that it is turtles all the way down."[35] Geertz tells the story to close an essay in which he considers the anthropologist's task as the interpretation of culture. For him, the tale speaks to the unending task of cultural analysis—"the more deeply it goes, the less complete it is." It is never enough to look for surface explanations for an event, Geertz warns, nor a solitary motivation for people's actions such as their economic interest. Nor should an anthropologist look only for deep-lying turtles and risk losing touch with the political and economic realities of life. Instead, Geertz reminds us that the closer we look at one turtle in a stack, the more we will be drawn into a need to understand the next turtle down. Meaning is always layered. Ethnographers should strive for "thick description" of cultures, even while knowing they will never truly get to the bottom of things. Like Stephen Hawking, Geertz argues that the best route to knowledge is to hang on to our own uncertainty. We have to keep asking questions about the things we take to be solid ground.

Geertz's turtle story is a reminder of the layered meanings of charismatic animals like Lonesome George. Galápagos giant tortoises are flagship species who sit at the crux of an alliance uniting science, conservation, and tourism.

For biologists, the animals are an "umbrella species"—protecting them means indirectly protecting many other species in their ecological communities. For broader publics, George and other giants play an important role as "boundary objects" that translate scientific understanding of nature to those concerned about endangered species.[36] Yet such translations are not merely a matter of education; they reflect the power relations that permeate assertions about how best to understand the world.

To begin to answer Stengers's question about with whom the tortoises live, we need to pair Geertz with work on political ecology and multispecies ethnography that focuses not only on the meanings that people ascribe to animals, but also on the ways that human and nonhuman lives are tightly entangled.[37] These perspectives can help us flesh out how multiple forms of knowledge and political economies mediate differently situated actors' relationships with nature.[38] They will also help to show us that a tortoise is not just a tortoise, is not just a tortoise. The animals in our stack change because of the ways that they become entangled in social life—and they change history too. Understanding these entanglements is an essential task for conservation because, as Ecuadorian historian Teodoro Bustamante has argued, it is the limits of understanding the dynamics of social life surrounding protected areas that become the limits of our ability to understand, and conserve, nature.[39]

Take for example the giant tortoises' roles as charismatic conservation icons who sit in the thick of a thriving economy. The evolutionary Eden is a political economic production as much as it is a biological and geological one—the national park is, to use political ecologist Roderick Neumann's phrase, a "landscape of consumption."[40] Tourism to this bucket-list destination generates much of the funding for conservation. George and other animals are valuable commodities for an industry that sells visitors the opportunity to get close to wildlife known for being unfazed by the presence of humans. Many Galápagos residents—who are mostly Ecuadorian but also include European and North American expatriates—see the tortoises as representing the prosperity that nature tourism brings to the islands. Indeed, more than two hundred thousand tourists visit the Galápagos every year. The local economy is booming—from the late 1990s to the 2000s, the islands' annual economic growth rate was an astounding 78 percent, making them one of the fastest growing economies in the world and a significant source of revenue for the Ecuadorian state.[41]

But this alliance between tourism and conservation is uneasy. There are downsides to iconic charisma: the booming industry also brings with it a host of issues that threaten conservation, including a rapidly growing human population that has increased from about two thousand in 1960 to more than thirty thousand today. These island residents live in the 3 percent of the islands not included in the national park. Supporting them—and tourists— means importing from the continent everything from food to air conditioners to fuel. With the increased transportation that brings people and goods to the archipelago also come foreign species and diseases. Conservationists and journalists alike fret about whether the Galápagos archipelago has become too successful for its own good, creating a crisis of overdevelopment in a place best known as an isolated sanctuary of nature. Anthropologist Diego Quiroga calls this the "Galápagos paradox." He writes, "the appeal of the Galápagos is based on its marketing as a pristine, people-free 'natural laboratory,' but at the same time the success of tourism is creating unpredictable flows of people and other organisms that undermine the unique terrestrial and coastal environments."[42]

Largely in response to these tensions, in 2007, Ecuadorian president Rafael Correa declared the islands to be "at risk," and the United Nations Educational, Scientific, and Cultural Organization (UNESCO) put them on its list of World Heritage Sites "in danger." These declarations fundamentally shaped my research, as it was in their midst that I made my first trip to the Galápagos in May 2007. As I conducted interviews with a research team that summer, I little could have imagined that I would spend the next decade of my life thinking through what locals told us about these "crisis" declarations. The park guards, municipal officials, foreign conservationists, farmers, fishers, and tourism workers we talked to all had differing perspectives, but they made it very clear that the status quo—the particular alliance between conservation, science, and tourism that had made George a world-famous tortoise (with his own biography even)—would not do for the future. The islands were taken off the "in danger" list a few years later following pledges of institutional reform, but deep questions remained about the islands being at a "crossroads": conservation or development, a sustainable future or continued crisis.[43] This book is my attempt to think through how the Galápagos Islands got into this situation, and how their conservation might be approached differently.[44]

One of the things I learned that first summer and came to better understand over the following years was that deep tensions exist among various populations in the islands about the meaning of iconic species. As charismatic icons, the giant tortoises' umbrella casts a broad shadow across the Galápagos. But not everyone there fits underneath, or wants to. While tourists congregated around George's corral and cooed over the baby tortoises at the breeding center, not everyone in the Galápagos was charmed. Among *galapagueños*—island residents whose identity is bound up with the archipelago's namesake animals—opinions of the tortoises vary greatly. Many, perhaps most, see the tortoises and their conservation as a source of pride. Yet others told me that Galápagos is not only tortoises. Residents have at times gone to great lengths to assert themselves outside the shadow of their umbrella. Fishers, who often have been excluded from the wealth of tourism and seen their livelihoods as threatened by conservation measures, have targeted the tortoises in political protests. As one told a group of students, "Lonesome George here, solitary George there, that is the best business that the conservationists have invented."[45] In the mid-1990s striking fishers called for Lonesome George's death; in 2004 another strike closed the road to the headquarters of the Galápagos National Park Service, including the breeding center where George and other tortoises lived, prompting international headlines that cried out "tortoises held hostage."[46]

Such actions can be difficult to comprehend for those concerned about the fate of wildlife. But the charisma of iconic wildlife is not innate. Instead it is a historical product and culturally situated.[47] Not everyone grows up as I did learning about endangered species by watching nature documentaries or through childhood class projects on saving the whales. Because of their fame, iconic animals can become lightning rods of tension among people who understand and interact with nature in different ways.[48] Conflicts surrounding iconic wildlife—from white rhinoceroses to Yellowstone wolves—are common and show that it is not the same thing to visit national parks and care about wildlife from afar as it is to live with these animals. Political ecologists push Geertz's analysis of the turtle story a step farther, helping us see that while conflicts about wildlife are of course about wildlife, they are not *just* about wildlife. They are about the validity of deep-seated questions about identity, belonging, and access to resources. They reflect the frictions produced as competing ways of life, economic systems, and goals for the future rub up against one another.[49] These frictions are at the core of an environmental politics that has transformed the landscapes of the Galápagos, both culturally and ecologically, as well as the lives and identities of the people who inhabit them.

Despite their sturdy appearances, the giant tortoises have not been a stable and unchanging foundation on whose backs these environmental politics have played out. The species have transformed along with the flows of history. The environmental philosopher Thom van Dooren sees species as "incredible *achievements*: intergenerational lineages stretched across millions of years of evolutionary history" and "interwoven in rich patterns of co-becoming with others"—both nonhuman and human.[50] Both symbolically and materially, the giant tortoises have coevolved with human histories in the archipelago. While the tortoises may be living fossils that have astonishingly survived since an earlier, reptilian age, they are far from *pre*historic beings—they are not isolated remnants of a former, fallen world, but rather are inseparable from global histories that have brought human beings to the Galápagos. The animals have changed as they have been caught up in histories of imperial exploration and natural history collection, the development of a postcolonial Latin American state, conservationist concern for endangered species, and nature tourism. The past five hundred years have produced quite a tall stack of giant tortoises who have been everything from soup and steaks to natural history specimens, resources for economic development, endangered objects of conservation, celebrated breeding studs and unnamed "harems" of dams, subjects of tourists' photo-shoots, and symbols of protest.[51] Giant tortoises have no single inherent value. As I have learned over the past decade, the tortoises have much to teach us not only about histories of evolution, but also about relationships between humans and nature.

According to his biographer, Lonesome George was an allegory who "embodie[d] the practical, philosophical and ethical challenges of preserving our fragile planet."[52] While for some locals George may have been a target of contestation over whose lives matter in the Galápagos, for conservationists he was a testament to a history of exploitation and a living plea to protect wildlife. His story puts into stark relief the ways dominant human conceptions of nature have changed over the past centuries from God-given resources for human consumption to beings with whom we share an earth-bound history of evolution. Yet his story also makes clear just how difficult it is to save wildlife, let alone an entire archipelago, in a world that little resembles a prehistoric paradise.

The history of the Galápagos has been forged through the intersection of various processes—geological, climatic, and ecological as well as political, economic, and scientific—that, though they span different geographies and operate on different temporal scales, are not so very distinct but come together on the backs of tortoises.[53] Giant tortoises are both objects and agents of a socionatural history in which people and animals—and plants and all nature—together make history, though not on equal terms or in conditions of any of our own choosing. How, then, did this place and these animals that so captivate us come into being?[54] To answer that question, we need to query our way through the stack of giant tortoises that comprise Galápagos history. To begin, let's turn to the Galápagos visitor who, unbeknownst to him, has done the most to shape modern understandings of the archipelago as a particular kind of solid ground. Whatever problems there may be with the vision of the islands now ascribed to Charles Darwin, we can hardly help but visit in his shadow.

2 • In Darwin's Footsteps

In which the tortoises are soup, and souvenirs

Going to the Galápagos can be like stepping into a nature documentary. Take a cruise of the archipelago and you will spend a week at sea stopping at some of the thirteen large islands and dozens of islets. As naturalist guides explain histories of evolutionary adaptation, you can tiptoe around blue-footed boobies protecting their nests, watch pairs of waved albatross tap their beaks as they dance together, and listen as piles of marine iguanas snort out saltwater as they sun themselves on lava rocks. You can see newborn sea lions snuggle against their mothers and be chased off a beach by aggressive, barking male sea lions. You can kayak among a school of sea turtles and snorkel with penguins, dolphins, octopi, sharks, and countless fish. Galápagos tourism is based on the remarkable ability the archipelago affords for visitors to routinely encounter rare and endangered animals in their natural habitats.

What you will not see are many other humans—aside from your shipmates and crew and those on a couple of other boats. This is not chance, but deliberate. It is what scholars call an "ecotourism bubble"—the carefully edited view of the islands presented to tourists who cruise the archipelago.[1] This edited Galápagos is the world presented in nature documentaries like the BBC's *Galápagos: The Islands That Changed the World*. It is a world of ostensibly pristine nature sold primarily to North American and European tourists as a place where they can walk deserted beaches, imagining themselves in Darwin's shoes and exploring the islands much as he did. Sporting outdoor adventure gear and high-end cameras, they can relive the experiences of a naturalist who was delighted by birds who landed on his water jug and puzzled by iguanas who dove into the sea to feed. Might they hike the same trails Darwin did as he followed giant tortoises lumbering toward pools of water in the islands' highlands? If they are lucky—just maybe—might they see the very same tortoises Darwin saw, still alive and gulping water from muddy pools more than 150 years later?

Of course, not everyone goes to the islands dreaming of Darwin. But in a place where Darwin statues now seem as ubiquitous as the animals he described, it is hard to miss the idea that visiting the Galápagos is about walking in his footsteps. The statues remind visitors that Darwin is the reason we understand the archipelago's strange creatures as evidence of millions of years of evolution. They also remind us that it was in his name that conservation was undertaken in the mid-twentieth century. On Santa Cruz Island, a monument arch at the end of Avenida Darwin, which terminates at the entrance to the national park, depicts Darwin as a bearded sage who crafted the concept of evolution, gazing Godlike from the heavens over the tortoises and other Galápagos creatures (fig. 2.1). This Darwin is the patron saint of Galápagos science and conservation—someone who, after centuries of human predation of wildlife, taught us that nature is not a God-given resource for human consumption. As two conservationists put it recently, Darwin "lift[ed] the veil of ignorance that had cursed Galápagos . . . reclaim[ing] the islands and their inhabitants from human condemnation and bequeath[ing] them long-lasting fame and a life-saving future."[2] The graybeard Darwin is the secular deity in whose steps tourists follow when they visit the Galápagos National Park. But he is not the Darwin who actually walked the islands' hills.

On a sunny day in October 1835, a twenty-six-year-old Darwin hiked from the parched coast of Santiago Island (then called James) to the island's green, damp highlands, where he was taken with the giant tortoises he found. After a long walk, he sat in the shade and watched them as they ambled along broad roads that had been trodden by their elephant-like feet over countless generations. He timed their gait (faster than he supposed), measured their carapaces (six and eight feet in circumference), and tried to lift them (some weigh well over five hundred pounds). Finding their weight too much, he climbed aboard, rapping on their shells and trying not to slip off their backs as they trudged along.[3] Riding tortoises is today, of course, not among the approved ways of following in Darwin's footsteps. Tourists may observe the animals Darwin saw, but they are not permitted to do as he did. Re-creating the experiences of the Darwin who actually walked the islands would mean doing a host of things now prohibited by national park rules—trying to net birds with your hat or catching iguanas by the tail and dissecting their stomachs to see what they eat.

Figure 2.1 Darwin watches over native Galápagos species as if a God in heaven, in an arch at the end of Avenida Darwin, Puerto Ayora, Santa Cruz. A blue-footed booby sits in the foreground, in the traditional position of the dove of the Holy Spirit. (Photo by the author)

Perhaps worst of all would be doing what Darwin did after a day spent observing tortoises: feasting on them. In the highlands of Santiago, Darwin spent two days camping with local tortoise hunters, during which time he subsisted on nothing but the animals. The "Spaniard" hunters, as Darwin called them, were political exiles from the new state of Ecuador who were collecting meat for a colony on another island. For dinner, they made tortoise *carne con cuero* in the style of Argentinian *gauchos*—roasting meat on the breastplate and frying it in tortoise fat. Jotting down his impression of the meal afterward, Darwin deemed it "very good," although he preferred the "capital soup" made of young tortoises.[4]

This young Darwin would be abhorred by the supposedly conservationist graybeard Darwin who watches over the islands. But even as an old man, Darwin did not see the Galápagos through the eyes of an environmentalist intent on protecting nature. The first Darwin statue was a product of mid-twentieth-century campaigns to preserve the islands, when international conservationists heroized the naturalist to protect the archipelago as what they saw as a natural laboratory of evolution. Others are more recent additions designed to appeal to tourists. The actual Darwin who hiked the islands' shores was instead a product of his own time, when eating tortoises was not a cultural and legal taboo, but the main reason people visited the Galápagos. After weary sailors had been months at sea, the tortoises were a sought-after prize; and most sailors were more enthused about their taste than Darwin: as one British admiralty captain-turned-whaler wrote in 1798, giant tortoise meat "in whatever way it was dressed, was considered by all of us as the most delicious food we had ever tasted."[5]

When we walk in Darwin's footsteps today we follow the path of the nineteenth-century sailor as well as the older sage. Is it possible to reconcile these two Darwins? We need to take our own tour through the islands, comparing the experience of a tourist cruise to see an evolutionary wonderland with Darwin's own exploits as one of thousands of sailors who hiked the islands' tortoise trails to find their dinner. Doing so will show that the consumption of tortoises has not changed quite as much since Darwin's day as conservationists would like to believe.

Cruising the Galápagos

A cruise around the Galápagos Islands today appears to offer the opportunity to return to the world Darwin saw—a place where nature had gone unchanged for more than a century, if not millennia. Much of a Galápagos tour is dedicated to viewing wildlife and hiking, snorkeling, and kayaking in a landscape where it looks as if few have tread before you. Mid-twentieth-century celebrations of Darwin's "discovery" of evolution in the Galápagos can make it seem as if he was the first to set foot on the islands, or at least to explore them. But this speaks more to the power of mythology than to the reality of Darwin's experience. To understand what the islands were like when he arrived, we need to go back to the seas that the *Beagle* sailed.

When Darwin's captain, Robert FitzRoy, headed west from the coast of South America toward the Galápagos, he followed a well-traveled route. It was September 1835, and the Pacific churned in the wake of ships that routinely crossed its waters. Since 1513, when Vasco Núñez de Balboa first spied the deep-blue waters of the Mar del Sur after crossing the Isthmus of Panama, the Spanish crown had claimed sovereignty over the Pacific. From the mid-sixteenth century to the early nineteenth, the profitable Manila Galleon trade connected Nueva España to the riches of the East. Lying far south of the trade winds that connected Acapulco to the Philippines and west of the route between Peru and Panama, the Galápagos remained a peripheral location in early histories of world travel—a stopover rather than a destination. Spanish explorers' requests to the crown for permission to establish a colony in the archipelago in the late sixteenth century went unanswered. They were dismissed just as the Galápagos's first Spanish visitor, Fray Tomás de Berlanga, the bishop of Panama, had himself dismissed the islands. In 1535, Berlanga was en route from Panama to Peru to resolve a dispute between warring conquistadors when his ships drifted into the archipelago, where they were becalmed for weeks. When eventually they were able to go ashore, he found the islands so dry and desolate—save for enormous tortoises and birds "so silly they didn't know to flee"—that it "looked like God had showered stones" upon the earth.[6] By the time his crews had found water and a breeze to carry them back toward the continent, two men and ten horses had died. Berlanga's letter to the crown sent a clear message: this was not a place worth returning to. (Looking back, it is ironic, perhaps, that it *is* now considered a landscape worth restoring—and that Berlanga's letter is a baseline for a return to Eden.)

Having been passed over by the Spanish, the islands became a useful refuge for seventeenth-century buccaneers who used them as a place to recuperate between sacks on Peruvian ports. These sailors, though, had little better opinion of the islands. English pirate-naturalist William Dampier managed to find springs of water and praised the abundance of tortoises—there were so many of this "extraordinarily large and fat" animal that he thought five hundred men could live on them for months.[7] But pirates generally found the islands too harsh for long-term stays, preferring to rendezvous much farther south on the more hospitable islands of Juan Fernández and Masafuera (now named Alejandro Selkirk, after the man said to be the inspiration for Daniel Defoe's *Robinson Crusoe*).

Views of the islands would change by the end of the eighteenth century as Spanish authority over its New World colonies, and the Pacific, slipped. The buccaneers who had challenged Spanish rule from the eastern Pacific were replaced by British and French explorers. In their wake came fleets of principally US traders who sold furs to the Chinese and whalers who hunted sperm whales for wax and oil to fuel industrial development. As stocks of Atlantic whales diminished after a century of predation, whalers moved to the Pacific in the late eighteenth century. By the time the *Beagle* entered the south sea, the Pacific whaling industry was well established and the Galápagos had become a popular crossroads for US, British, Dutch, French, and Peruvian ships that now sailed seas that had long been claimed by the Spanish.

When we sail the Galápagos today, therefore, we follow in the wake of not just the *Beagle*, but of centuries of ships that came before it. Tourists see a window onto this history during one of the common stops on a Galápagos cruise when they land at Post Office Bay on Floreana (Charles) Island. Today, the "post office" is a barrel on the beach where tourists drop unstamped letters for subsequent visitors to pick up and deliver when they return home. Tradition holds that tourists should deliver those letters by hand—as nineteenth-century sailors might have done. Back then, the post barrel was a landmark for ships—a dried tortoise shell served as a roof that stood out bright white against the landscape.[8] The post office facilitated communication among ships and provided a way for sailors to send letters home on vessels that would (they hoped) return before them. For tourists, the stop is a playful way of recognizing that the islands were once a common crossroads, and of learning a bit about what the archipelago was like in centuries past, when sailors came for quite different reasons than tourists do today.

For the post office to work, ships needed to pass regularly—not multiple times a day as tour boats now do, but at least monthly. This was possible only when the Galápagos became a regular provisioning stop for traders in the late eighteenth century, thanks in part to the travel writing of a former British naval captain, James Colnett. In 1794, Colnett visited the Galápagos at the helm of a ship backed by British whaling and sealing company Enderby & Sons. After a long career of Pacific exploration—Colnett had learned the seas under Captain James Cook and later played an integral role in the disintegration of Spanish claims on the Pacific—he returned to scout locations that would support British commercial development.[9] As he put it in the lengthy title of his 1798 travel narrative, he embarked on a *Voyage to the South Atlantic and Round Cape Horn into the Pacific Ocean, For the Purpose of Extending the Spermaceti Whale Fisheries and other Objects of Commerce, by Ascertaining the Ports, Bays, Harbours, and Anchoring Births, in Certain Islands and Coasts in those Seas at which the Ships of the British Merchants Might be Refitted*. His forthright title betrays the spirit of the age: one of unabashed exploration to support imperial exploitation of the seas' resources.

In the Galápagos, Colnett found just the kind of place he was searching for. Most early accounts emphasized the stark inhospitality of islands with little more to offer than giant reptiles. But Colnett saw riches in the archipelago and urged British navigators to pay attention. Off the westernmost islands of

Fernandina (Narborough) and Isabela (Albemarle) he had found no less a treasure than the calving grounds of sperm whales—the most desirable commercial species because of their wax and oil—which migrated to the islands from coasts along the eastern Pacific. Near this rich whaling ground was an island that was "in every respect, calculated for refreshment or relief for crews after a long and tedious voyage." It had wood to repair ships, a spring of water, and easy anchorages. It also offered food—the ever-essential provision for lengthy journeys. Colnett found the "best fishing grounds" but noted that "all the luxuries of the sea, yielded to that which the island afforded us in the land tortoise."[10]

In an age when travelers braved unfamiliar waters, accounts of previous voyagers served as crucial guides. To accompany his travel narrative, Colnett made a chart of the islands that FitzRoy later used as he navigated the archipelago. Colnett named the islands after British royalty and patrons, as was custom, and also identified useful places for finding water, which helped future travelers know where to stop. Travel writing like Colnett's is akin to the Lonely Planet and Frommer's guides that tourists consult today to plan their trips. Eighteenth- and nineteenth-century sailors depended on their own contemporary guides—maritime charts and the narrative maps of ships' logs and sailors' tales—in a similar way. They used them for "getting on," in the words of geographer Paul Carter, helping them find safe places to rest, repair their ships, and refresh their stores.[11] Narratives like Colnett's quite literally put the Galápagos Islands on the map.

Yet travel writing—whether centuries ago or today—does not simply help travelers plan where to go and how to get there; in a more fundamental way, it shapes our expectations of the places we encounter and our understanding of our own place in the world. Today, guidebooks and travel writing figure the Galápagos archipelago as a Darwinian paradise, a place where visitors from around the world are welcomed to see the animals who inspired Darwin and learn about the history of evolution. But earlier narratives described an almost unknown land: a place that, for Europeans journeying to the Americas, had no history other than the accounts of previous travelers. As these Europeans described the places they visited, they brought these new lands into historical being, creating what Carter calls their "spatial history."[12] That is, voyagers did not merely describe places that were already there; rather, by telling their travel stories, they made unknown places come alive in ways that would shape how readers, and future travelers, understood new worlds and their meaning.[13] Stories of the abundance of tropical islands created a fantasy world ripe for human consumption during an age when nature was widely understood in Western culture as a God-given resource for supporting human life. During a time in which the boundaries of the known world were continually expanding, travel writing made voyagers feel at home in what Mary Louise Pratt has called the "planetary project" of Enlightenment imperialism.[14] Indeed, the narratives of travelers who visited the Galápagos before the *Beagle* reflect an ethos of imperial expansion based on the exploitation of nature. They shaped geographical imaginations of the archipelago as a way station that would remain dominant for almost a century after Darwin visited there.

These early guides for "getting on" can also help us understand the context of the *Beagle*'s visit and how it was that Darwin found his way to the islands.

Fixing Enchanted Islands

Darwin's weeklong visit to Santiago, where he met the tortoise hunters, came about because of a common trouble in the Galápagos: the difficulty of finding freshwater. About a month into their five-week stay in the archipelago, FitzRoy sailed into a bay now called Buccaneer Cove on Santiago. Today, cruise ships stop here for a kayaking excursion or panga ride below the cliffs that line this protected harbor, but the site got its name from the broken liquor bottles and wooden seats Colnett had found—evidence that pirates had once frequented the beach. FitzRoy, though, was looking for water—Colnett's chart identified the place as Freshwater Valley. But when the *Beagle* arrived, Darwin reported that they found only a "miserable little spring of water."[15] FitzRoy would have to return to another island to water the ship. As he did, he left the perpetually seasick Darwin behind on Santiago to explore. Darwin was a rather weary passenger who took advantage of the *Beagle*'s stops to steady himself on solid ground.

This story illustrates the questionable reliability of early travel writing as well as the climatic variability that shaped the availability of water on these equatorial islands. Colnett had claimed that this island was an ideal place for restocking a ship, but he himself had found only a tiny spring at Freshwater Valley. He wrote that by the time he visited, the buccaneers' watering place "was entirely dried up, and there was only found a small rivulet between two hills running into the sea."[16] Some twenty years before the *Beagle* arrived, US captain David Porter also struggled to find water on the island. Porter sharply criticized Colnett's narrative, asking, "Where is the advantage of James' Island furnishing fresh water 'sufficient to supply a small ship,' if we are ignorant where it is to be found?" He ranted on, decrying Colnett's errors and the importance of good directions for men whose lives depended on them.

Indeed, it was the need for superior directions, rather than Darwin's interest in natural history, that took the Beagle to the Galápagos. Providing a better chart of the islands was FitzRoy's mission in the archipelago. At a time of expanding commercial interest in the eastern Pacific, when British whalers regularly stopped in the Galápagos Islands, the British crown had scant reliable knowledge of their position to assist its merchant fleet. After a quick visit to the islands in 1824, another British naturalist, Basil Hall, regretted not having time to make a proper survey, "a service much required, since few, if any, of them, are yet properly laid down on our charts."[17] How could merchants take advantage of safe harbor and the islands' resources if they did not well know where those resources were? "It is to be regretted," Hall continued in his 1824 travelogue, "that the true geographical position of these islands is still uncertain, and the hydrographical knowledge respecting them so exceedingly scanty."[18]

Despite the cartographic efforts of Colnett and other early visitors, the few previous charts of the Galápagos showed landmasses of inconsistent shape and location. Before ships started carrying chronometers in the late eighteenth century—giving surveyors a reliable method for finding their longitude—islands were notorious for never being quite in the same place.[19] This problem was confounded in the Galápagos by the irregularity of the winds—a feature of the archipelago that captured the attention of another famous visitor, Herman Melville, who arrived there as a sailor on a whaler from Massachusetts in 1841. "Nowhere is the wind so light, baffling, and every way unreliable, and so given to perplexing calms as at the Encantadas," Melville wrote in a series of fictional short sketches about the Galápagos. He called them Las Encantadas—enchanted—adopting a name used by early Spanish sailors. The capricious winds and tides that encircled the islands meant navigators produced inconsistent reckonings, which, Melville thought, might explain why early charts of the Pacific—including Gerardus Mercator's 1569 world map and Abraham Ortelius's 1570 atlas—often showed two sets of "Galapegos" (of various spellings) near the equator. The mysterious winds also accounted for the archipelago's alternate name: "This apparent fleetingness and unreality of the locality of the isles," Melville wrote, "was most probably one reason for the Spaniards calling them the Encantada, or Enchanted Group."[20]

The Beagle, though, was well equipped to fix the location of these enchanted isles, since it was outfitted as a hydrographical surveying ship with all the latest chronometrical tools. FitzRoy's mission was not to ferry an enterprising young naturalist on a journey of natural history collection. It was geopolitical. In the period of South American independence and British maritime expansion, improved cartographic knowledge was meant to further British industry and trade relations with nations newly independent from Spanish colonial control. FitzRoy had been instructed by the British Admiralty to visit the archipelago "if the season permits" after completing a survey of the southern coastline of South America, particularly fixing the location of Rio de Janeiro and charting Cape Horn to provide a safe route to the riches of the Pacific.[21]

Darwin was largely along for the ride—he secured his berth on this imperial mission not as its official naturalist, but as a gentleman to keep FitzRoy company for the long voyage. In fact, Darwin was not even FitzRoy's first choice but was recommended by a professor just as he graduated from Cambridge. He was a good candidate because he was fiercely eager to travel, having fallen for the romance of Alexander von Humboldt's tales, and would be able to pay his own way—after he persuaded his father to let him go. Darwin was himself something of a tourist, enjoying the privilege of his family's wealth and his nation's imperial might.

These details about the purpose of the *Beagle*'s voyage are important reminders that Darwin's theory of evolution emerged—like much natural history knowledge—from histories of imperial exploration aimed at commercial development of the world's natural resources.[22] When tourists follow Darwin to the Galápagos Islands today, they travel within the legacy of imperial exploration.

Island Encounters

Today, one of the many Darwin statues in the Galápagos—indeed, the largest one—towers over the bay on San Cristóbal (Chatham) Island, where the *Beagle* crew first anchored. This statue depicts a young Darwin clutching his Galápagos notebook and flanked by a giant tortoise, sea lion, and iguana. Tourists can make a short trek from town to have their pictures taken at his feet—this Darwin is so much larger than life that adults barely reach his thighs. The location of the monument speaks to first impressions, of the enduring myth that Darwin was struck by inspiration in the islands—that in this notebook, he would have first recorded the secret of evolution, scribbled down as he encountered the tortoises, finches, and mockingbirds.

When tourists go to the Galápagos today, this is the experience many want to relive. Even if our own intellects do not permit the jolt of understanding we imagine Darwin to have had, it *is* easy to get close to animals in the Galápagos;

visitors since Berlanga have been awed by their now-legendary indifference to human presence. Tourists can creep up to piles of hundreds of marine iguanas sunning themselves, tiptoe along paths where blue-footed boobies nest, study the wrinkles of giant tortoises, and swim alongside sea lions and sea turtles. Travel writers gush about losing themselves in a romanticized world of wild nature. In 1995, one tourist described his first outing: "every time our little group of 10 came upon a bird or a sea lion, we'd all gather around as if it were the last living specimen in the world." This was probably an overreaction, he conceded, but an understandable one: "after all, this was not just a visit to an exotic locale, but a communing with nature on a level we'd never experienced before."[23]

It is tempting to think that Darwin would have understood his own experiences in much the same way—that his theory of evolution emerged from a communion with nature in the Galápagos. But Darwin did not have an epiphany there. It was many months after his visit before he began to piece together the archipelago's evolutionary significance. Buying into the myth of a flash of inspiration means misunderstanding the slow process through which Darwin came to believe in evolution back in England, as he and other naturalists worked over the specimens he collected during his voyage.[24] It also means missing a nuanced understanding of how Darwin and his contemporaries would have seen the animals they encountered in the islands.

Today, a short hike from the bay where the *Beagle* crew first landed is the town of Puerto Baquerizo Moreno. Not far from the waterfront is an intersection where another Darwin Avenue crosses a street named after Herman Melville. The roads head in opposite directions—appropriate signposting for these two famous Galápagos visitors who are often said to have had diverging interpretations of the islands. Darwin supposedly found them an enchanting evolutionary wonderland, as we know them today, but Melville interpreted their enchantment much more darkly.[25] For Melville, who visited six years after Darwin, the tortoises were the dominant inhabitants of a "fallen world" that was striking for its "emphatic uninhabitableness"; he called the islands "evilly enchanted ground," where "the chief sound of life . . . is the hiss" of reptiles.[26] The tortoises, with their heavy, dented, even moss-covered shells, appeared self-condemned: "Lasting sorrow and penal hopelessness are in no animal form so suppliantly expressed as in theirs; while the thought of their wonderful longevity does not fail to enhance the impression." Melville told of a common superstition sailors held about the tortoises—that they were the reincarnated forms of cruel ship captains destined to spend eternity in this hot, dusty purgatory where they would continue their reign as "sole solitary lords of asphaltum."[27] Melville's "sardonic travel narrative" showcases some of his most vivid descriptions, presenting a haunting image of the islands and dark glimpse into the mystical worldview of sailors.[28]

It is appropriate, too, that Melville and Darwin Avenues intersect on San Cristóbal, for Darwin's first accounts of landing on the island have more in common with the world Melville described than the one we now associate with Darwin. When Darwin first came ashore, he was dismayed by the island's desolation. "Nothing," he wrote, "could be less inviting on first appearance"; the air felt like a stove, and even the "bushes smelt unpleasantly."[29] Darwin, who so loved Milton's *Paradise Lost,* saw the islands somewhat like Melville did: not as "a Tartarus of clinkers," as the novelist put it, but as "what we might imagine the cultivated parts of the Infernal regions to be." The island was dry, barren, and hot—although a "paradise . . . for the whole family of Reptiles," Darwin noted.[30] When he encountered his first giant tortoises, he did not think them evil reincarnations, but did think them well-suited to the harsh landscape. This was not because of any evolutionary fitness, but because both land and reptile seemed surreal: "Surrounded by the black Lava, the leafless shrubs & large Cacti, they appeared most old-fashioned antediluvian animals or rather inhabitants of some other planet."[31]

Although Darwin most loved the verdure of lush tropical landscapes, the otherworldly Galápagos volcanoes also appealed to him. Indeed, though we remember him today for his contributions to biology, when he was young he was primarily interested in "geologizing."[32] He was eager to reach the Galápagos so that he could indulge this passion in a dynamic volcanic landscape—islands made, as Melville put it, from "heaps of cinders."[33] On the *Beagle,* Darwin had been reading English naval captain George Byron's account of the Galápagos as a place where volcanoes were "burning around us on either hand," flows of lava making the surrounding sea boil. On Fernandina Island, Byron reported that "as far as the eye could reach we saw nothing but rough fields of lava" like hardened waves. Life amidst these lava flows was "like a new creation": "the birds and beasts do not get out of our way; the pelicans and sea-lions look in our faces as if we had no right to intrude on their solitude; the small birds are so tame that they hop upon our feet. . . . Altogether it is as wild and desolate a scene as imagination can picture."[34] Even before Darwin stepped onto one of the islands, their landscapes spoke of origins.

Darwin made careful geological observations in the Galápagos, particularly as he hiked crater-pocked Santiago with the tortoise hunters. But the animals he found living among the islands' craters and jagged cliffs proved charming enough to pull him from hammering away at the ground. On a rocky point, he found "most disgusting, clumsy Lizards" as black as the lava they crawled on. "Somebody calls them 'imps of darkness,'" Darwin recalled from his shipboard reading; "they assuredly well become the land they inhabit."[35] Darwin was intrigued by these creatures, which fed in the sea but otherwise spent their time on land. He pitched one repeatedly into the water, only to have it climb back to his feet each time: "though possessed of such perfect powers of diving and swimming, nothing would induce it to enter the water."[36] Darwin was perplexed by this behavior, but unlike Melville, who might have found it further evidence of an enchanted purgatory—Darwin surmised that the seaweed-eater must have fewer predators on shore than in the sea.[37]

Darwin was a curious and dedicated observer of the natural world—characteristics that would allow him to eventually piece together evidence of evolution—but his travel stories also reveal him to be like many explorer-naturalists who were awed by the unusual creatures they encountered far from home. Explorers often struggled to conjure a picture for readers of odd animals—like platypuses and kangaroos—that stretched, even violated, the known order of the natural world.[38] In the Galápagos, the oddities were mainly reptilian. In 1712, English privateer Woodes Rogers described the tortoises by comparing them to more familiar creatures: they had feet like elephants and a long neck and small head like a snake—a combination he thought made them "the ugliest in Nature."[39] First encounters with such strange animals could be startling—would they be aggressive? dangerous? As sailors walked unknown lands, they did not know what they would find and often played up the sense of adventure in their writing. When US sealer Amasa Delano first saw a giant tortoise, the animal stretched out its snakelike neck, mouth wide as if to attack. Delano readied his musket but allowed the animal to advance close enough that he could lean forward and reach it with the tip of his gun. When he did, he realized a gun was superfluous with these slow, shy beasts: the tortoise "dropt himself upon the ground and instantly secured all his limbs within his shell."[40]

The tortoises' propensity to withdraw made it easy for sailors to climb on their backs—the animals, with their built-in saddles, seemed made to be ridden. To express the immense size of these homely creatures, travelers judged the tortoises' girth and weight by how difficult it was to lift them and how many men the beasts could carry. The animals Tomás de Berlanga saw in 1535 were so large that "each could carry a man on top of himself."[41] More than a century later, Rogers reported a huge tortoise who lumbered along easily with two men on its back. It must, they thought, have weighed at least seven hundred pounds. For Rogers, the great size of some of the tortoises resembled another common form of transportation of the day: he compared their shells to the "Top of an old Hackney Coach, as black as Jet, and so is the outside Skin, but shrivel'd and very rough."[42] Whether saddle or coach, the imagery well captures the time's dominant understanding of the relationship between human and beast—beings meant to support human life. Indeed, the tortoises supported a way of life that made Pacific industry possible.

A "wild nightmare" Melville had after his first encounter with three giant tortoises that had been brought aboard the *Acushnet* makes this clear.[43] He wrote of losing himself in "volcanic mazes, brush[ing] away endless boughs of rotting thickets," until he found himself sitting on a tripod of tortoises that "seemed newly crawled forth from beneath the foundations of the world."[44] They were upholding "the universal cope," he wrote. Yet this stack of tortoises also upheld the ship-borne industry Melville would chronicle in *Moby Dick*. The next evening, his dream state cleared, Melville sat down to a dinner of tortoise steaks and stews with his shipmates. After eating his fill, he pulled out his knife and helped make soup tureens from the leftover carapaces and "gorgeous salvers" from the flat breastplates.[45] Although Melville may have dreamed of meeting the very tortoises who held the world on their backs, his consumption shows that in a more material sense the animals also upheld the strength of thousands of sailors, nourishing them during an era of shipboard merchant capitalism.

In the Bellies of Whalers

Nearly all the cruises that circle the western edge of the Galápagos today stop at Tagus Cove, the safest natural harbor on the coast of Isabela. The *Beagle* stopped here too; while FitzRoy's survey crews took their measurements, Darwin went ashore and hiked up a volcano that now bears his name. From the top, he looked down into the cove, which was made from the volcano's partially submerged caldera, a geological formation that delighted him. Today, visitors are often more interested in a wall of graffiti where for centuries sailors carved the names of their ships into the cliff. Much of the cliff is now covered with the white paint of graffiti tags from the 1970s and 1980s; but behind the paint, you can still make out the worn engraving of dates from the mid-1800s, when hundreds of ships stopped here because this was the best place from which to hunt whales. The cove itself is named for a British warship that visited in 1814. Like the post office, this tourist stop is a tangible re-minder of the political economy that once enmeshed the archipelago.

By the time Darwin and Melville visited, the Galápagos had become the supply station Colnett had envisioned nearly fifty years before.[46] Whaling was in its golden age, dominated by the United States, whose peak fleet included more than seven hundred ships—the overwhelming majority of a total global whaling fleet of nine hundred. Tens of thousands of men—and a few dis-guised women—crewed these ships and killed hundreds of thousands of sperm, right, gray, and humpback whales across the world's oceans. Whales, as one historian has written, were "swimming profit centers," and whale ships were floating factories for rendering leviathans into strips of baleen that struc-tured the corseted waists and full skirts of women's fashions, waxy spermaceti that made clean-burning candles, ambergris used as a base for perfumes, and tens of thousands of barrels of oil that lubricated the Industrial Revolution.[47] In 1853, the industry's most profitable year, whalers sold more than 100,000 barrels of sperm oil, 260,000 barrels of whale oil, and 5.7 million pounds of baleen—for some $11 million. The industry was a backbone of the New En-gland economy, a "source of national wealth," in the words of one senator.[48]

Keeping this fount of wealth flowing required sustaining the thousands of sailors who worked whale ships and commonly spent three or four years at a time in the Pacific. After rounding Cape Horn, whalers sailed north along the coast of South America, often stopping in Chilean and Peruvian ports to re-plenish supplies, depending on the era's shifting political alliances. Like Fitz-Roy, whaling captains also relied on travel narratives from earlier sailors, which guided them to the Galápagos. There, sailors collected water, chopped wood to repair their vessels, traded with local colonists for vegetables, and, above all, hunted tortoises. After months of standard sailors' fare—which, as historian Nancy Shoemaker has written, included provisions that were often "scant, rotten, and worm-riddled" and relied heavily on salt pork and beef, various carbohydrates, molasses, and coffee—sailors needed fresh meat.[49] Perhaps the most obvious source during an era of industrial whaling would have been whale; but although US whalers did eat some fresh whale meat, they tossed most of it overboard with the carcass after cutting out blubber and baleen.[50] They did so because they had other options—ships often carried poultry, pigs, goats, and even cattle on board. Sailors fished when they could and salted catch to preserve it. They also sampled all kinds of exotic animals they encountered. In the Galápagos, US naval captain David Porter recalled clubbing hundreds of "guanas," which despite being of "the most hideous appearance imaginable" were "excellent eating, and many preferred them greatly to the [sea] turtle."[51]

But owing to somewhat mysterious cultural histories of taste, these meats were never esteemed as widely as that of giant tortoises, whose praise was an often-repeated refrain in sailors' narratives. "The finest green turtle is no more to be compared to them, in point of excellence," wrote Porter, who helmed the USS *Essex* during the War of 1812, "than the coarsest beef is to the finest veal, and after once tasting the Gallapagos tortoise, every other animal food fell greatly in our estimation." Not only was tortoise tasty, but Porter found the meat to be "the easiest of digestion," reporting that "a quantity of it, exceeding that of any other food, can be eaten without experiencing the slightest inconvenience."[52] Tortoise was meat meant for a feast.

Sailors held Galápagos tortoise meat in the highest esteem because of its taste, but they likely found tortoise and turtle so delicious in part because they provided a break from the monotony of shipboard fare and because they provided much-needed nutrients. Fresh meat was essential for keeping crews healthy. Tortoises were "excellent food," wrote one captain, "and have no doubt saved the lives of thousands of seamen employed in the whale-fishery in those seas, both American and Englishmen."[53] A host of diseases afflicted sailors on crowded ships; typhus, malaria, tetanus, dysentery, tuberculosis, influenza, and pneumonia were all common. But the most widespread malady on long voyages was scurvy, which caused fatigue, sore and bleeding gums, and, for many, death. At the time, sailors did not know scurvy was caused by a lack of vitamin C—Porter thought the disease was brought on by harsh weather, changes in climate, bad provisions, and unclean water. But a good captain, as Porter thought himself, did know how to avoid the disease. He knew a good diet was essential for health and insisted in having on board fresh limes, oranges, lemons, vegetables, fish, and the best water he could get. Unlike most captains of the time, Porter bragged that his men "were never affected by the scurvy."[54]

To keep up their stores of essential provisions, Porter and other captains took advantage of the Galápagos Islands' strategic location as a whaling grounds and supply station. Porter was more aggressive than most—he made himself into one of the few American heroes of the War of 1812 by chasing enemy British ships. He excelled at this, capturing twelve ships from which he stockpiled not only prestige and prodigious quantities of whale oil he planned to sell, but also "prizes" of the "greatest importance": "cordage, canvas, paints, tar," as well as dry food stores and tortoises other ships had recently collected.[55]

Ship captains often left behind their own livestock—goats, sheep, cattle, and pigs—to propagate on the islands and stock them for future hunting. By the 1830s, ships could also call on the small colony of political exiles on Floreana (Charles) Island and trade goods, and liquor, for fresh vegetables and meat. FitzRoy did this, feasting with the islands' acting vice governor, Nicholas Lawson, who showed off his colony's plantations of bananas, sugarcane, maize, and sweet potatoes.[56] Lawson also sold some dried tortoise meat and oil to passing ships; indeed, he had sent the tortoise hunters Darwin hired to be his guides on Santiago partially for this reason.[57]

By the mid-nineteenth century, the islands had become not only profitable grounds for whaling and the collection of fur seal pelts, but also a center of subsistence that fed thousands of sailors.[58] These men relied on colonists who provided them with vegetables, and both groups hunted giant tortoises to sustain them. That the *Beagle* was an integral part of this set of commercial and political exchanges is a story that is not often told on island tours today.

TURPININ'

When the *Beagle* men first went ashore in the Galápagos, on San Cristóbal, they found a handbarrow that had been left at the base of a trail. It was evidence of previous visitors and also of the presence of giant tortoises. On that first outing, FitzRoy's crew returned to the ship without having found any tortoises to haul, but they would later bring at least forty-five on board—not as natural history specimens, but as food.[59] Like many ships before and after, the *Beagle* was well stocked with "turpin," as sailors called the tortoises.[60] Darwin would have eaten tortoises not just while camping on Santiago, but as the *Beagle* crossed the Pacific as well. It is tempting to look back on Darwin's experience hunting and eating tortoises with the exiles on Santiago as a personal anomaly—the forgivable actions of a curious young naturalist. But Darwin's experience was not an anomaly; he walked among men whose practices conservationists now condemn.

Darwin was a gentleman, however, not a laboring sailor who would have been sent to shore to hunt. He paid the tortoise hunters on Santiago to be his guides and feed him. The *Beagle* sailors who stocked the ship with tortoises followed in the footsteps of hundreds of others for whom "turpinin'" was a day's work—and a rite of passage. It was a storied enough experience to have been the subject of an account published in a sailors' newspaper in 1847.[61] The author, identified only as Camillas, was not a typical sailor, but a writer who tagged along to record his experience. His story, supplemented with oral histories from Nantucket whalers collected by an early conservationist in 1924, shows how Darwin's contemporaries, and the *Beagle*'s crew, would have experienced the Galápagos and understood the nature of their most famous inhabitants.

The labor of tortoise hunting—and it was, by all accounts, difficult labor—shaped how sailors understood both the animals and the islands. A tortoise hunt started as men pulled skiffs they had rowed to shore up onto a beach. From there, they headed inland along paths first worn by tortoises. Not all of them were broad trails through open land like the one Darwin hiked on Santiago. On San Cristóbal, Camillas and his hiking partner, the "old man" Capt. A, followed a path into dense thickets of wild cotton and other bushes with briars that scraped their hands and ankles. As they marched on, the dusty path turned to shards of broken lava rocks that pierced their shoes. After a day of trekking through a landscape better suited to armored animals a third of their height, hunters returned exhausted, their clothes torn, shoes worn through, and hands streaked with bloody scratches.

After hours of searching, Camillas and Capt. A found several tortoises near the rim of a mile-wide crater, nestled comfortably under a cactus tree, where they imagined the animals might have lain for the previous five hundred years. Noting how the tortoises fed on cactus pads that had fallen in front of them, Camillas saw the work of a divine creator who had provided just enough for the tortoises' survival: "how wonderful are the arrangements of Him who provides feed but few facilities for obtaining food, neither being able to climb trees or to kill animals, so slow is their motion." It was a common way of understanding the world in an age of natural theology—the islands were not created with human inhabitants in mind, but instead were well suited for giant reptiles.[62]

These old tortoises, though, were too big to carry, so the men hiked farther until they found two of just fifty pounds. Smaller tortoises were said to taste better than large ones, and they were a feasible weight for "backing down" to the shore, as sailors called the work of hiking with a tortoise rigged up like a backpack.[63] "With two bits of canvass, as straps," Camillas wrote, "we fastened their legs and placed them on our backs, precisely as a soldier does his knapsack, the strings passing under our arms." Other sailors did take larger tortoises, using either handbarrows like the one FitzRoy found, or lashing tortoises to poles they brought ashore—one man carrying each end of a pole with a tortoise in between, as if for roasting on a spit.[64] One sailor even recalled a boy trying to ride an exceptionally large tortoise to shore—a tedious process on an unbroken animal.[65]

With their tortoises secured on their backs, Camillas and Capt. A stumbled back along the path, soon losing their way in a thicket under blazing afternoon sun. Swatting at a swarm of flies that had been biting at their lips since they came ashore, they stopped for swigs of water and caught glimpses of their ship still miles off. Eventually, men on their second descent to the ship found them, gave them a keg of water, and helped with their tortoises as they made their way back to the shore.

Not all turpin hunters were so fortunate as to have someone bring them extra water. Losing sailors forever in such thickets was not unheard of (it continues today with unfortunate frequency among tourists who hike alone)—at least one sailor reported others perishing for lack of water and recalled his own shipmates refreshing themselves by drinking the blood of tortoises and water the animals store internally like camels.[66] This water, Porter reported, was "perfectly fresh and sweet."[67] Darwin drank from the tortoises too, out of experiment rather than necessity, and found their fluid "only slightly bitter—The water in the Pericardium," he noted, "is described as being more limpid & pure."[68] On dry islands where freshwater was difficult to find, the tortoises not only provided meat and fat, but also water to help sustain human life—they were, as one US whaling captain reported, "the living spring of this wilderness." He boasted that one huge tortoise his men decapitated would feed a hundred men and carried three gallons of water that, "wonders of wonders! . . . was cool." These tortoises, he wrote, were "Nature's own water-bottle. Such is one of God's providences for man in dry places."[69] His delight in the affordances of giant tortoises demonstrates what was a common understanding of nature as supplies sent by God to rescue men from a barren landscape. After a difficult hike, parched men could rely on these animals to sustain them—a way of knowing, and valuing, nature conditioned not through a naturalist's sensibility, but through the labor of the hunt.

Today we do not think of following Darwin as also following the tortoise hunters who led him around Santiago. Indeed, most of the tortoise trails that hunters once followed are now closed to tourists. But some tours have featured an environmentalist version of a tortoise hunt. Until the mid-1990s, tourists could hike to the rim of Volcán Alcedo on Isabela Island. One Berkeley tourist described the experience of visiting the giant tortoises who live at the rim of the volcano.[70] She let on no awareness of the thousands who hunted tortoises before her, but her description echoes their experiences. She too wrote of the ardor of hiking the "brutal" trails up the side of the volcano—indeed, she probably hiked farther than most nineteenth-century hunters, who did not often visit this volcano where the tortoises live far from shore. Back then they could find the animals more easily elsewhere. She also decried the difficulty of carrying a backpack full of provisions—although for her this was not a tortoise slung on her shoulders, of course, but food and water she hauled *up* the volcano. This change in provisioning betrays the most striking shift in visitors' ways of interacting with the islands—tourists no longer rely on local resources for food.

Her other experiences, though, resonate with the past. After five hours of hiking, her tour group, covered in grime, reached the rim of the caldera to set up tents in a "campsite from hell" where they were greeted by hundreds of chiggers. Finding the giant tortoises, though, was worth the discomfort. After a "meager" dinner, the group slept and then set out for a hike the following morning as a fogfall spilled into the volcano's caldera. They followed tortoise trails around the rim and down into the caldera. Along the way, they spotted 258 tortoises (eventually getting too tired to keep count). The animals plodded along the trails with them, stopping to eat leaves and grass or wallow in the mud. Like Camillas, this author also imagined that the tortoises had been alone in this quiet scene for quite some time; they were unaccustomed to human visitors, and when the hikers would get close, they would retract into their shells, exhaling with a hiss.

The sounds of reptile life that had spooked Melville were interrupted by the brays of donkeys galloping across the caldera ridge. These animals were an unwelcome intrusion. They threatened the survival of tortoises by trampling nests and crushing eggs. Other species—feral pigs, dogs, goats, and rats—competed with tortoises for food, or even preyed on juveniles. These animals had all been introduced, intentionally or not, by sailors or colonists. Today, conservationists see these animals not as useful food stores and laborers, but, like rats, as scourges on the landscape that necessitate control.

After an exhausting two days, the tour group hobbled back down the volcano with lighter packs. Back aboard their cruise ship, they settled in for a feast, much of which would have been imported from the continent. Days later, at home in California, "where the refrigerator is full and the water flows from gleaming chrome," the author began to make sense of her experience. She imagined the tortoises as gentle animals in a romantic scene, moving slowly along the rim of the caldera, surrounded by mist and illuminated by moonlight. The volcano, she concluded, echoing Melville and other early visitors, was unfit for human life—it was a "hostile place for mankind, and we have no business there, except for short visits to see these wondrous animals." Where sailors once made brief visits to collect tortoises they saw as resources designed for their consumption, she, like many tourists, took an opposite message—that these animals should be protected, and other than being admired by visitors, left alone to live in their remote world. Although the labor of a hike in pursuit of tortoises is much the same as it was for Darwin's contemporaries, the meaning attached to knowing nature in this way has changed dramatically since the days when visitors counted not how many wild tortoises they saw, but the number of animals added to ships' stores.

The Takeaway

In Darwin's day, crews filled their ships with tortoises after a turpin hunt—many taking far more than the *Beagle* did. Camillas's ship collected 37 animals the day he went out, for a total of 130 over four days.[71] Porter reported taking fourteen tons worth of tortoises—at least four hundred animals.[72] Crews stowed them wherever they would fit—on deck or below, stacked on their backs among the casks. Today tourists can get an image of what this might have looked like at the visitor interpretation center on San Cristóbal, where the *Beagle* first anchored. The center displays a re-creation of a ship's storage hull stacked with wooden barrels and tortoises. The scene allows tourists to peer back into a different time when the tortoises were not exemplars of evolution to be respected and conserved, but, as one sailor put it, an "almost invaluable" resource.[73]

Packed away in ships' hulls, rendered into barrels of oil, used as soup tureens and serving platters, giant tortoises were crucial provisions on hundreds of ships as they crisscrossed the Pacific. After the difficult work of hunting, sailors were rewarded with rich meals. "No animal," Porter wrote, "can possibly afford a more wholesome, luscious, and delicate food than they do."[74] Sailors ate boiled and roasted tortoise meat, considered fried terrapin liver a delicacy, and made them into a stew called "sea pie." Camillas and his shipmates lived on meals of terrapin soup three times a day, every other day for four months. These huge animals also provided plenty of fat for frying their own meat and doughboys and for adding flavor to dry ships' biscuits. Their fat was delicious—better than olive oil one sailor reported, and "as yellow as our best butter, and of a sweeter flavour than hog's lard" another said.[75]

In the days before refrigeration, the tortoises embodied an ideal, low-maintenance way to transport fresh meat and fat. Numerous accounts echoed the animals' most useful quality: they could stay alive for months—some said

six, others up to eighteen—without food or water, and apparently not dimin-
ish in tastiness. Some sailors did feed them—scraps of bananas and potato
sprouts and peels. US sealer Amasa Delano gave them more "suitable" grass
and prickly pear cactus pads he collected on the islands, which would draw
them from all corners of the ship's deck. They learned to eat on board "as well
as any domestic animal."[76] But unlike domestic animals, tortoises did not
compete with sailors for limited stores of freshwater. Their hardy longevity
was a life-saving gift—truly a feat of nature that was celebrated in many awe-
flecked and grateful narratives.

Today, of course, we look back on this history with chagrin. We no longer
read records of how many tortoises were taken as a measure of stores to be celebrated. Ship captains once carefully recorded those numbers in ships' logs to
keep an accurate count of provisions. Read from a contemporary perspective,
these logs provide records of the whaling industry's ecological impact. In the
1920s, decades after Pacific whaling declined following the outbreak of the US
Civil War and the development of the petroleum industry in the late nineteenth
century, the director of the New York Aquarium, Charles Haskin Townsend,
pored over old whaling logs in New England and collected oral histories from
aging sailors. He published a study in 1924 that estimated the cumulative effect
of whalers' tortoise hunting. From seventy-nine ships' logs, he found that whalers had made 189 visits to the Galápagos between 1831 and 1868 and recorded
taking no fewer than 13,013 tortoises. If just under eighty ships of a total fleet of
more than seven hundred had taken that many tortoises, he reasoned, then the
total sum must have been much higher—more than 100,000 animals since
1830.[77] Conservationists today have expanded on Townsend's study, considering
not just the take of US whalers, but also that of earlier explorers, natural history
collectors who came after Darwin, and island settlers. Ecologists now believe
that as many as 200,000 tortoises were taken over the past four centuries—
leaving just 20,000 tortoises.[78]

During the twentieth century, these take records became a core justifica-
tion for conservation efforts aimed to save tortoise species after centuries of
consumption. We tend to credit Darwin with inspiring these efforts by enlightening our understanding of nature. Yet he had little to say about tortoise
hunting besides his preference for soup over steaks. FitzRoy, however, was
more concerned. When settlers in the Floreana (Charles) Island colony hunted
tortoises, FitzRoy noted, they killed them where they found them, stuffing
useful meat and fat in a bag to carry back to the settlement. This method left
evidence on the landscape of tortoise kills. "The quantity of tortoise shells ly-

ing about the ground," FitzRoy wrote, "shows what havoc has been made among these helpless animals."[79] Sailors who took tortoises on board effectively erased the evidence of their take, throwing tortoise remains into inky seas. Vice Governor Nicholas Lawson was concerned as well—not for the animals' sake, but for the state of his provisions. He thought there were enough tortoises left on the island to last twenty years, but he also sent the tortoise hunters to Santiago to augment his stores.[80] As it turned out, Lawson and FitzRoy were right to be concerned—the Floreana tortoises were among the first island populations to become extinct.

Today, tourists learn about the effects of past tortoise hunting when they stop at one of the archipelago's tortoise breeding centers. Most Galápagos tours make their final stop at the center on Santa Cruz, between the headquarters of the national park and the Charles Darwin Research Station. Several dozen adult tortoises live there—survivors from the age of hunting that drove four island populations to extinction and decimated others. Tour guides explain that these tortoises are kept in captivity in an effort to restore populations diminished over hundreds of years of hunting. Visiting is an experience that encapsulates the mourning for a past world of untouched nature that is at the heart of nature tourism.[81] The chance to see tortoises up close captivates tourists, but the message of the visit can be depressing. Tour guides do their best to buoy tourists' spirits by showing them the tiny baby tortoises born in captivity, and by telling stories about the great lengths to which conservationists have gone to encourage the adult tortoises to mate.

Tourists generally spend the hour at the breeding center with their faces behind their cameras, listening to the guides while tinkering with their lenses to be sure to capture the tortoises in the only way permissible today—in photographs. Tourist interactions with the tortoises are mediated by the distance the camera puts between subject and object.[82] At the breeding center, waist-high corral walls also separate people from the protected animals—gone are the days when visitors could reach out and touch the animals or hop on their backs. Tourists used to be able to wander down from the boardwalk that snakes between the corrals and into some of the tortoise enclosures to get closer; however, the national park closed access in 2011 after a photograph of two tourists sitting on the back of one of the captive tortoises was published in the local newspaper, alongside an editorial criticizing the park's lack of control. National park rules—and the mandatory tour guides who enforce them—prohibit visitors from coming within about six feet of endemic animals, even when the animals are the ones making the approach.

These rules are as much about separating humans from nature as they are about separating the present Galápagos from the islands' past.[83] The meaning tourists ascribe to seeing tortoises today has changed considerably since the days of sailors' hunting. Tourists still marvel at the tortoises' great size and wrinkly skin, but the animals now convey a mournful conservationist message rather than one of gastronomical indulgence. Much as early travelers' tales about the Galápagos shaped voyagers' understanding of the islands, contemporary travel writing, as well as messages in interpretation centers, now reworks the kind of place we understand the Galápagos archipelago to be. Rather than serving as storehouses intended to sustain human life, as they did during Darwin's time, the islands are a fragile place endangered by histories of understanding nature in this very way.

What tourists now take away from the experience of hunting giant tortoises with their cameras is a moral sense of concern and mourning. Where visitors once delighted in trying to ride these animals, the same pastime now invokes pity and chagrin about the past: "As children, we saw them in zoos; maybe, God forgive us, we rode them around a pen. And something remained in the mind: Somewhere these creatures are born free, and live wild," the Berkeley tourist wrote. Going to the Galápagos is about making amends for the sins of the past. We credit Darwin with inspiring this change of perspective, and imagine that like us, he also learned in the Galápagos a new way of understanding nature. "The truth is," one guidebook put it, "that a visit to these islands does change you, just like it changed the great Charles Darwin."[84]

The Galápagos Islands might have changed Darwin, but they did not make him an environmentalist. Understandings of giant tortoises have changed enormously over the past century. These animals, once a favorite soup, are now more frequently captured in photographs or collected as plush souvenirs. These mementos carry an implicit message: they renounce the history Darwin participated in—that of taking actual tortoises and other natural life from the islands. Yet there is a risk in making too much of this change: of believing that the consumption of tortoises is a thing long since past.

Soup and Souvenirs

At the end of their tours, after an hour spent photographing tortoises in the breeding center, tourists stroll back into town along Avenida Darwin to buy souvenirs. The young Darwin, of course, never walked this road—the town did not yet exist when he visited, nor did he set foot on this island. But, like the trails he hiked on Santiago, this one is busy with tortoises. Here you'll see them on signs for tour companies, on T-shirts, as souvenir tchotchkes, even on bottles of Galápagos-branded water.

Darwin is here as well—gray-bearded Darwin, that is—watching tourists pass by from the arch near the entrance of the park. He is also on T-shirts and tote bags with slogans like "evolution revolution" that remind tourists of the transformation in understandings of nature he inspired. One summer when I visited, a coffee and frozen yogurt shop had even set up a life-size papier-mâché Darwin at a café table. You cannot join Darwin for a dinner of tortoise soup, but you can share a table and a latte with him. He is an icon emphasized by tourism companies and merchants in the Galápagos in an effort to meet the expectations of foreign visitors.

After collecting souvenirs, at the end of a week's vacation in the Galápagos, tourists file back onto planes to the continent, their cameras and smartphones full of digital photographs and videos, and their suitcases packed with trinkets, T-shirts, and stuffed animals. These remembrances symbolize the incredible experience of a week spent getting close—but not too close—to wild animals. They symbolize the moral message of learning to respect and protect nature rather than exploit it.

But we should not overemphasize the distinction between tortoises as soup and as souvenirs. When we do, we risk hiding from sight the similarities between contemporary tourism and the consumption of young Darwin's age. Remembering that the younger Darwin ate tortoise soup reminds us that he was part of an industry that rested on the tortoises' backs—an industry that, like today's tourism, brought thousands of visitors to the islands, connecting them to global economies that continue to envelop the islands, despite conservationists' attempts to ensure their isolation. By focusing so intently through their viewfinders on an imagined world of pristine nature, tourists miss the fact that they themselves are also participants in an industry that rests on the backs of tortoises. Today the animals are primarily seen not as ambulatory feasts but, as one Galápagos resident told me, as "oro," gold, valuable not for their meat, but for their role in supporting the tourism economy. Plush tortoise souvenirs—and sharks, blue-footed boobies, iguanas, and sea lions—are reminders that animal encounters are big business in the Galápagos. The cruises that continuously circle the archipelago to show tourists wild animals, and a bit of the islands' history, are the mainstay of the Galápagos economy. They drive demand for local industry based on selling hotel rooms, day trips, dinners out, and souvenirs to people who are passing through. This is the economy that comes into view when we burst the bubble of pristine nature tourism. Looking more closely at Darwin's own history in the islands shows how the seeds for contemporary tourism were sown long ago as nascent island colonies developed economies based on serving visiting sailors.

The floating army of whale and tortoise hunters of Darwin's time has been replaced by a flying consumer army of tourists who come to see Darwin's Galápagos. Darwin is credited with transforming the islands from a world enchanted because of its inhabitability to one enchanted because it allows close encounters with evolutionary nature. But he is also associated, then, with transforming the Galápagos from a busy crossroads for Pacific whalers to one of the world's most popular nature tourism destinations. Yet the conceit that the Galápagos are "Darwin's islands" is a vast overstatement of the role the archipelago played in his thinking, and a grave underplaying of the many people who have shaped the history of the islands. This then is what it means to think through the relationship between the two Darwins—to see that the way we celebrate history sometimes bears little resemblance to the past. The view of Darwin presented in the BBC documentary series does a dual disservice to our ability to understand the archipelago: it invents a false past of timeless nature and also obscures the present reality. The idea of traveling to the Galápagos to walk in Darwin's footsteps is, as Ecuadorians might say, a gringo construct—a story told by North Americans and Europeans that celebrates a familiar northern hero and downplays, if it does not erase completely, the far more complicated entanglements of humans and other species that have shaped the history of the islands. Visiting Post Office Bay, Buccaneer Cove, and Tagus Cove provides a window onto a past world of whaling and tortoise hunting, but we need to get outside the ecotourism bubble and visit other sites that will tell us about Galápagos histories that have little to do with Charles Darwin. The tortoises too provide a route into this history. To begin, if Darwin ate tortoise steaks and did not collect the animals as scientific specimens, how did they become examples of evolution?

3 • What's in a Name?

In which the tortoises become living fossils

The scutes, or plates, that make up a giant tortoise's carapace grow in rings, made of ridges that increase in number and size as the animal grows. They are similar to tree rings, although they grow so unevenly and become so worn over the decades that they are not a reliable means for measuring the passage of time. Nonetheless, the giant tortoises' shells, marked with the many scrapes and dents they acquire during a lifetime of plowing into lava boulders, as well as their deeply wrinkled, leathery skin, do make it seem like we could read history off their backs. The animals are like time machines, beings who seem to take us from our own time back to an earlier, prehuman epoch. They are exemplary "living fossils," a term Darwin coined to describe species that had survived a former age.[1] But he never used the phrase to refer to the Galápagos giants. The animals became seemingly prehistoric evidence of evolution only in the decades after Darwin's *Origin of Species*.

While Darwin was in the Galápagos Islands he missed what in retrospect was a huge clue to the puzzle of evolution he was just beginning to work through. When he visited the Floreana colony, the acting vice governor, Nicolas Lawson, told Darwin that the tortoises on each island differed and that he could "with certainty tell from which island any one was brought."[2] On small, dry islands like Pinta and Española, the tortoises are smaller and their shells curve up at the back of their necks, making their carapaces look like saddles— an adaptation to island environments that do not have much vegetative ground cover that allows the animals to stretch their necks up to reach leafy trees and juicy cactus pads. On larger, higher islands like Floreana, Isabela, and Santa Cruz, where volcanic calderas are often enshrouded in misty clouds, the tortoises are much larger when full grown and their carapaces are dome-shaped, without the upside-down *U* at the back of their necks. These animals live in places where the ground is blanketed in greenery year-round, giving them a steady food supply. In addition to these two dominant morphotypes, the tortoises also differ slightly in coloring—some are deeper black in coloring, others have yellow marking under their chins and on their throats. Lawson would not have understood these differences as evolutionary adaptations, but they were just the kind of significant details that hinted at the diversification of species.

Weeks later, as Darwin camped with the tortoise hunters on Santiago, they told him the same thing Lawson had. But, "I did not for some time pay sufficient attention to this statement," Darwin lamented back in England years later. "I never dreamed that islands, about fifty or sixty miles apart, and most of them in sight of each other, formed of precisely the same rocks, placed under a quite similar climate, rising to a nearly equal height, would have been differently tenanted."[3] When Darwin was in the Galápagos, he was not yet thinking about evolution as we now understand it. Although he made notes about the tortoises' behavior, gait, hearing, and size, he did not make collections of the animals, as he did with island birds, plants, rocks, lizards, and insects. In addition to the tortoises the *Beagle* men took as food, Darwin and FitzRoy brought aboard only four tiny living tortoises to keep as pets. This was perhaps because at the time naturalists thought the Galápagos tortoises were part of the same species of the giants that sailors had long reported finding on islands in the Indian Ocean. Shortly before the *Beagle* voyage, British taxonomist John Gray had named the animals *Testudo indica*, following the system of Latin binomial names for genus and species that eighteenth-century Swedish naturalist Carl von Linné—better known himself by his Latin name, Linnaeus—had made the basis for natural knowledge.[4] But when Darwin returned from the *Beagle* voyage, he began to doubt this name was an apt description for all the world's giant tortoises.

Back in London in late 1836, Darwin quickly distributed his collections to well-established British taxonomic experts. John Gould, curator of the Zoological Society of London, examined his birds and found that what Darwin had thought were three different types of birds were in fact all ground finches with such curiously divergent forms that they represented twelve new species. But when Darwin had labeled these birds in the Galápagos, he had not included their islands of origin, so despite this exciting news, the specimens did not make for reliable evidence of evolution. Two months later, in what has become one of the most famous meetings in the history of biology, Gould told Darwin that the mockingbirds he had collected, and labeled by island, were not just varieties but differed enough to constitute separate species—related to, but distinct from, South American mockingbirds. Darwin realized the import of this news and filled the sheet of paper he had brought to the meeting with excited scribbles. Back at home, he soon started his first notebook on "transmutation," as evolution was often called at the time.[5]

Darwin realized that the news about the finches and mockingbirds was parallel to Lawson's observation about the tortoises and asked Gray to look at the four small animals the *Beagle* men had brought home. But they were too young to show significant differentiation. Darwin did not have the evidence to confirm his hunch that the Galápagos tortoise, as he wrote in his *Journal of Researches,* was an "aboriginal inhabitant" of the islands, much less that the tortoises differed by island as the mockingbirds did.[6] It would be nearly eighty years after his Galápagos visit before naturalists proved this hypothesis.

How did the tortoises become evidence of evolution? The answer lies in the stories of how the species got their names. Today, the Galápagos giant tortoises are not considered *Testudo indica* but have Latin names that reflect their endemism—the fact that they evolved only on the different islands in the Galápagos archipelago and cannot be found elsewhere in the world. Several of the fifteen different species are named for their islands of origin—such as *Chelonoidis hoodensis* and *C. chathamensis* (Latinized versions of the English names for islands Hood [Española] and Chatham [San Cristóbal] used by scientists a century ago). Others are named for the morphological characteristics that make them unique—for example, the saddle-backed tortoises from Pinzón Island were once called *C. ephippium,* Latin for saddle, and the smallest species, from Volcán Darwin on Isabela, is *C. microphyes.*

But four of the tortoise species are named, following Linnaean tradition, for natural historians who were closely involved in their collection, naming, and classification: Darwin (*C. darwini*); Albert Günther (*C. guntheri*), keeper of zoology at the British Museum at the end of the nineteenth century; Rollo H. Beck (*C. becki*), a collector who made four trips to the islands hunting specimens at the turn of the twentieth century; and John Van Denburgh (*C. vandenburghi*), curator of herpetology at the California Academy of Sciences who, in 1914, wrote the monograph that established the fifteen different kinds of tortoises, each on its own island or volcano, as Darwin had supposed. Over the twentieth century, despite debates about whether the tortoise populations should be considered species or subspecies, and changes in the genus name from *Testudo* to *Geochelone* to *Chelonoidis,* Van Denburgh's monograph served as the authority on the species and the basis for conservation breeding.[7]

These species names recognize natural variation among tortoise populations, but they also reflect how, and to what effect, these species came into
being as the tortoises became entangled in the social world and (neo)colonial
politics of natural history. Species names, as historian Gordon McOuat has
written, "are about power, about possession, about capital and the status of
naturalists."[8] Both the tortoises and the men for whom they were named were
part of a world of imperial science, major investment, collecting mania, personal obsessions, globe-spanning expeditions, and excruciating fieldwork.
Tracing the history of the names of these four species will tell us about their
origins—not in a prehistoric world, but in the world of nineteenth- and early-
twentieth-century natural science as it transformed natural history into evolutionary biology. The tortoises were collected from remote islands, sent on
oceanic voyages, and circulated through major centers of scientific study in
London, San Francisco, New York, and Paris, among other northern cities.[9]
These are the places where naturalists spent hours meticulously measuring
and comparing the shape of tortoises' carapaces, and where taxonomists wrote
and published the essays that formally brought species into being. More than
a century later, I found Galápagos tortoises mounted on display in natural history museums, preserved as specimens sitting on museums' backroom
shelves, and represented in photographs in the brittle, cream-colored pages of
aging taxonomic monographs. It was in natural history museums rather than
in the Galápagos that the tortoises first became evidence of evolution.

What was at stake in the mania to collect and name species were the careers
of naturalists, the veracity of Darwin's ideas, and the continued existence of
the tortoises themselves. Through these socio-natural histories of scientific
study, the tortoises became understood not only as discrete species and evidence of evolution, but also as proxies for debate about the nature of life.

Darwini: History of a Type Specimen

Today, the tortoises Darwin ate on Santiago as he camped with the tortoise
hunters are named for him; they are known formally to biologists as *Chelonoidis darwini*. They are one of several species in the archipelago named for Darwin. Darwin's finches are the best-known example, although this is a colloquial
name, not a formal taxonomic designation. Galápagos cotton (*Gossypium darwinii*), coral (*Pacifigorgia darwinii*), red sheephead fish (*Semicossyphus darwini*),
and even two kinds of ticks (*Ornithodoros darwini* and *Amblyomma darwini*)
have been formally designated in Darwin's name. Today, hundreds of other

Figure 3.1 Specimen 8109. Photographic plate of holotype of *Testudo darwini*, from John Van Denburgh, 1914. (© California Academy of Sciences)

species around the world, even entire genera, have been named for Darwin, an homage to someone whose work now serves as the foundation of biological understandings of what species are.

The Santiago tortoises were named for Darwin in 1907 by John Van Denburgh. Darwin had made detailed observations about these tortoises in 1835, but the animals Van Denburgh studied were quite different from the living creatures who Darwin had watched drinking from muddy pools. The Santiago tortoises Van Denburgh studied—five of them—were dead research specimens, sitting on tables at the California Academy of Sciences in San Francisco. Of these five, Van Denburgh chose one, Specimen 8109, to represent the species he called *Testudo darwini* (fig. 3.1).[10] By formally describing this individual tortoise—its desiccated head and neck resting on a table, clearly lifeless and only part of the animal who once existed—Van Denburgh made it the holotype, or "type," the single specimen that a naturalist describes to represent a particular species, tying the name to a whole population of animals with shared characteristics. For biologists, type specimens serve as physical proof of a species, a referent that makes a taxonomic name meaningful.

Darwin might have ridden on the backs of these tortoises, but *T. darwini* could no longer move of its own accord. While its carapace, at thirty-eight inches long, would have been large enough to carry a man, Van Denburgh's tortoise had been carried on men's backs and in a ship's hull from the Galápagos to California in the interest of scientific study.

The animal who became Specimen 8109 was one of 266 giant tortoises captured during a Galápagos collecting expedition launched by the California Academy of Sciences (Cal Academy) in 1905. A team of eight natural history collectors set off with the mission of "collecting evolution"; when they returned after spending 366 days in the archipelago, their specimens would provide the basis, as Matthew James has written, for "vindicating" Darwin.[11] Nine months into their stay, in late July 1906, the team caught five tortoises on Santiago. During a previous visit to the island in January, Rollo Beck had found the remains of the camp once used by tortoise hunters but found no living tortoises there, only scattered bones. Determined to find tortoises, he hiked farther inland. Arthur Slevin, the ship's herpetologist and Van Denburgh's assistant at the Cal Academy (who would succeed him as curator of herpetology in the 1920s), kept the log of their activities. His description of tortoise hunting echoes that of earlier sailors in its complaints about the harsh terrain: "The country is extremely rough—the worst we have encountered since we arrived in the islands. The lava-flows are all comparatively recent, and many places have no vegetation whatever. . . . There is no earth whatever here; everything is lava, and it is impossible to do any trailing" of the tortoises.[12]

The following day, though, Beck and another collector, Ernest King, found two large male tortoises in better "tortoise country" where the lava had eroded into dirt and there were cacti for tortoises to feed on. Slevin helped King prepare the animals, skinning them by dissecting them and cleaning out their internal organs, scraping meat and fat from their skin and bones (fig. 3.2). For the largest tortoises, which were too heavy to carry alive and whole, the men often "fixed" them on land; smaller tortoises they could take aboard alive to live in a pen on the ship's deck until Slevin had time to prepare them. That day, while Slevin and King skinned, Beck continued searching, eventually finding three more. The men hauled these five tortoises back to shore in the same ways earlier sailors had—"backing down" smaller tortoises on their shoulders or strapping large tortoises to poles shouldered by two or more men (fig. 3.3). Slevin wrote of having to leave one tortoise for the night: "The country was so rough and hard to get over that our shoulders became so

Figure 3.2 Members of a Rothschild expedition skinning tortoises on grass, mixed brush and Bursera beyond. Photo by Rollo Howard Beck, 1901. (© California Academy of Sciences)

sore that we could not hold the tortoise up any longer." These Santiago tortoises, Slevin wrote, had "the heaviest shells and bones of any taken by us" and were "very fat."[13]

As he skinned tortoises with help from Beck, the ship's captain, and other collectors, Slevin made notes about the animals. He recorded where the specimens were collected and where they were killed; he also took measurements of the length of tortoises' legs, tails, and necks. He compared their body temperature to the temperature of the air, noted their sex (and whether females were producing eggs), and recorded the contents of their stomachs and, often, the state of the liver, which the men regularly ate. Echoing sailors' earlier reports, he noted that the fat was a "rich yellow color and looks almost like butter." But unlike earlier tortoise hunters, Slevin did not write about the tortoises' taste. The fat was noteworthy for the Cal Academy men because they had to scrub carapaces dry of copious amounts of oil to prepare them for storage or mounting—something for which they relied on arsenic soap and a long soak in pickling solution while on board the ship.[14]

Figure 3.3 "Packing out" a tortoise strapped to a pole. Photo by Rollo Howard Beck,
1902. (© California Academy of Sciences)

Transporting specimens from far-flung locations like the Galápagos to metropolitan museums was always difficult; damp sea air, mold, bugs, and bacteria
ate away at organic matter, and there was always the risk of simply losing collections during transport. Animals were more difficult to preserve than plants,
and thus many collectors predominantly saved animals with hard features—a
tortoise's shell was an ideal material because of its durability, if not its size and
weight.[15] Indeed, many museums' tortoise specimens were just cleaned shells,
the animal's internal organs, bones, and soft tissue having been discarded or
rotted away. The specimens Slevin and Beck preserved for Van Denburgh,
though, were more complete: some they fully stuffed in lifelike positions,
whereas others included the animals' neck and head, which had been cleaned,
stuffed, and dried and were dusted with arsenic to keep insects away.

In this form, Specimen 8109 was—and remains—no longer an ecological
being, but a still life. It is not an animal to be observed in its natural habitat,
nor a tasty delicacy for consumption. Instead, specimens were consumed as
"working objects" of scientific inquiry that provided a standard reference,
standing in for "unrefined natural objects [that] are too quirkily particular to

cooperate in generalizations and comparisons."[16] Natural history classification depended on the consumption of specimens of plants and animals and minerals plucked, captured, or chiseled out of their natural environments and preserved in special boxes, botanical presses, and jars of alcohol and formalin or as skins or taxidermied mounts.

Metropolitan museums like the Cal Academy in San Francisco and the British Museum in London were storehouses for these specimens, "centers of calculation" where specialized taxonomists, like Van Denburgh, made use of them.[17] These centers were hubs of scientific study as well as of imperial or state power; they housed vast collections acquired through global exploration and reinforced that power by concentrating knowledge of the world's natural resources.[18] For centuries, giant turtle and tortoise shells had been popular items in curiosity cabinets. But the wonderment of curios shifted in the eighteenth century toward a focus on systematized knowledge and classification made prominent by Linnaeus. In the early nineteenth century, natural history museums emerged as state-funded institutions designed to collect, house, study, and display examples of all the world's species. Eighteenth- and nineteenth-century natural history came to be structured around an institutional division of labor in which collectors traveled the world, often for months or years at a time, and museum workers put order to the natural wonders they received. It was an arrangement Linnaeus made prominent, sending his students on early expeditions to collect in remote regions of the globe and providing a uniform system for ordering what they found.[19]

Species classification is inherently a comparative science, one that depends on systematic analysis of multiple specimens from a given set of plants or animals as well as closely related species. In the halls of natural history museums, as Rachel Poliquin has written, taxidermy and taxonomy became inseparable—the "twin soldiers in the quest for a comprehensive catalogue of nature's diversity."[20] Through the work of ordering, specimens became the basis for knowledge of nature. Taxonomists took meticulous measurements of specimens, examined their anatomy, compared them with other similar creatures, and wrote formal nomenclatural designations. Van Denburgh's 1914 monograph, for example, includes dozens of plates and illustrations of tortoise specimens. Animals who on first glance appear to be multiples of the same species reveal, on closer inspection, differences in morphology that for natural historians were the factors that distinguished one species from another. What features were important varied by kinds of animals. For Darwin's finches, differences in the shapes of beaks are a defining characteristic;

Linnaeus had based his botanical classifications on reproductive anatomy. For the Galápagos tortoises, the taxonomically relevant features included the shape of their carapaces—dome-shaped or saddle-shaped—as well as the shape of their bones and the number and placement of scutes. From these basic distinguishing features, Van Denburgh formally described *T. darwini* on the basis of meticulous measurement of carapace shape: "No nuchal [scute]; gulars paired; fourth cervical vertebra biconvex; carapace high, elongate, somewhat dome-shaped but high in front; posterior declivity beginning about middle of third vertebral; height at nuchal notch more than 41% (45%) of straight length . . . carapace not saddle-shaped . . . straight length 38 inches; shell heavy; . . . jaws and throat black." These tortoises, in other words, were in between saddle and dome shapes and were also a quite "large, heavy, thick-shelled species."[21]

In contrast to museum taxonomists' dry accounts, nineteenth-century sailors' tales of tortoises were full of sensory observations that reflected their hunting encounters: the tortoises were "remarkably quick-sighted and timid, drawing their head into their shell on the slightest motion of any object; but they are entirely destitute of hearing, as the loudest noise, even the firing of a gun, does not seem to alarm them."[22] Captain David Porter was the first to write about differences between the islands' species, noting that the animals on Santiago appeared to be "a species entirely distinct" from the tortoises on Española and Floreana. The latter two species, he noted, had longer, brown, saddle-shaped shells and were "very disagreeable to the sight, but far superior to those of James' island in point of fatness, and their livers are considered the greatest delicacy." Yet the Santiago tortoises were "round, plump, and black as ebony." While not as displeasing to the eye, their "liver is black, hard when cooked, and the flesh altogether not so esteemed as the others."[23] (Perhaps this was why Darwin was less than enthused about the taste of tortoises.) Though Porter used the language of species, his classification was not a matter of origins, but rather focused on the animals' shape and taste.

Van Denburgh focused on technical description rather than subjective opinions, but his classification echoes Porter's description of different shell shapes. To compare variations in shape among the individual tortoises, Van Denburgh took close measurements, each of which he translated into a percentage of the straight length of the carapace so that he could compare tortoises of different sizes.[24] These measurements, repeated on each of the 266 tortoises the Cal Academy collectors took home, were the basis for his classification. The shell shape and overall size are the most noticeable distinctions among the tortoise species—the kind of differences Lawson was likely alluding to when he told Darwin he could tell what island a tortoise was from by looking at her.

Van Denburgh's formal presentation of the tortoises as fifteen different species, or "races," as he called them, included the measurements and charts, photographic plates (like fig. 3.1), and—crucially according to taxonomic rules—his formal description, part of which is quoted above. Each of these elements is based on representation of the specimens in themselves, an ostensibly objective presentation of nature's forms. It is the kind of description that Linnaeus had instituted, writing that "the first step in wisdom is to know the things themselves" in the tenth edition of his *Systemae Naturae*.[25] It also reflected the institutional structure of science at a time when studying nature meant collecting plants and animals from their natural habitats and examining specimens in far-away museums. There is no ecological description of the animals' behavior in their environs, as Darwin had given in his *Journal of Researches*. The plate of Specimen 8109 is the image of a species stripped of its ecological role and the human social history through which it became evidence of evolution.

Scholars have long criticized the methods of natural history for erasing the social relationships through which collectors found and extracted specimens from particular places.[26] Measuring hollowed-out carapaces may provide insight into species variation, but it tells little if anything about a species' ecological role or cultural significance in its native environment. Yet natural history during this period was not devoid of social history; it was just that its social history was self-referential. Van Denburgh, for example, included Slevin's stories about capturing giant tortoises in his description of each species. Doing so served to situate the Cal Academy expedition in a long history of explorer-naturalists. Taxonomists often started their monographs with long quotations from previous explorers describing their encounters with the animals. Van Denburgh's tortoise monograph, for example, includes the stories of Dampier, Delano, Porter, and Darwin as a collection of previous knowledge about the animals. It was a literary convention that established a canon of Western knowledge rooted in heroic adventure tales of European and North American exploration—implicitly determining whose stories counted in the history of a species. "Local knowledge," like that of Lawson and the other Galápagos residents, appears only as translated through a naturalist's story, and even then only when deemed valuable enough to contribute to the naturalist's project.[27] By reciting these stories, naturalists placed themselves within a history of progressive, positivist Western knowledge production—providing the latest, and most complete, chapter.

But Slevin's stories about capturing tortoises also played another role in Van Denburgh's monograph: they were proof of the specimen's geographical origins.[28] Despite pretensions to the contrary, tortoise carapaces and other specimens did not speak for themselves. What made Van Denburgh's classification of the different species possible was a classification of their shell shapes according to their island of origin. As Darwin realized when he returned to England, geographical knowledge of the animals' origins was essential to understanding patterns of evolution. Today such information would be recorded by taking satellite coordinates, but a hundred years ago, as Darwin had found, proof of geographical origin depended on a collector's notes. That Van Denburgh had access to this information, as well as a reliable record of each animal, was a testament to the organization of museum-based collecting expeditions and the growing emphasis on field biology in the early twentieth century. In the 1870s, when zoologist Albert Günther issued the first major work of classification on giant tortoises, based on the collections at the British Natural History Museum and specimens loaned out from other European museums, he complained that "the majority of specimens are young, or fragmentary, or without any history; and there will be found scarcely one with an indication of the particular island from which it came!"[29] He set forth a call for more expeditions and specimen extraction to improve the museum and the quality of knowledge it could produce. Thirty years later, before Van Denburgh sent Slevin to the field, he taught him how to find, capture, and preserve specimens, as well as what information to record.

This was the socio-natural history that brought the species *T. darwini* into being and vindicated Darwin's hunch that each of the islands had its own species. The tortoises had become evidence of evolution. But between Darwin's trip to the Galápagos in 1835 and Van Denburgh's research, it was not at all clear just what giant tortoise specimens would reveal about the nature of life.

Guntheri: "How the Deuce" Did Tortoises Get to the Galápagos?

Darwin's work on the question of origins recast nineteenth-century debates about natural history, but it was another naturalist, Albert Günther, who would "set off a craze" for giant tortoises that made the animals proxies for understanding evolution.[30] Between Darwin's Galápagos trip and Van Denburgh's monograph, the tortoises became "prehistoric" beings whose past could shed light on the origins of life. The mystery of the tortoises' origins centered on one question in particular: as paleontologist George Baur, who named *T. guntheri*, put it in 1891, "How did these large *land*-animals come to the *islands*?"[31] Darwin suggested that the Galápagos fauna had come from South America, but he could not explain how. He had told Günther in 1877 that giant tortoises had likely once existed over most of the world, but "how the deuce they got to volcanic islands I cannot pretend to say."[32]

In the decades after Darwin and Alfred Russel Wallace published their theory of evolution by natural selection, their contemporaries offered many competing hypotheses about how the tortoises got to the islands: Had they been brought by previous sailors? Created there specially by God? Had they walked to the archipelago over a land bridge that once had connected the islands to South America? Or, perhaps most far-fetched of all, had they been swept to sea by storms on the mainland and floated six hundred miles to the islands? These questions reflect the changing world of natural history in the mid-nineteenth century, when theories of transmutation upended widely accepted providentialist views about the history of the earth. Just how the tortoises got to the islands depended on one's take on the nature of the islands, the role of the divine in shaping the world, and how old the tortoises, and the earth, actually were. At stake was not only the veracity of Darwin's theory, but just what one could read off the backs of tortoises.

Günther took the position keeper of zoology at the British Museum in 1875 after more than a decade as an assistant keeper. Soon after he started, he presented a paper at the Royal Society in London—the elite club of British natural scientists—on giant tortoises in the Galápagos and Mascarene islands of the Indian Ocean. It was the first cut of his monograph, although he deplored that it was based on the "scanty and incomplete" collections available at the museum.[33] Nonetheless, even fragmentary remains, he wrote, "revealed important differences unmistakably pointing to a multiplicity of species. The results of these researches were startling . . . [they] bring us face to face with the mystery of the birth and life of an animal type."[34]

The paper was the first to convincingly establish that the Galápagos giants were not part of the *T. indica* species but were a distinct lineage. Günther analyzed Indian Ocean specimens and found different species on the islands of Rodrigues and Mauritius, both of which had gone extinct along with the dodo.[35] But on the uninhabited island of Aldabra, tortoises still survived. These animals showed "great affinity" to the Galápagos tortoises, but they were distinct species. Indeed, the Galápagos Islands had five different tortoise species,

Günther thought, a finding that "fully [bore] out" statements from Porter and Darwin that each of the islands was inhabited by a distinct species.[36]

The Galápagos animals had not been introduced by sailors. It was common practice in the eighteenth and nineteenth centuries for sailors to move tortoises from island to island or to introduce them in new places, as they did with goats and pigs.[37] Captain FitzRoy thought the Galápagos tortoises might have been so introduced by earlier "aborigine" sailors from the South American coast, for "in simple truth," he wrote, "there is no other animal in the whole creation so easily caught, so portable . . . so likely to have been carried, for food" in balsa rafts. But Günther did not think the "agency of man" could account for the presence of giant tortoises at "two so distant stations," the Galápagos and the Mascarene islands. The dispersed geography meant that each group "must be regarded as indigenous."[38] Darwin had thought so as well, writing in 1845 that the animals must be "aboriginal inhabitant[s]" since they were, he believed, found on all the islands in the Galápagos archipelago. "Had it been an imported species," he noted, "this would not have been the case in a group which has been so seldom frequented."[39] Although sailors moved tortoises among the islands, they were too widely distributed and had come to vary too much to be recent introductions.

As original and unique inhabitants, the tortoises, Günther demonstrated, were what biologists now call endemic species—not only native to the islands, but existing only in one particular place. They were not recent arrivals but had lived in the islands for quite some time. Darwin had mused in his *Journal of Researches* that the tortoises seemed like "antediluvian" animals, implying that they had outlasted a biblical flood. But as Günther and others worked up tortoise specimens, the animals might better have been called "living fossils," a phrase Darwin first used to describe other such species that had survived a former age.[40] The tortoises had become evidence of a new way of understanding the origins of life on earth, one based not on divine creation but on earthly change.

A NEW CREATION STORY

Günther remained ambivalent about evolution (in part perhaps because he worked for Darwin's archrival, Richard Owen, who was curator of the British Natural History Museum and used his position to vociferously challenge Darwin's interpretation of what caused evolutionary change). But his work on the tortoises positioned them as evidence of a new kind of origin story. Giant tortoises have long been part of origin stories; in several Eastern and indigenous

American traditions, the world was born from a tortoise egg or rested on a tortoise's back. But early scientific species classifications stripped out what were deemed to be merely mythical stories about tortoises who grew to be as big as islands or held up the earth. Instead, for many of Darwin's predecessors, scientific description of giant tortoises, like other species, positioned them as evidence for a providentialist origin story. Linnaeus, for example, believed that species were created by God as discrete, unchanging kinds that were ordered hierarchically. Collecting and naming God's creations was a form of religious study and devotion that would reveal the divine order of the universe God had designed.[41] Darwin's chief opponent, Louis Agassiz, also saw species as a "thought of God."[42] As historian Edward Larson has written, for Agassiz taxonomic work was a way of uncovering the divine: "In classifying organisms into species and higher groups, naturalists 'only translate into human language the Divine thoughts expressed in nature in living things.' Agassiz stressed that such manifested thoughts—found in the structure, relationships and geographical distribution of organisms—reflected design: 'They show the omnipresence of the Creator.' "[43]

Theories of transmutation, however, suggested that species were not a "thought of God" but evidence of earthly origins and slow material change. As living fossils that had evidently survived from an earlier age, the tortoises became proxies in debates about the history of life, and the earth, that gripped nineteenth-century scientists. Early in the century, geologists had become convinced that the earth was not merely six thousand years old, as seventeenth-century Irish archbishop James Ussher had calculated. Instead, the earth had a history that stretched back far before the history of human beings. Making sense of the fossilized records of past life and geological change convinced them of the idea of "deep time."[44] But the nature of historical change was a topic of considerable debate when Darwin was a young man. One of Darwin's key influences, Charles Lyell, argued against invocations of deluges or other catastrophic events to explain geological change. Instead, he put forth a theory of gradual change based only on presently observable processes: the uprising of mountain chains, the drying of former seas, and the volcanic creation of new islands were not special feats of nature but the outcomes of steady processes that would have taken nearly infinite stretches of time.[45] The assumptions of such deep history and gradual change were foundational for Darwin and Wallace's theory of evolution by natural selection. But they were also among the more controversial aspects of this idea.

When Darwin wrote of the Galápagos as a place of "new Creation," he meant not divine origins, but volcanic ones; islands newly arisen from the sea were a glimpse into what life on much older continental landforms might once have been like. But Agassiz read this evidence differently. No fan of transmutation, he designed a South America expedition in part to refute Darwin's views. Like Darwin, Agassiz thought that the Galápagos Islands were a fascinating case where "the mystery of change . . . is only increased, and brought to a level with that of creation." Agassiz stressed the youth of these volcanic islands that seemed "of most recent origin" as a way of challenging Darwin—the tortoises must be providential because they would not have had enough time to evolve on these young islands.[46] It was a good point. Even for the physicists of Darwin's day, who estimated the earth to be one hundred million years old, natural selection would have taken far too long. The mechanism Darwin and Wallace proposed to explain how evolution worked did not appear at the time to be a plausible explanation for the variation naturalists witnessed across animal forms. But despite this challenge, belief in evolution of some form was widespread by the end of the nineteenth century. Agassiz, who died in 1873, was the last prominent natural historian to believe in special creation.

One of the consequences of eighteenth- and nineteenth-century study of the history of the earth was a broad fascination with prehistoric life. As scientific understanding shifted from belief in a young, divinely created earth to one millions of years old, naturalists and the public alike were riveted by the idea that creatures such as dinosaurs, mammoths, and giant reptiles had roamed the earth for millennia before humans. It was a truly revolutionary idea for a society that had believed for centuries that God had created humans at the center of the world. Fossilized dinosaur bones and enormous mammoth teeth fascinated museum visitors; giant tortoises provided a further sense of awe, for these living fossils linked contemporary life with this prior world. But evolutionary origin stories about the tortoises required their own leaps of faith, particularly about how these lumbering animals might have first arrived on the islands—how they managed to navigate, for example, the six hundred miles of open ocean that separates the Galápagos from continental Ecuador.

SWIMMING TORTOISES?

To explain how distinct tortoise populations came to live in the Mascarenes and the Galápagos, Günther offered two hypotheses: either they walked there over land bridges that once connected the islands to the continents, or they drifted by chance in ocean currents. If one believed in a common origin—that the species at some past time were related, as Darwin's theory suggested—then Günther supposed that a "former continuity of land" between Africa and the Mascarenes and South America and the Galápagos would have been necessary. By this hypothesis, the tortoises once would have been widely distributed on the earth's continents, "without being able to survive long the arrival of man," who must have hunted continental species to extinction since they were delicious and easy to capture.[47] A deepening ocean would have closed the land bridges, effectively isolating and protecting the island tortoises. Günther acknowledged, though, that this hypothesis depended on assuming that the tortoises had lived "over an enormous period of time, of which the period required for the loss of power of flight in the Dodo or Solitaire is but a fraction."[48]

Another possibility, Günther believed, was "equally justified" if one thought the Mascarene and Galápagos tortoise species had originated independently: "Without overstepping too far the limits of probability," he wrote, "we may assume that some Land-Tortoises were carried by stream and current from the American Continent to the Galápagos, and that others from Madagascar or Africa, found in a similar manner a new home in the Mascarene islands." The tortoises would have arrived by chance, and although they might have originated from dissimilar species on the two continents, they became more similar to one another as they adapted to similar island environments.[49]

Günther did not offer his opinion on the likelihood of either of these options, though he did think the land bridge supported Darwinian theory. Darwin, Wallace, and Agassiz, however, all thought the oceanic islands had never been connected to the continent. As Wallace described in *Island Life*, published in 1880, the tortoises, like all life-forms on oceanic islands, had most likely arrived by chance: "Considering the well-known tenacity of life of these animals, and the large number of allied forms which have aquatic or subaquatic habits, it is not a very extravagant supposition that some ancestral form, carried out to sea by a flood, was once or twice safely drifted as far as the Galápagos, and thus originated the races which now inhabit them."[50]

But in Wallace's day, belief in such dispersal necessitated extraordinary faith in chance. A decade later, paleontologist George Baur argued that the tortoises, whom he believed could not swim, clearly demonstrated the implausibility of the Galápagos's oceanic origins. If one believed the islands had been elevated out of the sea, then "by a peculiar accident" a tortoise might have arrived ashore. But, Baur noted, "alone it could not propagate." Reproduction would be possible only "after a similar accident imported another specimen of *the same species, of the other sex, to the same island.*" Although improbable, that could account for one island population. But fifteen? "To explain this," Baur wrote, "we would have to invoke a thousand accidents."[51] Yet Baur's skepticism reveals more about his gendered preconceptions of seafaring exploration and colonization than about the islands' origins: only a pregnant female tortoise was necessary to start a new island population—surely a feat of remarkable chance, but one apparently too outlandish for Baur to consider.

For Baur, the "harmonious" distribution of different, but closely related, species on different islands meant the archipelago could not possibly be of oceanic origin.[52] Instead, he believed the islands were the tops of former mountains once connected to the mainland. Baur was the chief proponent of the idea of a land bridge, himself making a trip to the Galápagos to collect specimens of tortoises and other animals that would, he thought, prove his theory that the tortoises were "not introduced but left there; the Galápagos originated through subsidence of a larger area of land; they do not represent oceanic islands, as generally believed, but are continental islands."[53]

It was an opinion, as Baur himself noted, that was intensely ridiculed. "Every geologist will stand up against Dr. Baur's hypothesis," wrote German naturalist Theodor Wolf, who spent much of his career in Ecuador and conducted geological surveys of the archipelago.[54] For Wolf and most others, the islands' active volcanoes meant they were obviously of oceanic origin and never connected to the continent.

But this did leave the problem of the "thousand accidents." Could the giant animals really have survived an oceanic swim to an archipelago six hundred miles from the continent? Darwin had done experiments with seeds in saltwater to see whether they might have been able to germinate after crossing an ocean.[55] He found that some varieties could, but whether a giant tortoise could survive such a crossing was a different matter entirely. Baur gave no evidence for his assertion that the tortoises could not swim but had certainly read Porter's account of finding tortoises floating a few days after one of the ships he captured had thrown them overboard: they had been "lying in the same place where they had been thrown over, incapable of any exertion in that element, except that of stretching out their long necks."[56] Baur was convincing enough for Günther and Van Denburgh, who concurred that the tortoises were "absolutely helpless in the water" and were "at the mercy of the winds and currents."[57] At the turn of the twentieth century, a land bridge was the most prominent explanation for how the tortoises got to the islands.

Decades later, though, another zoologist argued that Porter's story was "proof only of great buoyancy, a quiet sea and absence of land in sight to direct and stimulate exertions."[58] Whether the tortoises could exert themselves would be tested by another naturalist, William Beebe, during a cruise of the archipelago in 1923. Beebe conducted his own tortoise swim test by tossing the only tortoise he found on Pinzón over the side of a luxury yacht and filming it float miles from land. He bobbed quite well, Beebe found, and could raise or lower his head below the water. But most surprising "was the ease and excellence of its swimming ability. The reptile would swim toward the row-boat which I occupied, and finding it too high, would turn and swim over to the *Noma*, stretching its head high along the waterline. Then it steered its way to the companionway. This was with, across, and against the very appreciable current at will." Eventually Beebe plucked the animal from the sea, but a week later the tortoise died suddenly. His lungs and small intestine were "heavily congested"—Beebe thought ingesting saltwater must have caused his death. He did not comment on the ethics of his experiment but concluded that it "would negate any possibility of the tortoises being able to make their way over wide expanses of water, either from the mainland or from island to island, in spite of their unusual swimming ability."[59]

The unlikely probability of swimming tortoises, however, was for another Cal Academy collector trumped by the implausibility of bridges stretching across the vast Pacific: "If we must invoke the aid of geographical changes to account for all the anomalous biological conditions that we find, truly the Pacific Ocean would look like a spider-web if the ghosts of all the supposed land-bridges should suddenly rise up to confront us."[60]

It was not until the early twenty-first century that direct evidence of a giant tortoise ocean crossing turned up, when in 2006 a giant tortoise from Aldabra drifted some 460 miles across the Indian Ocean before walking up a beach in Kimbiji, Tanzania, emaciated and covered in barnacles, but otherwise no worse for the voyage.[61] By then, the land bridge hypothesis had long since been disproved by mid-twentieth-century research on plate tectonics. Yet the debates Günther inspired about what giant tortoises could tell naturalists about the history of the species and the nature of the islands have had lasting resonances.[62] For more than a century, scientists have echoed what Baur once wrote about the value of the archipelago for studies of evolution: "There is no other place on the whole earth," he asserted, "which affords better opportunities for such a work than the Galápagos. Here we have the original natural conditions, hardly influenced by man."[63] The relative youth of the islands once posed a problem for Darwin but now underscores both his and Agassiz's assertions that the islands brought them nearer to the mysteries of creation. It is the geological youth of the islands that makes them seem prehistoric; because they are a more recent creation than continents, they seem like a portal through time for understanding earlier forms of life. It was in this period that the tortoises became understood as species with prehistoric lineages—if not "antediluvian," as Darwin had mused in the *Journal of Researches,* then island escapees who had survived from an earlier era when giant tortoises were common throughout the world. The idea that the tortoises were living fossils, holdovers from a previous world protected on remote islands, has become a central part of conservationist thinking. It also inspired a mania for collecting complete "sets of examples" from the islands to determine the origin of the species.[64] As Samuel Garman wrote after examining the specimens Baur collected in 1892, "a most important contribution to the scientific history of the region might be made by one who is able to gather from each of the islands series large enough to supply the now-lacking means for comparisons."[65]

Becki: Saving Tortoises for Science

During the decade surrounding the turn of the twentieth century, seven expeditions sponsored by major scientific institutions visited the Galápagos.[66] Inspired by Günther's and Baur's calls for complete collections, their naturalists scoured the archipelago and gathered nearly one hundred thousand specimens of various species.[67] What naturalists found, though, was not the origin of these species, but their demise. Intense field collecting at the turn of the twentieth century was intended to save species for museum study. What emerged was also a greater understanding of the species' ecology and the reality of extinction.

Natural history collector Rollo Beck, for whom the species *C. becki* is named, was at the center of these efforts. Beck was a California naturalist called by one scientist *"the* field collector of his generation."[68] He made his name as a skilled skinner of birds and an intrepid explorer who made ornithological collections around the world. But early in his career, he made four expeditions to the Galápagos—in 1897, 1901, 1902, and 1905–1906. With the haul he and his collecting colleagues made from these trips, Beck helped stock museums and zoos in London and San Francisco with Galápagos specimens, making them the largest centers of Galápagos research.

The species *C. becki* was named in 1901 by Lionel Walter Rothschild, who hired Beck to make the first three trips to the islands. Walter was the oldest son of British banking magnate Lord Nathaniel Rothschild and was, to his father's chagrin, an avid natural historian. Walter was socially awkward—painfully shy, with a tendency, as his niece wrote in her biography of him, for "pathological secrecy" and a "crippling lack of voice control that made normal conversation virtually impossible."[69] But he was inspired from a young age to be a naturalist. He acquired his first specimen at age seven and opened his first "museum" a year later in a shed on the family property at Tring, outside of London. Visiting the Natural History Museum in London as a boy, he also befriended Albert Günther, who became a mentor and eventually a close friend and collaborator.

Over the course of his life, Rothschild used his fortune and social standing to turn his shed museum into the largest private natural history collection in the world. Although Rothschild lived at the tail end of the popularity of natural history as a scientific endeavor—as systematic classification was superseded by laboratory genetics, field ecology, and eventually evolutionary biology—no individual better exemplifies the mania for collecting that made natural history a common pastime during the nineteenth century. When he died in 1937, he left the British Museum more than 3 million specimens, including 2 million butterflies and moths, 300,000 bird skins, a collection of 200,000 bird eggs, and some 4,000 animals stuffed for display, including 62 cassowaries, 40 apes, and 144 giant tortoises. Today, many of them remain on display at the Tring Museum (no longer housed in a shed), which still stands just as Rothschild had arranged it in the 1910s. Many of his specimens started out in the menagerie he made of the Tring grounds, which were once home to cassowaries, rheas, kangaroos, and giant tortoises who roamed the gardens in the summer and wintered in the orchid house. When Rothschild went to study at Cambridge, he took a flock of kiwis with him; as an adult, he trained zebras to pull a carriage and drove them to Buckingham Palace. He acquired these animals by sending expeditions around the world and purchasing collections from others. Once they died—few survived long in the damp English climate—they became specimens to add to his museum. To manage his collections, Rothschild, whose day job was ostensibly in the family bank, hired two professional naturalists—Karl Jordan and Ernst Hartert—to run his museum and publish its journal, *Novitates Zoologicae*. Together, the three men named more than five thousand new species and published twelve hundred papers and monographs.[70]

Rothschild was a man of many obsessions, with an inordinate fondness for giant tortoises among them. The first one he acquired was from the governor of Mauritius—he speculated that perhaps the last living *T. indica* roamed the gardens at Tring.[71] He would later have more than fifty living tortoises from

Figure 3.4 Walter Rothschild rides a *C. darwini* giant
tortoise from Santiago Island, Galápagos. (© The
Trustees of the Natural History Museum, London)

Aldabra and the Galápagos at Tring and at the London zoo, where he paid to
have a special house and yard built, complete with wading pool. Something of
a social misfit, he took pleasure in his animal collections and delighted in seeing them. Perhaps the most famous portrait of Rothschild is of him in a top
hat riding on a tortoise (fig. 3.4).

Rothschild could write to British colonial officials to purchase rare tortoises
from Aldabra, but acquiring Galápagos animals was a more involved affair. In
1897, he hired Massachusetts collector and taxidermist Frank Webster to organize an expedition to the Galápagos with "orders to send every single bird
they can shoot except Flamingoes & to bring alive & dead . . . 150 tortoises
from any island & to collect every bone, skull and carapace of dead tortoises to
be found in the whole of the Galápagos."[72] Rothschild was keen to gather the
complete collection of Galápagos tortoises that Günther had said would be

necessary to understand the origins of variation. This mania reflected a will to know and possess the world's nature.[73] The grand quantities of specimens also were necessary for the detailed work of classifying species into trinomals (names of varieties in addition to genera and species) that Rothschild and Karl Jordan promoted.[74] Rothschild had bought much of the collection of Galápagos specimens that Baur had made in 1892, but wanted more. He entrusted Webster to acquire specimens of excellent quality with which, he wrote to Günther, "We ought to be able to do something this time, as both skinners & tortoise hunters are trained American taxidermists & collectors."[75]

Rothschild, though, was no evolutionary theorist. His interest and abilities lay in amassing a complete collection and sorting through species names under Günther's close direction. He was motivated by fears of the disappearance of tortoises. Günther had been quite concerned in the 1870s about the fate of Indian Ocean tortoises, and Rothschild was clearly spurred to action by Baur's call to study Galápagos fauna *before it is too late. I repeat, before it is too late! Or it may happen that the natural history of the Galápagos will be lost, lost forever, irreparably!"*[76] In 1897, Rothschild wrote to Günther explaining the rationale for his first Galápagos expedition: "My chief reason for telling Dr. Harris to bring away every tortoise they saw big or little alive & dead was that the Orchilla moss hunters had already reduced them by more than half since Baur was there in 1892 & they would have eaten them all in 2 or 3 years more & I wanted to save them for science."[77] For Rothschild, saving tortoises meant stockpiling them in a museum (preferably his), where they could be studied before they were lost to the bellies of Ecuadorian colonists.[78]

Saving species for science depended on reliable collectors, their skill at preserving specimens, and luck. Although Rothschild hired skilled naturalists, his first attempt at the Galápagos ended in disaster. Webster hired collector C. M. Harris to lead the first expeditions, but once in Panama, Harris had trouble finding a ship to charter to the islands. He was forced to dismiss one of the crew for drunkenness and then lost three men, including the captain, to yellow fever, as well as another to "desertion" after he fled home in fear of the disease. He turned back before they reached the archipelago. A few months later, Harris tried again, hiring young collector Rollo Beck as part of his replacement team for $25 a month. This expedition would be far more successful.

Rothschild's collectors tried to oblige his desire for complete collections. They took twenty-nine tortoises from Pinzón, where one of the collectors thought there might be only two or three left on the island.[79] This was the island where Beck first saw a giant tortoise in its native habitat, after climbing fifteen hundred feet up a mountain on an island notorious among park guards today for its difficult terrain. Beck recalled the experience several years later as if he had traveled back in time. Standing inside a crater, its walls towering around him, Beck wrote that "the intense stillness of the place was broken only by the drone of a cricket." It was easy to imagine that "we were back in a bygone age." When a tortoise appeared on the scene, Beck "felt that it would not be surprising to see a pterodactyl come flying over the rim of the crater, or a megalosaurus rise out of the bushes near by."[80]

After four months in the archipelago, the Webster-Harris expedition returned to San Francisco with sixty crates of specimens, including more than 3,000 bird skins, 150 iguanas, and 65 tortoises, many of them alive.[81] Harris had to babysit the tortoises for weeks in San Francisco as they, and he, regained their health from weeks at sea before shipping them across the Atlantic. When he did finally send them in February, he annoyed Rothschild by dooming the living tortoises because he had not waited for warmer weather.

Harris's team had explored thirteen islands but did not manage to get to the northern part of Isabela, to Rothschild's disappointment. Knowing this, before he had even received Harris's tortoises, Rothschild was writing to Beck to plan a return voyage. After a failed start in 1899, Beck returned to the Galápagos in December 1900 for a trip during which he collected the tortoise that is now named for him. Rothschild was pleased to have this specimen from the northern part of Isabela where Harris had not gone because it proved the existence of a third species from the large island—one that was even quite distinct from the others, having a more pronounced saddle shape.[82]

Beck was concerned, though, about the poor quality of some of his tortoise preps. For collectors, nature preservation was about how best to keep skins and mounts. But the damp climate during the rainy season and the difficulty of working belowdecks on a cramped ship challenged his ability to preserve life: "The moisture in the air on board the vessel seemed to effectually prevent the successful curing of the skins. They were poisoned with arsenical soap and alum to the inside of skin, but nearly all had more or less loose skin when dried." Even when dried and placed on a shelf for storage, the specimens were not safe from the frustration of captured tortoises awaiting similar fates: "one day our largest tortoise got on a 'rampage,' tore down the shelf and took 2 or 3 of bites from legs and heads of the skins."[83]

These challenges were worrisome for Beck because preserving specimens well also meant preserving his own earnings from the trip, and his future career. Beck's prospects for work depended on his reputation as both an able fieldworker who could capture animals in difficult environments like the Galápagos and a careful preparer of specimens. Rothschild, however, was apparently pleased with Beck's work. After Beck took the 1901 animals to the London zoo—including several living tortoises, an owl, and half a dozen land iguanas—he noted that Rothschild paid more than he and other collectors could have earned in California.[84] Beck saw this as an opportunity. He lamented to Hartert the incompleteness of his collection and pitched a return trip to several islands in the eastern Pacific that might serve both his and Rothschild's desires: "I should very much have liked to work the entire group more thoroly, & could I see my way to come out ahead financially would be tempted to try it again. . . . If you think you could use a $1000 or more worth of material from Galápagos, Revillagigedo, Cocos or Clipperton, I may go down again in a few months." He repeated the offer again in September, promising to be sure to dry tortoise skins ashore, rather than do it belowdecks in the ship, where the dampness of the rainy season would not be so consuming.[85] Rothschild consented and Beck went for a third trip, gathering a five-year total of 150 tortoises, including what he thought was the largest tortoise ever taken from the islands.

AN IMAGE OF EXTINCTION

Saving tortoises at the turn of the twentieth century meant killing animals to preserve them in museums, but Beck was not unconcerned with the tortoises' fate in the wild. Like Baur, he worried about what he thought was their looming extirpation, particularly as colonists established themselves in the islands: "It is only within the last two years that the home of these very large tortoises has been invaded by man, but the rapidity with which they are being killed, and the reason for their destruction, leaves us but little hope that they will survive any longer than did the American bison after the hide hunters began their work of extermination."[86]

The colonists Beck was concerned about were hunters for whom the tortoises were not only a source of meat for subsistence, but also a source of fat rendered into oil to be sold to merchants in Guayaquil. As he described,

> The outfit of the oil hunter is very simple, consisting merely of a can or pot in which to try out the oil, and three or four burros for carrying the five- or ten-gallon kegs in which it is transported to the settlement. After making a camp near a water-hole, and killing the tortoises there, the

collector brings up a burro, throws a couple of sacks over the pack-saddle, and starts out to look for more tortoises, killing them wherever found. A few strokes of the machete separates the plastron from the body, and 10 minutes' work will clear the fat from the sides. The fat is then thrown into the sack, and the outfit moves on.

When the burro is well laden, man and beast travel back to camp, where the oil is tried out. Each large tortoise yields from one to three gallons of oil. The small ones are seldom killed, because they have but little fat. By daily visits to the few water-holes during the driest season, in the course of a month the hunters get practically all the tortoises that live on the upper part of the mountain.[87]

Beck drew a distinction between such hunting and his own, a double standard that served to justify scientific collection. Yet he not only criticized local hunters—he also relied on them.

During the 1902 trip, he spent two weeks camping at the highlands cattle ranch of colonists from the Ecuadorian coast at a settlement called Santo Tomás. Nearby were more tortoises than he ever thought he would see in their "native state." In every open field along the trail tortoises "could be seen feeding, walking about, or quietly sleeping with their heads against the base of some bush or tree where they had dug a form in which to lie."[88] But this paradise for tortoises, just discovered, was imminently threatened: "While at the ranch we were amazed at the reckless and heart-less manner in which some of the natives destroyed the tortoises. The proprietor informed us that only the males were killed, but we noticed that the working people made little distinction in the sexes when killing for food." Beck was appalled at how little meat hunters took from each tortoise. "Some evenings, two or three men coming in from different directions would each carry in his hand a small piece of tortoise meat, and a pound or so of fat with which to cook it. Of each tortoise killed not over five pounds of meat would be taken, the remainder being left for the wild dogs that swarmed about."[89] At this rate, the animals would disappear quickly, he thought: "it will require but a few years to clear this entire mountain of tortoises, and when we see the methods pursued by the proprietor in getting tortoise oil for shipment to the mainland, we know that the large tortoises can last but a few months after the work of the oil hunter begins in earnest."[90] The remarkable abundance of tortoises that sailors had routinely described during the eighteenth and nineteenth centuries had shifted to a state of pending extinction.[91]

Figure 3.5 Tortoise remains left by oil hunters on Sierra Negra, Isabela, Galápagos,
1902. Photo by Rollo Howard Beck. (© California Academy of Sciences)

To document local overhunting of the tortoises, Beck snapped photographs.
One showed tortoises not yet attacked by hunters as they waded and rested in
a clearing with muddy pools; it served as a basis for comparison with another
photograph that depicted a watering hole where Beck estimated oil hunters
had killed 150 tortoises, leaving behind the animals' macheted carapaces to
bleach and weather in the sun (fig. 3.5). Another watering hole a half mile
away looked much the same, Beck reported, with the remains of one hundred
slaughtered animals.

At the time, Beck's photograph became an icon of impending extinction,
evidence of human destruction of nature that was a core concern for turn-of-
the-century naturalists. It also provided evidence of the "profligate native"—a
common trope of colonial-era exploration that justified naturalists' collection
and conservation.[92] Today, the photograph is used repeatedly in conservationist publications as an alarming depiction of wanton destruction. When it is
presented today, although Beck is credited, it is often done without much comment about the conditions of its taking. The imperial will to possess that drove
natural history collection and further contributed to the species' endangerment is easily edited out of the way that conservationists remember the past.

But at the turn of the twentieth century, the photograph motivated not conservation, but more collection. Indeed, early-twentieth-century collectors repeatedly warned of the extinction of tortoises and claimed they had collected the last specimens, only to have someone else later find additional animals.[93] Beck returned to the archipelago in 1905 for a year as the leader of the Cal Academy expedition, the longest and most thorough collecting trip in the archipelago's history. Under the direction of Everett Mills Loomis, director of the California Academy of Sciences, Beck and seven other collectors, plus a crew of three, scoured the islands collecting specimens. They took more than seventy-eight thousand specimens, including hundreds of bird skins that would become the basis for work on the evolution of Darwin's finches and the 266 giant tortoises that Van Denburgh would work up.[94]

Vandenburghi: Saving Tortoises from Extinction

After a year of extensive collecting in the archipelago, the Cal Academy crew returned to San Francisco on Thanksgiving Day 1906, with the largest and most complete collection of Galápagos specimens ever made. In addition to scores of birds, insects, fish, and lizards, they had found previously unknown populations of tortoises on Santa Fé Island and Volcán Alcedo on Isabela and brought home what for more than a hundred years was the only tortoise specimen ever found on Fernandina. Their search, Van Denburgh wrote in 1907, had "met with far greater success than I had dared anticipate."[95] But for a century, it appeared that this success had come at the expense of the extinction of a species just at the moment of its discovery.[96] It was also the last expedition of its kind. Over the following decades saving tortoises would come to mean saving them alive rather than preserving museum specimens.

In 1930 a new breed of tortoise collector, one who was concerned with protecting living tortoises, named a new species *T. vandenburghi*. The name identified animals from a population on Volcán Alcedo that Van Denburgh had described on the basis of Beck and Slevin's collections but did not feel he had enough information to formally name. The collector, Ralph DeSola, came across the Alcedo tortoises during a 1928 expedition that took 180 animals from Isabela to distribute to zoos across the southern United States, with the goal of creating breeding colonies that would save the species.

Van Denburgh's 1914 analysis of the tortoises was a turning point in the history of tortoise collection. What was new about his monograph, aside from the breadth of its coverage—describing what he thought were fifteen different

races—was that he made the first chart showing the extinction status of Galápagos giant tortoises.

Van Denburgh's status chart was a representation of concern about the
tortoises, but it also provided good news. Although Beck had sounded an
alarm about the tortoises' pending extinction, his collections showed Van
Denburgh that the species were doing better than naturalists had thought. All
the species previously thought to be extinct were still living, with the exception
of the Floreana and Santa Fé tortoises. What the chart showed was not so
much a report of extinction losses, but the idea of populations at risk. Although the distinction seems subtle, the difference between imminent extinction and endangerment created a space for thinking differently about saving
tortoises. Coupled with greater knowledge about the tortoises' ecology—
gleaned from Rothschild's and others' attempts to keep them alive in London
and Slevin's detailed notes about Galápagos climate and tortoise diet and nesting behavior—this was news that would reshape efforts to save the animals
over the following decades.

DeSola's description of *T. vandenburghi* was not a normal species designation
like the ones Van Denburgh and Rothschild had made on the basis of morphological descriptions of type specimens. Instead, DeSola wrote of his experiences
on Isabela watching the tortoises' *"liebespiel."* "In the midst of volcano scarred
tortoise country" north of the Santo Tomás colony on Isabela, DeSola stumbled
across mating tortoises. After trekking past the nearest peak, "Villamil Mountain" (Sierra Negra), and passing "whitened shells, bleached bones, and . . .
dried dung" along old trails, DeSola and his local guides found living tortoises
just south of the Perry Isthmus. He described watching their slow, crashing
mating—punctuated by guttural grunts audible from hundreds of yards away.[97]

DeSola's paper on tortoise mating was far from a typical taxonomic description. Nonetheless, it was published in a well-respected herpetology journal,
and the name he gave the "specimens"—though they were living animals, not
preps—has persisted, being recognized by subsequent naturalists. The greater
attention DeSola paid to descriptions of mating practices than measuring variation is indicative of an emerging new phase of attempts to save living giant
tortoises and encourage their reproduction.

DeSola had found himself a voyeur of tortoise *liebespiel* as part of a 1928
expedition led by Charles H. Townsend, director of the New York Aquarium.
Townsend had first become enamored of the Galápagos tortoises as a young
naturalist when he accompanied Alexander Agassiz's deep-sea dredging
expedition, which visited the Galápagos aboard the US Fish Commission's

Albatross, on which he was posted. Townsend was one of several naturalists of his generation who started their careers as taxidermists but became involved in efforts to protect living animals in zoos and aquariums after the turn of the twentieth century. In 1887, he had been publicly rebuked for collecting what were believed to be the last living northern elephant seals, although he later led an expedition that would encounter another colony of the animals.[98] The experience was perhaps instrumental in shaping his later conservation efforts. During the 1920s, Townsend spent time in New Bedford, Nantucket, and Salem, Massachusetts, examining the logbooks of whaling records and collecting stories from retired whalers about their experiences hunting tortoises.[99] Townsend's stories, unlike Beck's description of the oil hunters, did not condemn the whalers who hunted tortoises, perhaps because these hunters were not his contemporaries, but of an earlier generation. Nonetheless, his research on their take spurred new concerns about the animals' "impending extinction," albeit with a different mission than the one Rothschild and Loomis had given Beck. What was now important, Townsend wrote, was "not the collecting of more specimens for Museum purposes, but the preservation of such species as may be living."[100]

Although Günther had worked to protect Indian Ocean tortoises, the Galápagos animals remained, in the late nineteenth century, on the far side of the Americas and out of reach of the colonizing instincts of the British government. It was not until the late 1920s that elite American naturalists tried to save living giant tortoises—something facilitated by the opening of the Panama Canal in 1914 because it took men like Townsend, associated with the New York Zoological Society, within cruising distance and thus facilitated the southern reach of neocolonial US science.[101]

Saving tortoises alive was the task Townsend set out for himself, and in March 1928 he led an expedition to the archipelago aboard the US Fish Commission's USS *Albatross II.* After searching Pinzón in vain, they went to the Villamil settlement and hired locals to help them capture tortoises (fig 3.6). Although none of Townsend's crew had been there before, the settlers were familiar with northern tortoise hunters. It was the same place the Cal Academy crew had spent two weeks more than twenty years previously, where colonists had helped Beck hunt during his visits. Beck had even relied on them to send him tortoises after his 1902 trip. In 1903 Beck wrote to the leader of the Isabela colony requesting that his hunters send him more tortoises in San Francisco so that he might sell them. His letter had strict instructions for keeping the animals alive: "If they die I will lose the money [and not be able to pay] . . . you

Figure 3.6 Settlers on Floreana surround a pen with giant tortoises, 1905. The Floreana species was extinct by this time, so these tortoises would have been brought to the island for subsistence and to sell to visiting ships and naturalists. Photo by Rollo Howard Beck. (© California Academy of Sciences)

would have to handle them more carefully than those other big ones you brought down. Two of them died before we reached San Francisco and another died after they got up to San Francisco. None of those we caught died. Their legs *must not be tied* too tight, and only one leg left tied each night and a different leg every night. Then probably none will die coming up in a steamer."[102]

Townsend hired a crew of twenty islanders for a week's hunting trip, during which time they managed to secure 180 living tortoises. DeSola went along as overseer, translator, and photographer. He learned from his guide how to differentiate male and female tortoises and had plenty of opportunity to watch the animals during mating season.[103] Once they returned, the ship's crew loaded the tortoises on deck and headed for the Panama Canal.

Based on Beck's and Rothschild's earlier successes, Townsend was confident that the tortoises were "hardy animals" and could withstand being shipped abroad. Taking the tortoises home, he thought, was "the only hope of keeping the stock alive" because naturalists could not protect the animals'

safety in the far-flung Galápagos Islands where they were susceptible to colonists' hunting.[104] "The only hope," he stressed, was "to establish breeding
stocks under domestication. Otherwise the great tortoise will be listed with the
dodo, the great auk, and the passenger pigeon."[105] It was a goal that fit the
changing mission of zoos at the time, which sought to expand previous work
on science, education, and entertainment to also increase populations of wild
species, such as American bison.

To avoid the problems Rothschild had faced keeping the animals alive in
London, Townsend distributed the animals to zoos and botanical gardens in
the Panama Canal Zone, Bermuda, San Diego, Arizona, San Antonio, Houston, New Orleans, Miami, and Hawai'i. These were places where zookeepers
had applied to the New York Zoological Society to house and breed the animals
and that, he thought, would be climatically suitable—not as cold and damp as
the English climate that had claimed the lives of Rothschild's tortoises.

Townsend's goal was to save the animals from extinction but also to propagate tortoises as the British colonial government in Aldabra had. Since the
mid-1870s, thanks largely to pressure from British naturalists organized by
Albert Günther, the Mauritius governor had agreed to start a breeding colony
of giant tortoises to help protect animals jeopardized by a timber business on
uninhabited Aldabra Island. But the government had also permitted private
citizens to keep and even farm tortoises—they were popular pets and food
sources on the islands. Townsend's aim was similar—to save the tortoises by
breeding them in zoos and eventually to use them as a food supply: "in another decade," he wrote, we may be able to "add the great Galápagos tortoise
to the domestic animals available for the arid regions of our Southwestern
States."[106] The animals—once a favorite and plentiful food source for generations of sailors—had become endangered and in need of saving, but that did
not imply that they should not be eaten. (That was a taboo that would begin
with in situ conservation in the 1960s.)

As with Rothschild, Townsend's efforts to bring living tortoises home ful-
filled goals both public and personal, scientific and amusing. Whereas Rothschild was only minimally involved in caring for living tortoises at the London
zoo, Townsend took an active role in directing zookeepers on the proper methods of tortoise care. Keeping tortoises alive required a different set of skills
than making museum preps. Townsend wrote to keepers with advice on diet
(tortoises loved cactus) and how to keep the animals warm at night; he even
visited the colonies and moved tortoises out of zoos that did not have ade-
quately warm housing.

Despite his efforts, a decade later Townsend had to report that his zoo-based breeding colonies had not been successful. Although many of the animals grew rapidly, in just two years more than fifty had died. Those who did survive did not breed within Townsend's lifetime. The amorous tortoises De-Sola described were not so amorous in zoos. A near-tropical latitude was not enough to ensure the effective acclimatization of animals to new environments. It would be nearly thirty years before zookeepers became attuned enough to the tortoises' needs to allow for successful captive breeding.[107] The will to save the animals alive was only marginally successful outside of the islands, and in the 1930s, emphasis among foreign naturalists turned to protecting them in situ.

What's in a Name?

The Latin names for species may seem like scientific minutiae, but these names, assigned by naturalists during a period in which the nature of species changed from God-given, fixed kinds to evidence of evolution, reflect the very foundation of Western understandings of nature and conservation. At the turn of the twentieth century, when naturalists named these tortoise species, they were ostensibly making objective descriptions about nature. But the epithets they chose to distinguish different populations are telling paths into the worlds that shaped the conditions of natural history. These were worlds shaped by imperial exploration, a will to possess, and the assumed authority of Western naturalists to take and to name the nature they saw as their own.

What emerged from this period was a will to save the species that would power conservation for the next century. Through histories of collection and classification, scientists identified giant tortoises as individual species and also became aware of the fragility of their existence. As the tortoises became understood as endemic to the archipelago, so too emerged an understanding of them as being in danger of extinction. This story is not unique to the Galápagos. But the particular history of the Galápagos, with its Darwinian overtones and charismatic megafauna, makes for an illustrative case that demonstrates how the foundation of modern understandings of nature rests not only on the backs of species, but on the institutions and practices of natural history through which evolution was created rather than discovered.

Indeed, species names are not representative, but constitutive. They are a telling part of the history of species, epithets that tie together the evolutionary history of the animals themselves and the ways that animals have come to

embody deep meaning through entanglement with histories of human knowledge production. Examining these entwined histories shows that species are
not just biological organisms, or conceptual representations of them, but beings that embody big ideas—the proper relationship between humans and
animals, cosmology, the origins of life, and the biological and geological processes that have shaped the history of the earth. What we save, when we save
species today, are not only animals, but specific ways of understanding them
forged through histories of scientific exploration and collection. The legacy of
this work is not only an evolutionary understanding of the tortoises, but a will
to save them that positioned particular actors and forms of knowledge as the
appropriate trustees of nature.[108] Over the next century, these living fossils
would be at the center of a desire to save the Galápagos archipelago as an evolutionary Eden. The understanding of species as both endemic and endangered laid the groundwork for movements to protect the Galápagos as a
natural laboratory. But as Beck, Townsend, and other naturalists well knew,
their understanding of the islands as a Darwinian world was not the only way
of seeing and engaging life in the archipelago.

4 • The Many Worlds at World's End

In which the tortoises are cursed, and curse settler colonies

There is a story in the Galápagos, at least according to Ecuadorian historian Octavio Latorre, of the curse of the giant tortoise. As he tells it, if you look into the eyes of a giant tortoise, you will be met by "a gaze that is both mysterious and piercing." The animals' long stares make it seem as if they are reflecting on the purpose of life. They "presumably have stood witness for at least a century," gathering experience that "enables them to perceive the motives and ambitions of those visiting the islands: weather [sic] they come to destroy this peaceful refuge or simply to admire it."[1]

The tortoises, observing the world with their small, black eyes, draw on the wisdom of their many years to judge the intentions of the humans who come to their islands. They have ostensibly seen thousands of sailors arrive to hunt their brethren, whether for food and oil or for scientific study. Today they watch as thousands more tourists come to see the islands and to see them, to stare into the inky abyss of their eyes and confront a nonhuman being whose consciousness we can never fully understand but who nonetheless holds and returns our gaze. "The slow, deep stare of the giant tortoise," Latorre wrote, "marks the approval of the visit or the announcement of the newcomer's death; it is a safe welcome or a curse which will be carried out without fail in the most varied of circumstances." For him, the message of the tortoises' gaze was clear: "ALL ATTEMPTS AT BLIND EXPLOITATION, WITH NO REGARD FOR THE ENVIRONMENT OF THE ISLANDS, WILL END IN FAILURE OR DEATH."[2]

This ominous warning was the lesson Latorre drew from looking back at the fraught history of island colonization. The tortoises were not passive victims of centuries of exploitation, he warns, but rather had the power to determine the fate of island visitors. The curse resonates because of the emotional pull of asking readers to consider the tortoises' perspective. This is a common move among conservationist writers who combine the mystery of the romantic sublime and the charisma of nonhuman animals to make an ethical plea, to ask publics to care about the fate of wildlife. It is certainly an effective ploy, and it is tempting to look back on histories of sailors' hunting, natural history collection, and colonization from the tortoises' perspective and to tally the cumulative effect of hunting associated with these histories on tortoise populations. But this is not a very helpful lens through which to understand histories of island colonization.

I say this not because I doubt the curse—I actually quite like how it asks us to reimagine the world from a position not our own. Rather, I say this because Latorre's conservationist values, his stated desire to protect "the enchanted Islands of Galápagos" as a place apart from "urban mass production or commerce" that "should only be visited, admire[d], remembered," have seeped into the worldview he ascribes to the abstract figure of the tortoise.[3] The judging tortoise looks back on history and condemns those who exploited his kind. While we might find ourselves sympathetic to this reading of history, it does not help us understand who came to settle on the islands, what their motivations were, or what their lives were like. Understanding these earlier perspectives is essential for understanding the plight of the giant tortoises, both then and today. We must therefore look back at the history of the islands through the eyes of those who, along with the tortoises, lived there.

From the 1830s to the 1940s, a sense of just what kind of place the Galápagos Islands were was very much unsettled. Over the century after Darwin's visit, while naturalists were debating the evolutionary history of the giant tortoises, the Galápagos hosted various settler colonies based on disparate, and conflicting, understandings of nature. The archipelago became a meeting place for people who arrived with many different visions for the kind of world they wanted to create at world's end: agricultural colonies, remote prisons, and desert island retreats. Each of these geographical imaginations is quite different from the natural laboratory for science, conservation, and tourist enjoyment that many people now understand the Galápagos to be. Yet these older imaginations of the archipelago did not become things of the past when conservation campaigns began in the 1930s. The natural laboratory ideal emerged from scientific understanding of the islands' species, but so too did it evolve out of these other imaginations. Over the past two centuries, these disparate visions have become entangled in ways that have remade, and continue to remake, the nature of the islands. Before we turn to the older worlds at world's end, however, let us begin with the history of the archipelago as a natural laboratory.

Scientific Adventure at World's End

On 21 September 1935, one hundred years after Darwin first came ashore in the Galápagos—"precisely to the day, month, and year"—German-American writer and naturalist Victor Wolfgang von Hagen erected a bust of the scientific hero at the bay where he had first landed on San Cristóbal.[4] At least, that was von Hagen's intention. He got both the day and the place slightly wrong.[5] Nonetheless, he saw his Darwin monument as an important marker that would draw international attention to the need to "conserv[e] the irreplaceable natural phenomena of the archipelago, and to save from extinction this living laboratory for the study of evolutionary processes."[6] It was the first tangible instantiation of what are now well-rehearsed calls to conserve the Galápagos as a Darwinian laboratory.

During the 1920s and 1930s, a new generation of naturalists, including Charles Townsend and Ralph DeSola, visited the Galápagos in Darwin's wake. Once the islands became more easily accessible from the North Atlantic with the 1914 opening of the Panama Canal, New Yorkers joined Californians in cruising tropical waters to the archipelago. The peace and prosperity of the interwar years led to a surge in ocean voyaging among wealthy northerners. The islands were a popular stop for yachters, including US capitalists J. P. Morgan, Andrew W. Mellon, George Vanderbilt, Vincent Astor, George Putnam, and California oilman G. Allan Hancock, among others. These men's private yachts stopped at Tagus Cove, the former rendezvous of whalers, where their passengers painted the cliff walls with the names of their ships, many of which reflect a desire for adventure—the *Corsair* (Morgan), *Vagabondia* (Mellon), *Pioneer* and *Ara* (Vanderbilt), and *Nourmahal* (Astor, fig. 4.1).[7] These luxury adventures made the Galápagos into a vacation spot, a place where guests could cruise and do some sport fishing in tropical waters. But many of these pleasure cruises also had a scientific purpose, often carrying well-connected young naturalists along for the chance to make observations in Darwin's "classic ground" and to take home collections for museums and zoos.[8]

The most famous of these elite, scientific pleasure cruises—and the inspiration for many others—was the 1923 voyage of the *Noma*, which was chronicled by the popular science writer William Beebe in the best-selling book

Figure 4.1 Members of Vincent Astor's 1930 expedition aboard his yacht the
Nourmahal with creatures to take home: six young giant tortoises from Santa Cruz,
three land iguanas, and two penguins. Several New York naturalists joined Astor
(wearing a tie) and friends to lend scientific credence to a sport-fishing trip. Charles
Townsend is third from the right. Photo by Kermit Roosevelt and Elwin Sandborn.
(© Wildlife Conservation Society. Reproduced by permission of the WCS Archives)

Galápagos: World's End. Beebe, the curator of birds at the New York Zoological
Society, led the cruise aboard New York electric power magnate Harrison Williams's 250-foot steam liner. Aboard was a party of twelve that included as
many socialites as scientific staff—in addition to the yacht's working crew of
fifty-eight. As the director of the American Museum of Natural History, Henry
Fairfield Osborn, wrote in the book's foreword, they took off "in the wake of
Charles Darwin!" for a two-and-a-half-month cruise from New York through
the Panama Canal to the Galápagos. Though the long cruise left the voyagers
only four days in the archipelago, the book did not suffer for want of adventurous tales.[9]

As Beebe narrated his first walk on Santa Cruz (which he called Indefatigable), he rejoiced in "the thrill of one's first desert island."[10] His vivid descriptions of wildlife were reminiscent of explorers' tales from centuries past—and

in case the similarity was not apparent, Beebe made the comparison to early adventurers explicit. When "[a] little duck flew down, paddled and waddled to our very feet, looked up into our faces, and quacked in curiosity and astonishment," Beebe "knew it for the fearlessness of the Garden of Eden, the old tales of Cook and Dampier come true again." Beebe and co-author Ruth Rose (an actress and "trip historian") quoted at length the tales of past explorers, pirates, and marooned sailors. In the Galápagos, the party copied Darwin's experiments, even throwing iguanas into the sea. The islands were a place where nature had "remained almost unchanged" since Darwin, the "immortal naturalist," had visited. "Generations of . . . creatures came and went without ever seeing a human being," Beebe concluded. In the following decades, *Galápagos: World's End* became a cornerstone for the construction of the Galápagos as both an evolutionary Eden and a destination for intrepid tourists. It would also spur European and North American colonization, despite Beebe's description of the human history of the Galápagos as a tale of "thirst, war, tortoise meat, and mystery."[11]

But Beebe also dramatized the islands as the setting for romantic stories of castaways, repeating a common trope of northern writing about tropical islands.[12] After first setting foot on Santa Cruz, he noted that "Robinson Crusoe was brought here by his buccaneer rescuers and must have rejoiced that his luck had not cast him upon these inhospitable shores."[13] Of course it was not the fictional Crusoe, but Alexander Selkirk, the sailor on whose life Defoe based *Robinson Crusoe,* whose rescuers stopped in the Galápagos after retrieving him from Juan Fernández; but Beebe's point about the difficulty of life on the dry, rugged islands was clear. The socialites aboard the *Noma* were grateful for the luxury accommodations, which were a far cry from the conditions castaways endured.

Upon returning from the trip, when the *Noma* docked in New York, Beebe met one of these castaways in person. A taxi driver approached him, having heard that the ship had just returned from the Galápagos. He told Beebe a firsthand account of being shipwrecked on Santa Cruz seventeen years before when the Norwegian bark running coal from New South Wales to Panama on which he was working was becalmed near the archipelago. After waiting for a breeze for weeks, the sailors abandoned ship and rowed a skiff loaded with food toward where they thought the islands were. Eventually they found land and rushed ashore to find water, but they found only brackish water in coastal pools. As they searched, their boat was smashed by the incoming tide. Stranded on the island, for two months they survived by drinking the blood of sea turtles

and eating their meat raw—before one man realized he had a box of matches
in his shirt pocket. They made shoes of sealskins. "One thickness," the driver
said, "didn't last long on the lava, so we'd put five or six thicknesses round our
feet and punch holes and lace them on with thongs made out of long strips of
skin. Our feet looked like ferry-boats, but it sure saved us a lot."[14] Two of the
ten men died before the others were rescued after five months on the island.[15]
This was not so much an Eden as a hell on earth.

Beebe's time in the islands had been much more pleasant, given the luxury
accommodations on the ship, and the *Noma*'s scientific crew's interactions
with wildlife focused not on their own survival, but on capturing animals for
museums, zoos, and aquariums. The ship had been converted into a floating
scientific station. The library was made into a "storeroom for fishnets, tents
and birdskins" and the "daintily appointed sun-parlor" into a dissection
room.[16] But the living animals they transported were what most captured passengers' attention. Animals had the run of the decks: three penguins were
"irresistible pets . . . so tame that it was almost impossible to keep them at
arm's length long enough to take their photographs." They waddled around,
"sometimes managing to hop over the high threshold into the smoking-room
where they carefully inspected the phonograph. With foolish little stumpy
wings outspread, they hurried from one person to another, looking up into
each face in an appealing, anxious way which meant that fish were ardently
desired." The phonograph was also popular with Benjamin, a pet sea lion pup:
"When allowed to roam, he always turned in that direction and after much
flopping, hoisted himself over the sill and fell with a thud into the smoking-
room. On one occasion he managed, by super-sea-lion efforts, to climb to a
leather-covered seat and from that to the top of the phonograph, where he was
discovered lying cozily in and on someone's new straw hat."[17] Back in New
York, the charming animals made their debuts into New York society at a party
on the lawn of the New York Zoological Society, before being placed in the zoo.
After they eventually died, they would be transformed into museum exhibits.[18]

As we have seen, for naturalists in the early twentieth century, taking animals to museums and zoos was a way of saving species from the threat of extinction.[19] Even if they were preserved as still life, at least museum specimens
would be available to scientists. But biologists of this era were not only interested in natural history collection. As the field of ecology gained prominence,
naturalists increasingly sought access to study species in their natural habitats.
But ensuring such in situ study necessitated conservation efforts.[20] This was
the origin of the geographical imagination of the islands as a natural laboratory.

During the 1930s, as US and European scientists returned from voyages to the Galápagos, they began calling for conservation measures.[21] For them, the Galápagos were a living laboratory in which to study evolution, as Darwin had before them. The importance of Darwinian ideas resurged with the modern synthesis of genetics and natural selection in the 1930s and 1940s. It was a time of changing methodologies for biologists, when laboratory-based work dominated research and its spokespeople belittled taxonomic museum-based study as less rigorously scientific than experimental research.[22] Following a Cal Academy expedition in 1932, ornithologist Harry Swarth (curator of birds at the Museum of Vertebrate Zoology at Berkeley) called on Ecuador to create an "outdoor biological laboratory" in the archipelago, which was "one of the most amazing natural laboratories of evolutionary processes on earth."[23] One of Swarth's shipmates, California botanist J. T. Howell, told an audience, "It would appear that there are few places in the world where [evolutionary] problems can be studied under such favourable and unusual conditions—a fact which has led me to call the archipelago Evolution's workshop and showcase."[24]

The idea of the Galápagos as a natural laboratory suggested a remote, isolated, and controlled space where scientists would have access to the last remnants of pristine nature and could study unimpeded natural processes. For these biologists, the metaphor was not an oxymoron that juxtaposed the unruliness of wild life with the enclosed sterility of a more traditional site of scientific experiment. Rather it reflected the scientific value of the archipelago as a place where *nature* ran "experiments" in evolution.[25] The Galápagos archipelago provided a unique landscape for field-based study where the simplicity of life on stark and scarcely populated volcanoes helped to make processes of evolution visible. The islands were an exceptional place where biologists could actually study evolution in situ. The giant tortoises were central objects of this laboratory-world, as were other native species, including finches, penguins, flamingos, albatross, and flightless cormorants. For biologists, protecting these animals within a nature reserve was the way to preserve the utility of the natural laboratory for scientific study. Rather than collect specimens for museums or zoos, for this next generation of field biologists, place-based conservation was crucial.

When foreign naturalists used the natural laboratory metaphor, they were arguing not just for the importance of fieldwork, but also for a specific vision for protecting the islands. Remote, tropical, scarcely inhabited, and of great scientific value, the Galápagos Islands were an ideal "natural monument" for preservation.[26] Although foreign naturalists at this time had multiple views on what form Galápagos conservation should take, their ideas reflected a romanticized view of pristine wilderness that was central to the creation of US national parks as well as European models of parks created for explicitly scientific use.[27] As von Hagen's monument made clear, Darwin was the heroic ancestor of a new, field-based "scientific" natural history.[28] In the Galápagos and elsewhere Darwin had worked outdoors with easily replicable experimental methods and used his sharp eye for observation as the basis for insightful analysis of the natural world. The most famous example of this new era of Galápagos research came in response to Swarth's and others' calls for more field-based attention to the study of Galápagos finches. In 1938, British intellectual and prominent Darwinist Julian Huxley—grandson of "Darwin's bulldog" Thomas Henry Huxley—who was then secretary of the Zoological Society of London, sent a young schoolteacher and amateur birder named David Lack to the Galápagos. Over the next decade, Lack's field- and museum-based research on patterns of evolution among Darwin's finches made both the species and himself stars of modern evolutionary biology.[29]

Yet studying the nature of evolution in Darwin's footsteps was not the only scientific motivation in the 1930s. Two years after von Hagen erected the Darwin monument, in 1937, the Ecuadorian National Scientific Commission sent a team of researchers to the archipelago.[30] Among them was geo-botanist Misael Acosta Solís, who was then the director of the state botanical institute and would become a leading conservationist in Ecuador, directing the state's first Institute of Forestry in 1948 and later its Institute of Natural Sciences.[31] Like the foreign naturalists, he also returned from the archipelago calling for the creation of reserves to protect rare species, but he was primarily concerned with reforestation and the possibilities the islands held for agricultural development. This perspective reflected long-running Ecuadorian desires to harness the value of natural resources, but it fit uneasily with foreign scientists' desire to preserve a natural laboratory.[32]

It was this desire to preserve a natural laboratory rather than to conserve natural resources for development that motivated an international campaign to protect the Galápagos in the 1930s. Acosta Solís and other Ecuadorian scientists and policymakers were involved, but foreign scientists and Ecuadorian diplomats in the United States and Europe had the greatest influence on protectionist legislation.[33] Their work was a precursor of the powerful influence of international organizations in the proliferation of conservation territories across Central and South America during the late twentieth century. Acosta Solís vociferously campaigned for conservation and the creation of a state forestry department beginning in 1936, but he lamented that little action was taken domestically on conservation until interventions by foreign scientists in the 1940s.[34]

Campaigns to study and protect the Galápagos were part of a rush among US biologists in the early twentieth century to establish research stations across the tropical Americas—what historian Megan Raby called a "scramble for the Caribbean." To establish Caribbean stations, "U.S. biologists became embedded in the networks of empire," both formal, such as US control of the Canal Zone in Panama, and capitalist, in the form of private agricultural companies.[35] These biological stations, also established in Europe, the United States, and Africa, served a variety of scientific and political purposes but tried, as historian Raf De Bont wrote, "to provide 'universal' truths, while stressing that many of the great problems of zoology could not be solved in placeless urban sites, but only at specific locations of 'real' nature."[36] Although it was relatively unusual, some of these stations were established alongside nature reserves created for explicitly scientific purposes. One such reserve for "exclusively scientific aims" was the Prince Albert (now Virunga) National Park in the Belgian Congo, overseen by Belgian scientist Victor Van Straelen, who would play a central role in Galápagos conservation in the decades to come.[37]

Galápagos preservation, however, was organized not through formal or capitalist empires, but through neocolonial networks of transnational governance that extended the influence of northern scientists across much of the world.[38] One of these institutions was the American Committee for International Wild Life Protection. A spin-off of Teddy Roosevelt's Boone and Crocket Club of elite, male hunter-naturalists, the American Committee formed at a time when naturalists became increasingly convinced that national approaches alone could not protect endangered species, many of which were migratory or otherwise not confined to national political boundaries. The first effort to protect the Galápagos was led by US ornithologist Robert T. Moore, a researcher with the California Institute of Technology. Moore chaired a subcommittee of the American Committee on Galápagos protection, working with Ecuadorian officials and other northern institutions, including the British Association for the Advancement of Science, the Carnegie Institution, the Fifth Pacific Science Congress, and the Office international pour la protection de la nature in Belgium. These organizations were at the heart of an increase in international cooperation on conservation during the 1930s—the seeds of what would become, over the coming decades, major postwar international environmental organizations, such as the International Union for the Protection of Nature

(1948; later changed to the International Union for the Conservation of Nature) and the World Wildlife Fund (1961).[39]

Moore worked with Víctor Manuel Egas, the Ecuadorian consul in Los Angeles, as well as primatologist Harold Coolidge. Coolidge was particularly well connected: a cofounder of the American Committee and curator at Harvard's Museum of Comparative Zoology, he played a central role in conservation efforts in the Americas during the 1930s and 1940s and across much of the globe after World War II. Together, the men drafted legislation that provided the first protection for wildlife in the Galápagos. It prohibited capture of species threatened by industry and scientific collection, including tortoises, fur seals, sea lions, penguins, albatross, flamingoes, and flightless cormorants; made several of the islands "inviolate refuges for all forms of zoological life"; and called for the creation of a Darwin Memorial Zoological Laboratory. Ecuadorian president Abelardo Montalvo signed the decree with only minor changes in 1934 but did limit foreign participation in this would-be laboratory.[40]

In an article in *Science*, Moore praised this "first step" toward Galápagos conservation but noted that the American Committee's work would continue because "legislation without enforcement is usually of little value." They needed to finance, appoint, and supply a warden, otherwise "poaching may continue with more or less impunity."[41] The British concurred with the need for protection and a warden but hesitated on the wisdom of establishing a research station, lest a large scientific presence "lead to destruction of fauna rather than their preservation."[42] Even scientists themselves might have disturbed the nature of the evolutionary laboratory. At the time, a major concern was specimen collection—biologists debated how many samples of rare species they could ethically take home from fieldwork. The chief concern, thought Huxley and others on the British committee, was to appoint a scientifically trained warden who could both make studies and enforce protection laws. But just where this warden would come from, and who would pay him, remained open questions to these foreigners. Writing from Ecuador, however, Acosta Solís argued that there would be no need to import foreign experts; Ecuador had plenty of trained scientists to fill the post.[43]

The following year, under a new government, von Hagen, who had not been involved in the Moore/Egas decree, launched an effort to strengthen the legislation and build a research station. Working from Ecuador following a six-month exploration of the islands with his wife, he adopted the ground-up strategy that Coolidge advocated for conservation planning across national borders, recognizing that he needed local support and a personal touch. Von

Hagen pulled together a committee of naturalists, including Acosta Solís, to serve as a local political base from which to lobby the state. But he looked to major US and British institutions of natural science for further backing and funding. He wrote to officials at the American Museum of Natural History, the American Committee, and the British Natural History Museum, feeling that these institutions had a "noblesse oblige" toward the Galápagos because their collection efforts had "sacked the islands of the species." He wanted to open the islands to management by foreign scientists and assured Swarth that his work with the Ecuadorian committee would not jeopardize international direction of a research station: "there would be no fear of disturbance of the plans of the International Wild Life Association [American Committee], for these people will soon lose interest and the whole thing will eventually be worked and operated by outside interests."[44] Von Hagen's disparaging attitude toward his Ecuadorian peers reflects a common assumption of twentieth-century environmentalism: the belief that white northerners knew best how to manage nature.

Yet von Hagen found that opening the islands to foreign control was "a very ticklish political question in Ecuador."[45] Ecuador's sovereignty over the Galápagos had been a sensitive issue for more than a century. Historian of Ecuador Alfredo Luna Tobar argues that the period from 1832 to 1946 was one in which the country struggled to defend its claim on the archipelago in the face of multiple competing attempts by foreign nations—particularly the United States and Peru—to gain control of natural resources and the islands' strategic location near the Panama Canal. Like the Amazon region in eastern Ecuador, the Galápagos was a frontier of the new Ecuadorian nation-state, relatively unknown but not without the promise of future benefit.[46] Whether, and how, the Galápagos might become a productive part of the nation had been a matter of sustained, if never terribly prominent, debate for a century.

In addition to von Hagen hitting a sensitive nerve, the 1930s was also a time of exceptional political tumult in Ecuador; fifteen men occupied the position of president during the decade. The crash of the cacao economy, Ecuador's primary export crop during the first decades of the twentieth century, and the Great Depression left in their wake economic hardship that radicalized a politically fractured Ecuadorian citizenry dismayed by infighting among the dominant parties and their leaders' inability to address social and economic problems. Territorial sovereignty was also a core national concern: Ecuador lost Amazonian land to Colombia and Peru in the early 1920s and continued to guard against Peruvian claims to land along its southern border.

With the political instability and economic woes, the Galápagos Islands were not a central concern of 1930s politicians, but both they and elite capitalists were interested in projects that would secure national sovereignty over the archipelago.

Von Hagen found a sympathetic ear in Frederico Páez, who had been installed as chief executive after a military coup ousted populist *caudillo* José María Velasco Ibarra from his first of five stints as president.[47] Von Hagen worked with the conservative dictator, with whom he drafted legislation that in 1936 declared most of the islands to be a natural reserve and set up a provisional committee of directors to manage them. A permanent presence in the Galápagos would help substantiate Ecuadorian sovereignty. A research station also would provide—as David Lack and other foreign scientists argued to Ecuadorian politicians—a base from which to generate revenue from foreign researchers and tourists.[48]

With the Páez legislation, von Hagen began organizing a research station, but implementing the new law was a frustrating experience. On Santa Cruz, von Hagen had found a working farm belonging to a Norwegian settler that he thought could easily be converted to a station. He began recruiting to the committee prominent naturalists, including Julian Huxley and Leonard Darwin, Charles's grandson, to oversee the project. Both men were sympathetic to the cause but declined the invitation because of growing skepticism about von Hagen. His position outside the scientific establishment encouraged well-connected naturalists to judge him at best as a "promoter" and "amateur" naturalist.[49] Von Hagen's credibility further fractured when he positioned himself as the scientific authority for the islands, even informing foreign researchers that they would need a permit from him to conduct fieldwork. This claim prompted the Ecuadorian government to investigate him and frayed his already strained ties with US and British naturalists.[50] Stymied by his own political missteps, von Hagen left the Galápagos project behind in 1937, decrying to Huxley the need for conservation: "I cannot make people understand that it is not years, but months, days, when some yachtsmen shall remove, or some inhabitant kill the remaining species of a rare tortoise or bird."[51]

Von Hagen's departure, along with the outbreak of World War II, marked the end of the decade's foreign conservation interventions, but Ecuadorian scientists continued working in the islands. In 1937, Páez sent the Ecuadorian Scientific Commission to the islands with, in addition to geo-botanist Acosta Solís, a veterinarian and an agronomist—scientists whose fields of study reflected a priority on encouraging productive agricultural colonies.[52] When

Acosta Solís returned from the Scientific Commission's excursion, he concluded his botanical study with an extended discussion of possible agricultural
development—including fishing, cattle ranching and pork production, and
export-oriented plantations of coconut, coffee, sugarcane, oranges, and guava,
among others. (Guava, or *guayaba* in Spanish, he noted, was already a pernicious invasive species—locals considered it a *mala hierba* that had overtaken
fields of sugarcane. Acosta Solís recommended using the plant's fecundity as
the basis for an industry of sweets production.) This was not the same emphasis on preserving a pristine laboratory that had motivated foreign naturalists;
most of Acosta Solís's plans involved plants and animals that had previously
been introduced to the islands by settlers and sailors.[53]

Calls to protect the Galápagos as a natural laboratory posited a specific view
of nature to be protected that was based on evolutionary understandings of
belonging. Native and endemic species—including giant tortoises, Darwin's
finches, albatross, fur seals, and penguins—were the valuable nature of the
archipelagic laboratory. Northerners' understanding of the Galápagos as an
idealized scientific world emerged from histories of scientific adventure—
from Dampier to Darwin to Beebe—in which naturalists voyaged, toured the
tropical islands, and returned home wiser for their encounters with native
flora and fauna. At stake in saving the Galápagos-as-laboratory was preserving
this history and the opportunity for future biologists to continue its tradition.

But at the close of the 1930s, with the limited success of protectionist campaigns, it looked as if the Galápagos might be headed toward a different
future—one in which scientific study would support island colonists through
a focus on what we would now call sustainable agricultural development. This
vision emerged not from histories of scientific adventure, but rather from a
century of Ecuadorian history in which elites and politicians had sought to
make the islands a useful part of the new nation-state. For von Hagen and
other northern conservationists, island colonists' hunting and introduction of
foreign species were the chief threats to the protection of a laboratory of evolution. But the settlers had come with other geographical imaginations of life at
world's end.

A Lawless Frontier

When von Hagen arrived in the mid-1930s, about five hundred people lived
in the Galápagos. The archipelago had been populated for just over a hundred
years by a diverse set of colonists who did not understand themselves to be

living in a natural laboratory. Among them were elite Ecuadorians who attempted to establish extractive industries, penal laborers brought out to work on haciendas, government officials and police officers who managed island penal colonies, and European and North American expatriates who came seeking opportunity and new lives on tropical desert islands. These island residents had multiple, and competing, ways of understanding island nature.

The Ecuadorian history of the archipelago is by no means a local history, bound by place in remote tropical islands, to be counterposed to the global history of traveling northerners. Both are histories made of hemisphere- and globe-circling connections. The Ecuadorian history concerning the Galápagos, however, principally focused on a different set of concerns than those held by explorers and foreign naturalists. Chief among them was how to ensure, and best use, national sovereignty over the archipelago.

For early Ecuadorian elites, the Galápagos Islands were a frontier for development to be brought into the new state. Ecuador claimed the archipelago in 1832, shortly after its independence from Gran Colombia, sending elite landowners and traders to establish settler colonies to extract value from island resources.[54] Nineteenth-century Galápagos economies focused on cattle ranching, sugarcane production, and the collection of orchilla lichen. The giant tortoises were also a key part of this suite of valuable species. Over the nineteenth century, the tortoises who sailors had long relied on were a resource that fed settler colonies as well as a commodity in a new circuit of trade that connected the archipelago to the Ecuadorian mainland—where, for a time, tortoise oil lit the streets of Guayaquil. The animals were also useful for island homemaking, their disembodied shells used as garden pots and baby bassinettes. This was the nature of value to settler capitalist societies that searched their new homeland for useful resources and introduced new species, remaking nature to make it profitable and to support their lives.

Extractive colonies were powered largely by penal laborers who did not come to island frontiers by choice. Throughout the nineteenth and early twentieth centuries, the Galápagos was not only a field site for scientific endeavor and a frontier for development, but also a place for Ecuadorian politicians to rid the nation of undesirable populations—from the political opponents of *caudillos*, to criminals, debtors, and poor, often indigenous, people. These peons supplied cheap labor that made extraction possible and economically viable. The labor history of the Galápagos at this time is a story of hardship, cruel patronage, and revolt. Workers' discontent repeatedly caused the demise of settler colonies, echoing histories of marginalized peasants across the region whose rebellions against ruling classes have been a dominant force in shaping Latin American societies.

The archipelago's first formal settlement was established in 1832 by General José de Villamil, who petitioned the new state of Ecuador to take formal possession of the islands. Born in New Orleans in 1789, Villamil had lived in Ecuador since 1811 and became a hero during the war for independence. After a voyage to the Galápagos in 1830 to study the islands' economic potential, he organized a colonization company and encouraged Ecuador's first president, Juan José Flores, to stake a claim. On 12 February 1832, an Ecuadorian colonel presided over an official ceremony on the island that the English had called Charles, with the captains of two North American whalers serving as witnesses. As three rifle shots rang out, the colonel claimed for Ecuador the "fertile territory which for the first time was born to the light of society."[55] Villamil and Flores considered the islands *res nuillus*, unclaimed territory, and thus established state domain over the renamed Archipiélago del Ecuador.[56] Soon after, Villamil arrived on the island, renamed Floriana, after Flores (the spelling was later changed to "Floreana"). He brought poultry, cattle, and a labor force of political prisoners and dove into the work of establishing a colony with, as he put it, "an ardor inspired by the country's progress, which convinced me that by increasing [Ecuador's] territory and population, I was also bringing new wealth."[57]

Villamil's colony, and several others that followed, were modeled on haciendas, with an elite *patrón* landowner and a conscripted labor force. That October, during a trip to the continent, Villamil was named governor of the archipelago and returned with eighty-five political prisoners, each issued six pesos and a machete. Tempting fate, Villamil named the settlement Asilo de Paz, Asylum of Peace. Over the following years, many of the colonists who could returned to the mainland, deterred by the difficulty of life on an island covered, in some places, by dense vegetation and in others with fields of sharp lava. They were replaced by penal laborers—criminals, vagrants, prostitutes, and poor mestizo and indigenous men and women packed into Guayaquil's debtors' prisons.[58] Villamil, whom Latorre considered a utopian dreamer, sought to reform and educate the colonists to be men of honor through hard labor and daily routine—a common, if not commonly successful, strategy among liberals of the era.[59] But the future of the colony was ill-fated—not because of a curse from the tortoises they exploited for subsistence, but by lasting patterns of personal power politics and brutality that were legacies of Spanish colonial rule.

Over the next century, agricultural penal colonies became a common strategy for liberal republican-era governments in the region to establish sovereignty over frontier lands while ridding continental society of undesirable populations. Colonization with penal laborers not only populated the territory, but also provided a means for exploiting its natural wealth in the form of fisheries, agriculture, and mining sulfur and salt.[60] Laborers' uncompensated or poorly compensated work allowed *patrones* to eke profit out of the islands. Agricultural penal colonies reflected elite visions for social and political-economic order for the emerging Ecuadorian nation. A central part of this vision was what historian Nicola Foote has called "racialised nationalism"—a desire among elites, most of whom traced their roots to Spanish ancestry, to whiten the national population through *mestizaje,* an ideology of racial mixing informed by social Darwinist assertions about the superiority of light-skinned populations.[61] Penal colonies were a place in which to banish from society those who did not fit ideal constructs of racial citizenship while ostensibly offering opportunities for reformation such that particularly indigenous populations could learn to more appropriately integrate themselves into society.

Yet when the *Beagle* visited Asilo de Paz in 1835, then directed by Vice Governor Nicholas Lawson in Villamil's absence, Darwin saw not a prison but thought the three hundred or so inhabitants were living a "sort of Robinson Crusoe life."[62] He and FitzRoy were surprised by the verdure of Floreana's highlands. After a long, hot, and "dusty uphill walk through sun dried trees and over rugged lava stones," FitzRoy wrote, "our bodies were here refreshed by a cool breeze, while our eyes enjoyed the view of an extensive, fertile and cultivated plain." Sugarcane, bananas, corn, and sweet potatoes were "all luxuriantly flourishing." FitzRoy found it "hard to believe that any extent of sterile and apparently useless country could be close to land so fertile, and yet wear the most opposite appearance. Our eyes . . . were completely dazzled by a sight so new and unforeseen."[63]

Nearby, Darwin described a village of simple houses made of poles and thatched with grass situated near several springs, where giant tortoises once swarmed. Although the animals were no longer as abundant as they once were, Darwin reported enough were left that two days' hunting provided food for the rest of the week. The island was not too dry to be inhabitable, although Darwin noted that scarcity of water was the "main evil" that threatened the colony's survival.[64]

That Darwin interpreted these colonists—who he acknowledged were "far from contented"—as akin to heroic castaways is a testament to the power of the Crusoe story as a geographical imagination that framed how he saw the islands.[65] It reflects a worldview shaped by what literary scholars call the "Robinsonade," the "frenzy" of stories centered around desert island colonists that was, as Elizabeth DeLoughrey argues, "a powerful and repeated trope of empire building and of British literature of the eighteenth and nineteenth centuries."[66] James Joyce thought Crusoe was "the true symbol of the British Empire" and the "prototype of the British colonist"—a figure who embodied the "whole Anglo-Saxon spirit" and who proved industrious with only meager tools and his own ingenuity.[67] For Darwin, the Floreana inhabitants, like Crusoe, were able to thrive on a desert island by subduing nature. "Although complaining of poverty," he wrote, they "obtain, without much trouble, the means of subsistence."[68]

But it was not only for their own subsistence that the Floreana colonists labored. In addition to raising cattle and tanning hides for Villamil, they also sold agricultural provisions to passing ships. The colonists complained of lack of money, but Darwin presumed they had a "more essential want than that of mere Currency, namely want of sale of their produce." But he dismissed this concern as a problem that would "gradually be ameliorated," noting that sixty to seventy whale ships called on Floreana a year.[69] Darwin's understanding of the settlers' complaints reflects the interpretation of Crusoe held by the classical political economists, for whom he was the prototypical *Homo economicus* who produced goods through the exertion of his own labor. These proto-capitalists were simply in need of a market for their goods, something Darwin seemed to presume would naturally grow. But Darwin, though a critic of slavery, did not comment on the fact that unlike Crusoe, these men were not free. Their complaints, as the future of the colony would show, were not about finding a market so that they could profit.

The island life that Darwin romanticized soon devolved into violence. Villamil returned to the continent in 1836 to look for new investment after depleting his fortune. In his absence, the former North American naval captain Jaime Williams was named the new governor. But Williams was a tyrant; Villamil reported that he whipped an eight-year-old boy, caused the death of another child, and broke the arm of one laborer, among other injustices. In 1841, the colonists revolted, forcing Williams to flee to the continent—a struggle between labor and capitalists that would shape not only the future of the islands, but patterns of development on the continent as well.[70] When Villamil returned months later, he found his colony in ruins. Several of the remaining colonists had moved to San Cristóbal, so Villamil rounded up some of the cattle and took them there as well.

The violent demise of the first Floreana colony inspired one of Herman Melville's *Encantadas* sketches. The author spent several days ashore on San Cristóbal (Chatham) in November 1841, when tales of Williams's overthrow would have been fresh.[71] In his fictionalized story "Charles's Isle and the Dog-King," Floreana (Charles) was a "riotocracy" where the peons were not criminals and the poor collected from the streets of Guayaquil, but rather sailors who had deserted their ships and who "gloried in having no law but lawlessness."[72] Melville's Dog-King was a Cuban-born creole who had been awarded the island by the Peruvian government for his bravery during the independence wars. He ruled with a "disciplined cavalry company of large grim dogs" whose "terrific bayings prove[d] quite as serviceable as bayonets in keeping down the surgings of revolt."[73] Only they did not. The sailors mutinied against the ruthless king in a vicious fight that left three men and thirteen dogs dead. The Dog-King, like Williams, fled.

Melville warned of the "difficulty of colonising barren islands with unprincipled pilgrims."[74] And indeed, Galápagos historians have often highlighted the scandalous nature of the *malandrinos* who populated the islands as a central cause for the violence that marred the histories of several colonies. As one wrote, colonists included "all manner of lawbreakers, regular convicts—thieves, assassins and prostitutes."[75] In these narratives, it was the ungovernability of these men and women, coupled with the brutality of island *patrones*, that led to the colonies' demise.[76] Yet the rebellion of lower classes against elites is endemic to Latin American history of this period, when new states used liberal rhetoric as a foundational logic for government but strictly circumscribed just who was to be included in new nations.[77]

The history of Villamil's colony, like Melville's story, reflects the social landscape of a frontier political economy in the first decades of Latin American independence. Villamil, Williams, and the Cuban-born creole are figures who demonstrate both the transhemispheric personal histories of postrevolution elites as well as the accord between their status as heroic military men and powerful *patrones*. Recently, Melville scholars have read the story as an indictment of the *caudillos* installed in the wake of liberal revolutions in South America—a shift that replaced the patronage politics of the Spanish empire with those of a creole elite. These stories are a window into a system of economic development built on a model of political control that reflects the instability of the early

Ecuadorian state. They reflect the strength of *caudillos* and governments' willingness to supply them with forced labor to support a hacienda economy organized to ensure the wealth of the *patrón* through the exploitation of labor and natural resources. As a response to this political order, the success of Melville's riotocracy in overthrowing its *patrón* is not so much a commentary on the lawlessness of the dredges of society, but on the ubiquity of political insurrections during this period and the possibility of an alternative "outlandish politics" based on a more egalitarian social order.[78]

Melville's focus on postcolonial political order differentiated his travel writing from Beebe's and Darwin's apolitical and romanticized stories of life in the Galápagos.[79] The "Robinson Crusoe life" on a desert island that Darwin and many subsequent travel writers imagined in the Galápagos was far from the experience of most settlers. The story of the Galápagos's own real-life Crusoe is a morality tale with none of the celebration of a Protestant capitalist ethic that Defoe emphasized.[80]

The first known inhabitant of the Galápagos was an Irish sailor named Patrick Watkins, who survived on Floreana for several years at the beginning of the nineteenth century. Watkins is said to have deserted an English ship and subsisted by eating tortoises as well as potatoes and pumpkins he grew and sold to passing ships for rum or cash.[81] But although he was industrious, he was not civilized like Defoe's Crusoe. As Captain David Porter told the story, "The appearance of this man . . . was the most dreadful that can be imagined; ragged clothes, scarce sufficient to cover his nakedness, and covered with vermin; his red hair and beard matted, his skin much burnt, from constant exposure to the sun, and so wild and savage in his manner and appearance, that he struck everyone with horror." In case the point was not clear, Porter continued, "He appeared to be reduced to the lowest grade of which human nature is capable, and seemed to have no desire beyond the tortoises and other animals of the island, except that of getting drunk."[82]

What Porter saw as Watkins's descent into animal savagery was reflected in his actions as well as his appearance. Rather than saving Friday, the quintessential "noble savage" whom Crusoe rescued from death, Watkins captured and enslaved five sailors, including a man described only as a "negro" from an American ship, getting them so drunk when they came ashore for provisions that he was able to hide them until their ships abandoned them as lost. Eventually, Watkins escaped the island with his captives by tricking visitors away from their boat. He left behind a note signed "Fatherless Oberlus" in which he protested poor treatment by passing ships and their unwillingness to take him aboard. He claimed he was setting sail for the Marquesas but soon arrived in Guayaquil—without any of the others. They, apparently, had died—or been killed—when freshwater became scarce. Watkins continued on to Paita, on the Peruvian coast, where he was caught and jailed, and there remained as Porter wrote the story in 1812.

Watkins was the anti-Crusoe: trapped alone on a desert island, he was not a *Homo economicus* who applied his industriousness to construct a proto-civilization from nature, saving and enlightening a trusted companion like Friday. Rather, he was himself savage, using his talents and ambition to seize and subjugate his peers, expending their lives in his personal quest for freedom. Melville wrote of Watkins that "the sole superiority of Oberlus over the tortoises was his possession of a larger capacity of degradation, and, along with that, something like an intelligent wit to it."[83] Watkins crossed the murky divide between human and animal—Melville judged his nature as more base than what he presumed the tortoises' to be. Watkins's story ends not as Crusoe's does, with victory over indigenous populations that Defoe presented as savages and a return to Europe—the center of modernity—where Crusoe discovers his plantation has made him rich during his twenty-eight-year absence, marries, starts a family, and eventually takes again to the seas for more adventure. Watkins, however, remained on the periphery, incarcerated literally and in his own savagery. He was an anti-modern anti-hero. The story of his degradation into beastly animality—whether stereotypically racist commentary on Watkins's Irish heritage or the supposedly deleterious effects of tropical climates that concerned nineteenth-century travelers—reflects concerns about what makes us human that were so much a preoccupation in the decades after Darwin's *Origin of Species*.

The world of misery and discontent we can glimpse through stories of the earliest settlers on the Galápagos is a far cry from the remote wilderness nearly untouched except by heroic adventurers like himself that Beebe romanticized in *Galápagos: World's End*. Yet like Beebe's tales of desert islands, stories of castaways and ill-fated settler colonies were important cultural devices that constructed geographical imaginations of the islands as far removed from modernity, at the fringe of ordered society. But the islands were not a fringe: the stories Porter and Melville narrated reflect preoccupations that were at the center of modern life in the nineteenth and early twentieth centuries: what it meant to be human, the supposed superiority of rational economic actors over colonized populations, and the political possibilities of postcolonial societies. Indeed, just how best to bring the archipelago more fully into the life, and prosperity, of the new Ecuadorian nation-state was the main concern of the elite colonists who followed Villamil to the islands.

RESOURCE-BASED DEVELOPMENT

If for von Hagen and his scientific contemporaries in the 1930s the Galápagos archipelago was a place valued for its endemic species, then the colonists who proceeded them saw value in a different, though overlapping, set of species: orchilla, sugarcane, coffee, cattle—and giant tortoises, too. For the colonists on the Galápagos from the 1830s through the first decade of the twentieth century, the archipelago was a place where natural resources could be harnessed for subsistence and profit. How to develop these resources, who could access them, and who had the right to profit from them were the central questions of nineteenth-century Galápagos history.

Villamil had built his colony on cattle, but after the demise of Asilo de Paz, the next great hope for island prosperity was guano, the dried excrement of seabirds, which was then in demand as a valuable fertilizer for high-input, high-output agriculture. At the time, the Peruvian economy was thriving on the export of guano from its coastal islands to Great Britain.[84] Since the Galápagos Islands were also home to the cormorants, boobies, and pelicans who produced guano rich in nitrogen, phosphorus, and potassium, it seemed plausible that they too would have profitable deposits. After Villamil renounced his position as governor of the Galápagos, he served Ecuador as *encargado de negocios* (business representative) in the United States.[85] In 1852, while in New Orleans, Villamil met with Julius de Brissot, an entrepreneur and sailor who proposed a voyage to the Galápagos to investigate whether the islands had guano. Brissot returned in 1854 with the news that several of the islands did indeed have guano deposits that were at least as high in quality as those of the Peruvian islands. The largest deposit—fifteen miles long, two miles wide, and one hundred feet deep—was on Isabela.[86] An enthusiastic Villamil wrote home that Ecuador would soon be "in the first row of the most opulent nations" and promptly returned to the archipelago to confirm Brissot's findings.[87] With a concession from the government, Villamil and Brissot would share guano profits with the state.

Other events in the archipelago, however, threatened their plan. In 1852, the escapades of island prisoners on Floreana caused an international uproar when pirate Manuel Briones and a crew of other prisoners escaped the penal colony by capturing a whaler from New Bedford, Massachusetts. They killed several men and raped a woman before they were caught in the Gulf of Guayas.[88] On behalf of the ship's owners, a US diplomat requested an indemnization of $40,000 to recoup losses. That petition came to naught, but another suggestion made in the wake of the affair—that the United States take the opportunity to acquire the islands for their guano—had more influence.[89]

As Brissot returned with good news for Villamil in 1854, the US minister in Quito was negotiating a treaty with President José María Urbina for rights to mine the Galápagos in exchange for a loan of 3 million pesos to be paid from guano profits. The treaty came just two years before the US Congress passed the Guano Islands Act in 1856, enabling any citizen to take possession of any island with guano deposits as long as it was "not within the lawful jurisdiction of any other Government, and not occupied by the citizens of any other Government."[90] But for this negotiation, the United States would be obliged to respect Ecuadorian sovereignty in the archipelago and to protect the islands from invasion.[91] For Ecuadorian historians, the treaty has raised suspicions about the Louisiana-born Villamil's national allegiances: Was he serving the interests of the United States or the Ecuadorian government? Had he been duped by Brissot, who seemed to be working to further US acquisition of the islands? As it turned out, however, the archipelago's guano supply was so meager that it did not warrant the cost of extraction, and the United States never signed the treaty.[92] The Galápagos guano boom was one of short-lived speculation, but it foreshadowed future extractive projects on the islands.

The focus of natural resource exploitation in the Galápagos soon turned from guano to orchilla, a lichen used to make purple dye before the discovery of aniline. At the turn of the twentieth century, Scottish writer Alexander Mann described orchilla that "grew in abundance on the trees, rocks, and cactus plants around the coastlines. . . . [It] hung in elegant festoons from one to three feet long on the twigs, branches, etc., clothing them in a patriarchal grace, quaint and beautiful."[93] In 1869, a prominent Spanish merchant from the Santa Elena peninsula named José de Valdizán bought exclusive rights from Gabriel García Moreno's government to exploit orchilla for 4,000 pesos a year for twelve years. Valdizán and his wife moved to Floreana, establishing the second colony at Asilo de Paz with a mix of both free men and women and penal laborers.

After visits in 1875 and 1878, naturalist Theodor Wolf described Valdizán's Floreana colony as a lush agricultural paradise, much as Darwin and FitzRoy had seen Villamil's. Wolf, a German geologist living in Ecuador, was surprised by the great fertility of the soil and the ease with which temperate-zone vegetables had acclimated to the island. In plots surrounded by living fences and rows of lemon trees to protect crops from feral cattle grew the "most beautiful" sugarcane alongside yucca, camote, potatoes, indigo, cotton, and huge heads of lettuce and cabbage—one of which was more than two and a half feet in diameter. Under the shade of *plátano* and banana trees grew beets, radishes, carrots, and artichokes. "Ah!," Wolf wrote. "If the islands in all their extension were so favored by nature as this privileged point of the hacienda, they could be transformed into a true earthly paradise."[94]

Such productive development was a common motivation for state-sponsored natural science in Latin America at the time. As historian Stuart McCook has argued, Latin American natural science during the liberal period (1870–1930) responded to elite interest in the development of export commodities—in Ecuador particularly sugarcane, cacao, and later bananas—and sought to "nationalize the natural world" through the production of inventories and maps.[95] Knowledge of the natural properties of the territory was a central element of state-building that fueled the export boom of the late nineteenth and early twentieth centuries. Wolf, a Jesuit, had gone to Ecuador from Germany in 1870 to be a professor at the new Polytechnic University in Quito, part of conservative Catholic *caudillo* García Moreno's efforts to strengthen Ecuadorian science. He traveled extensively studying the state's natural history and geology. His mission was similar to other national surveys in North and South America at the time, which, as they mapped the state and described its resources, *produced* both the nation's territory and its nature in ways that were economically legible.[96]

Wolf, a Darwinist, was eager to explore the Galápagos.[97] After spending months in the archipelago, he returned with mixed news for state development. He was little surprised that the islands did not have much guano, given the damp climate of the highlands. And while he praised the productivity of Valdizán's colony, he warned of the limited prospects of agricultural colonization in the archipelago. Of the 240 square leagues of land across the major islands, he estimated that only 20 (about 60 square miles) were cultivable. Floreana, he thought, would accommodate one hacienda, and San Cristóbal and the other large islands perhaps two or three. But sending numerous agriculturalists would, he wrote, be "a utopic dream." He also cautioned that "the colonist who comes from far away land in search of a new homeland, perhaps accompanied by his family, would not be content with working like a peon in a hacienda." Where in the archipelago, he asked, was there enough cultivable land for numerous free, landowning agriculturalists to not only support a family, but also produce enough for export? Better, he thought, to limit the number of agriculturalists and to diversify potential colonies to include cattle raising, fishing, even viniculture. What would make the islands more prosperous than natural commodities, he thought, was their geographic position. With the opening of the Panama Canal—then under construction by the French—the islands could be an important naval station for refueling and supplying ships with fresh meat and produce.[98] It was a prophetic statement, but such prosperity was much further off than Wolf could have known. Instead, he must have been surprised to learn that just three days after he left Floreana in 1878, Valdizán and another worker were murdered by colonists during an uprising. Like Villamil, Valdizán had hoped to reform his penal laborers but instead was undone by them. Following the incident, the remaining colonists abandoned Floreana, many of them moving to San Cristóbal.[99]

The failures of early colonies jeopardized Ecuadorian sovereignty in the archipelago. In 1881, the US secretary of state sent geographer George Earl Church to Ecuador for three months with the mission of proving that the Galápagos "NO pertenecía al Ecuador," that they did not pertain to Ecuador. At least that is how it was later interpreted by Ecuadorian writer Gonzalo Ortiz Crespo.[100] Church reported on the state's political and economic context, partially for English bond holders interested in recouping debt from loans made during Ecuador's fight for independence and partially to inform US speculation about the archipelago's strategic position in relation to the Panama Canal. In a condescending report that Church acknowledged was "not entirely favorable to Ecuador," he deemed the state incapable of developing itself because of poisonous race relations, an autocratic presidency, the limiting ideology of the Catholic Church, and a lack of entrepreneurial spirit. For Church, Ecuador's insufficient attention to the Galápagos after Villamil's death, as well as the presence of US whale ships, and Villamil's New Orleans birth (even though it was before the Louisiana Purchase) were the basis for a US claim to the archipelago. Church concluded that "the occupation of the Galápagos by Ecuador does not exist to a sufficient extent to entitle it to the respect of other nations."[101] It was a point vigorously, and successfully, contested in the *New York Herald* by Antonio Flores Jijón, son of Ecuador's first president (who would himself become president in 1888).[102]

In response to these northern intimations, the Ecuadorian government encouraged colonization with tax exemptions, land grants, and exemptions from military service.[103] At the turn of the twentieth century, two new colonies took advantage of these measures, creating a lasting Ecuadorian presence in the archipelago. One was led by a coastal caballero, Antonio Gil, who had made a fortune selling *sombreros de paja toquilla*—the white straw "Panama hats" that remain one of the most recognizable Ecuadorian products. Gil, who was appointed *jefe territorial* (territorial chief) of the islands, went first to Floreana in 1897 but found no remaining tortoises there to support a colony and settled instead on Isabela. The island was uninhabited but, he boasted, was home to more than three hundred thousand head of feral cattle. Decades earlier, Villamil had stocked at least five islands with cattle, horses, and possibly goats to take advantage of their native pastures. By the time Gil arrived, the cattle had reproduced prolifically—although his estimate was likely exaggerated. His plan was to export cattle and their skins, as well as to collect tortoise oil and mine sulfur deposits.[104] He established the town of Puerto Villamil, named for his predecessor, on the coast and the settlement of Santo Tomás several miles up into the highlands. Gil also used penal labor, drawing on his previous post as the superintendent of prisons in Guayas province. His colony of about 150 people was more successful than those at Asilo de Paz—these were the people Rollo Beck hired to collect tortoises for the Rothschild and Cal Academy expeditions.[105]

The other, and more notorious, Galápagos colony was headed by a man Latorre called the *emperador de Galápagos,* Manuel Cobos, from 1879 until his death in 1904.[106] In the 1860s, Cobos and his business partner had obtained an agricultural concession for land on San Cristóbal. Cobos was a well-connected commercial trader, and his primary business appears to have been running gunpowder and other contraband between Chanduy on the Santa Elena peninsula, Galápagos, and Panama—something for which he was wanted by conservative García Moreno's government. He fled to Mexico, taking about three hundred Ecuadorian *braceros* with him to harvest orchilla in Baja California. But after García Moreno's assassination in 1875, Cobos returned to Ecuador and headed to the Galápagos, where he expanded a small sugarcane planation. He took advantage of an 1885 Special Law for the Galápagos passed by the Ecuadorian Congress that encouraged colonization with free land, exemption from military service, tax exemption for five years, and the ability to export goods without duty. By 1904, he had grown a "small empire" that encompassed much of the humid southwest portion of the island.[107]

Under Cobos's strict rule, the settlement of El Progreso, about four miles inland from Puerto Chico (renamed Puerto Baquerizo Moreno in 1917 for the first Ecuadorian president to visit the archipelago), became a productive agricultural center where colonists grew coffee, yucca, *plátano*, potatoes, maize, beans, and vegetables.[108] They raised cattle descended from the animals Villamil had brought from Floreana decades earlier. When Alexander Mann visited El Progreso in 1905 he saw vast plains cleared for grazing, for which Cobos had imported *janiero* grass.[109] A photograph of the settlement taken in 1888 by members of the US Fish Commission's *Albatross* expedition shows these plains extending from a village of about three dozen thatched cottages toward the volcanic hills in the distance. In the foreground is a tightly staked fence that surrounds a community garden, likely designed to protect produce from grazing livestock and tortoises (fig. 4.2). Colonists produced salted and smoked beef and cured hides. They also harvested tortoise oil, which Cobos sold in Guayaquil, where it was used to light the city's street lamps. Their exploitation of the tortoises was so thorough that when the Cal Academy expedition visited in 1905, expedition members reported that the island's tortoise population was "nearly extinct."[110] As tortoises became scarce on San Cristóbal, Cobos sent *aceiteros* to Isabela to collect tortoise oil—a move that angered Gil, with whom Cobos had previously had a more productive relationship, even trading cattle for laborers, an economy that revealed the low value these capitalists placed on the lives of their captive workers.[111]

The centerpiece of El Progreso, though, was neither tortoise oil nor cattle, but sugarcane. Mann estimated that the plantation included seventy-four hundred acres of sugarcane and one hundred thousand coffee plants. At its peak, the plantation produced five hundred tons of sugarcane a year, making El Progreso a significant element of Ecuador's economy. Cobos was among the elite landowners on the continental coast (he also owned a plantation near Yaguachi, the epicenter of sugarcane production) who dominated national politics during the epoch of cacao in Ecuador, when the country was the global leader in the production of the luxury commodity.[112] In 1889, he imported to the Galápagos a modern sugar mill, with elements from Glasgow in Scotland and Chattanooga and Buffalo in the United States, that allowed him to produce alcohol and *aguardiente* at scale, in addition to *panela* and anisette liquor. To move cane from the fields to the mill, he installed portable French Decauville railway tracks and train cars.[113] At the time, Cobos's island hacienda was among the nation's most industrialized.[114]

Today, though, the El Progreso plantation is remembered more for Cobos's cruelty than for bringing modern industry to the islands. Sugarcane production is notoriously labor intensive and physically demanding. To provide the necessary labor force, El Progreso operated as a de facto prison with the brutal labor conditions and insular economy common of colonial plantation systems.

Figure 4.2 Fenced garden and thatched homes at El Progreso on San Cristóbal, 1888,
with cleared fields in the background. Photo perhaps taken by Charles Townsend
when he was a young man and did much of the photography for expeditions of the US
Fish Commission's *Albatross*. (Courtesy US National Archives, photo no. 22-FA-88)

Cobos issued his own currency that was valid only at his general store, where
he sold goods at inflated prices so he could recoup value.[115] Laborers reportedly
worked eighteen-hour days with only three days off a year. Cobos regularly
whipped them, sometimes to the point of death, and abandoned others
who displeased him on uninhabited islands—the severity of the punishment
depending on whether the island had freshwater or not. In one extreme case,
he abandoned a man on Santa Cruz with a sign posted near the entrance to
a bay telling passing ships not to pick him up, no matter how he might beg,
for he was "twenty times a criminal." After three years on the island, the man
was recovered by Ecuadorian authorities—but not until after a revolt at El
Progreso.[116]

By 1904, workers had lived under Cobos's harsh discipline and network of
spies for decades. Several times they had conspired to assassinate him without

success. But early one January morning, a group of laborers confronted Cobos with a stolen revolver. A friend was due to receive five hundred lashings that day, a punishment that likely would have killed him. The laborers shot Cobos, and after a struggle, he fell from a second-story window and died. The conspirators fled to the continent, where they were apprehended and marched as prisoners down a main street in Guayaquil. Their story caused a sensation. After a trial that horrified the city with tales of abuse, two men were sentenced to death, but others involved were acquitted.[117]

Histories of cruelty and revolt that plagued Galápagos settlements were, for Latorre, the basis for the curse of the giant tortoise. The lawless frontier economy of the Galápagos in the nineteenth century—so remote from the lives of Ecuadorians on the continent—made the islands seem cursed, a place of hardship and misery, much as they had to Melville. The tortoises themselves also seem to have been cursed by this history—seen by both elite and marginalized colonists as akin to other resources provided by nature to serve human needs. They became, during the century of Ecuadorian colonization, a vanishing resource. While colonists likely would have regretted the lack of these valuable animals, their disappearance was no great cause for alarm, for their use was one among several strategies for making the islands profitable.

During the early decades of the twentieth century both the Gil and Cobos colonies declined, in part because the uproar over Cobos's cruelty obligated state authorities to better monitor the power of Galápagos *caudillos*.[118] Yet in the 1920s, a new wave of Ecuadorian colonists demanded land in the archipelago following an economic crisis on the continent associated with the crash of the cacao industry. The Ecuadorian government also encouraged foreign colonization of the archipelago. While the Ecuadorians who migrated were mostly rural agriculturalists, government officials, and police officers stationed at penal colonies, many of the European colonists who settled in the Galápagos arrived with quite different ideas about the kind of life the islands would afford.

Robinsons in Eden

Unlike the majority of Ecuadorians in the nineteenth century who were coerced into a life in the Galápagos, Europeans and North Americans went willingly, inspired by Beebe's stories of desert islands and abundant nature at world's end. Also in contrast to most of the Ecuadorians living in the archipelago, European immigrants of the era enjoyed the welcoming hand of the state, including paid passage, extensive land rights, and considerable municipal autonomy.[119] Liberal Ecuadorian governments welcomed them, driven by hopes that white Europeans would spur economic development while whitening, and thus civilizing, the national population. But European settlers sought independence and solitude, so most settled not on San Cristóbal, where Cobos's son continued the El Progresso settlement, nor on Isabela, but on Floreana and Santa Cruz, which were uninhabited at the time. Many of these settlers called themselves "Robinsons" in search of a quiet life in "Eden," where they could find solitude and test their mettle by depending only on themselves for survival.[120] They were replaying a long history of European colonization of tropical islands. Yet a romantic life of solitary independence was not what they found. Indeed, one wonders how well they took in the details of the castaway stories Beebe and Ruth Rose told.

The first group comprised 130 Norwegian fishers who arrived in 1926 with plans to start a tuna cannery. The Galápagos were to be a promised land, where the mild climate and abundant sea life would provide settlers with an easy and lucrative living—at least according to the promoter in Oslo, although he had never been to the islands. Families invested 2,500 kroner each in this new life. When the first contingent arrived, they built a large house they named Casa Matriz at Post Office Bay on Floreana. A few months later a second group arrived to settle on Santa Cruz. But they fared worse than previous Ecuadorian colonists had.

When the second group arrived, they were met with the news that the Floreana colony was already dissolving—undone by political infighting exacerbated by the realities of making a new life on a desert island. Many of the Santa Cruz contingent returned home, but several of these new settlers stayed on, determined to make a go of their island life. The US conservationist Gifford Pinchot visited in 1929 and reported that they had brought canning machinery as well as "their cows, their dogs, their chickens, a huge radio set, a quantity of good sensible Norwegian pine for house building, barrels of gasoline, and fishing gear, and farming tools." The settlers built "seven houses and made them habitable, laid out streets and named them, and cut a trail to the top of Indefatigable [Santa Cruz]," Pinchot observed. But waiting at the island's peak was not "as advertised, a lake of living water." Nor did the settlers find coal, diamonds, or silver to mine, as they had been led to believe. The fishing was indeed rich, but they had no regular means of transporting their catch to market.[121] By the time Pinchot arrived, the community had been reduced to three young men, three dogs, and a handful of chickens. The graves of three settlers stood nearby, an ominous reminder that the islands were no Eden.

In 1928 the colony's leader had written to the Ecuadorian government pleading to be bought out. They had been misled by the organizers' promises. "They painted such an easy and simple picture of everything in reference to plantations, fishing, whale hunting, etc.," he wrote, "that [a] hundred of good Norwegians were tricked and today find themselves with no means of subsistence. This is why the colonists have lost everything they had, realizing too late that they were victims of the organizers' hoax."[122] Ecuadorian officials, however, were less than sympathetic. They refused to purchase the tuna canning equipment, which had been damaged in an explosion, and also blocked its sale to the US Van Camp Company, whose fishing boats took huge hauls from Galápagos waters, paying only nominal fees to the Ecuadorian state.[123] Many of the settlers returned to Guayaquil, having invested their life savings in the failed scheme.[124]

The next Europeans, three families also inspired by Beebe's book, went separately to Floreana. They took seriously suggestions like Alexander Mann's that, with a proper water supply, Floreana would be "an ideal 'Robinson Crusoe' retreat for people tired of the artificiality of modern civilisaton."[125] But their fates too seem to have been cursed—they soon became embroiled in a series of mysterious deaths and disappearances.[126]

The first pair arrived in 1929: Friedrich Ritter, a doctor and philosopher who adhered to Friedrich Nietzsche's strident critiques of European civilization, and his "comrade," Dore Strauch, a former teacher who suffered from multiple sclerosis and had been a patient of Ritter's in Berlin (fig. 4.3). They left behind in Germany their spouses, their careers, and even Ritter's teeth—he had them pulled and replaced by steel dentures to prepare for a life without dentistry. (It was perhaps wise foresight—he would end up pulling Strauch's teeth on the island.) But mostly they sought to leave behind the pressures of society for a life of contemplation on an "almost desert island"—"turning our backs forever upon civilization and the society of our fellow men." Inspired by a childhood love of *Robinson Crusoe*, Ritter wanted to be not *Homo economicus*, but, as he wrote, "*Homo solitarius*." He found "organized society . . . to be a huge, impersonal monster forging ever-new chains with which to shackle the free development of its members." Strauch, who proclaimed herself deeply enamored of Ritter and his philosophy, was prepared to escape the life of a *hausfrau* for a life as Ritter's disciple. She set out "to find [her]self in self-abnegation" and to "subjugate the eternal feminine in [herself] as far as possible" through discipline and hard work. It was a relationship dynamic that would shape the fate of their lives on Floreana.[127]

Figure 4.3 Friedrich Ritter and Dore Strauch at home at "Friedo" on Floreana in the early 1930s. They had imported many of their supplies from Germany and bought additional provisions in Guayaquil, but they built the home themselves with lumber and stones collected on the island. (Courtesy of University of Southern California, on behalf of the USC Libraries Special Collections)

When they arrived on the island, they hiked inland to a valley near a small spring. It was "a veritable paradise of tropical vegetation" full of "bananas, oranges, pineapples, lemons, guavas," and other fruits Ritter did not recognize that would have been planted by previous colonists. They performed a small ceremony and "pronounced the land ours," rechristening the place "Friedo"—a combination of their names. It took them weeks to move all their things up from the shore over the rough terrain—a labor for which neither Strauch nor their old horse was physically capable, though Ritter showed little compassion for either and worked the horse to death. Strauch fumed at the animal's ill treatment, but the eccentric duo soon settled in to a life of clearing the land for a garden and building themselves a home—wearing nothing but "high hip boots" to protect them from jagged rocks. As Ritter explained to an audience of *Atlantic*

Monthly readers in 1931, a few days after arriving, the intense rays of the tropical sun forced them to discard their clothing, except when visitors called: "Our bodies beet-red from the sun, we were a toiling Adam and Eve in boots."[128]

A few years later, their isolated Eden was spoiled by the arrival of another German family, the Wittmers: Heinz, Margret, and Heinz's twelve-year-old son, Harry. They had read Ritter's dispatches home and become convinced that they too could make a farm "out of this wilderness by the labor of our hands." They also sought an Eden that would serve as a sanatorium in which to improve Harry's health. Although Ritter and Strauch were less than pleased to share the island, the Wittmers settled some two miles away in Villamil's old Asilo de Paz and tried to give their odd neighbors as much space as possible. Margret prided herself on being a successful *hausfrau*—establishing a comfortable home, re-creating German dishes for special holidays, and even giving birth on the island.[129]

The third group to arrive was led by a self-proclaimed Austrian "baroness," Eloise de Wagner-Bousquet, who promptly declared herself the "Empress of Galápagos." She was accompanied by three male companions—Robert Phillipson, a German whom she introduced as her husband; Rudolph Lorenz, an emaciated man they treated as a laborer, also German; and an Ecuadorian named Valdivieso, who was returning to the archipelago after having escaped life as a peon on Isabela by stowing away on a steamer bound for Europe.[130] The foursome had met in Paris and went to the Galápagos with plans to build a hotel, named Hacienda Paradiso, for the millionaire yachters who frequented the archipelago. Unlike the other settler families with patriarchal, if not traditional, gender relationships, this party was driven by Wagner-Bousquet's whims. Both of the other white women on the island wrote dismissively of this new woman who flirted with all the men, including their partners, visiting sailors, and Ecuadorian authorities (whom she persuaded to grant her a much larger tract of land than the other settlers). These new settlers grated on the nerves of the others—settling too close to the Wittmers, reportedly opening the others' mail, and stealing supplies left by visiting ships. But the baroness was a favorite with at least one US yachter, Captain Allen Hancock, who made several visits in the early 1930s and even filmed a movie in which the baroness starred as a "piratess" (fig. 4.4). The gentlemen naturalists who visited on luxury cruises during this period were as taken with the lives of European settlers on these nearly deserted islands as they were with the native fauna: they took home stories, as one San Francisco headline put it, of "Thrilling Discoveries of Strange Animals and Human Exiles" in this remote world.[131]

Figure 4.4 Eloise de Wagner-Bousquet, Viennese
colonist and self-proclaimed "baroness," poses as a
piratess with German colonist Robert Phillipson for
a film shot during the 1933 Hancock Pacific
expedition. They later disappeared, presumed
murdered, in what remains an island mystery.
(Courtesy of University of Southern California, on
behalf of the USC Libraries Special Collections)

Any semblance of peace among these settlers did not last long. In 1934,
the baroness and Phillipson disappeared, never to be heard from again. The
Wittmers—the only of these families that remains in the archipelago today—
claimed that sailors had taken them to Tahiti. But no one else saw a ship
arrive, something that was usually the cause for much notice. Shortly afterward, Lorenz fled the island with a Norwegian fisher from Santa Cruz, but
both men were later found dead, their bodies desiccated on the shore of an
island in the far north of the archipelago, where they had apparently drifted in

the current. Months later the tortoises' curse appeared to claim another life when Ritter died of poisoning—perhaps an accident caused when the supposed vegetarian ate some bad chicken, or perhaps the doing of his comrade, who could no longer cope with his strict philosophy.[132]

The mystery of what precisely happened on Floreana adds to the romantic history of the Galápagos. Tourists can visit the old site of Asilo de Paz and read firsthand accounts written by Margret Wittmer and Dore Strauch, although the women's conflicting memoirs add more to the mystery than they resolve. Nevertheless, the memoirs provide insight on why colonists came to the Galápagos and how they managed to survive (or not) on deserted islands. These accounts and those written by other educated white women settlers for publics in their home countries detail the ceaseless labor that social reproduction on the island required. The women's narratives also describe social life on the islands—often dramatizing tensions but also acknowledging the many other residents in the archipelago. Ritter's Adam and Eve stories in the *Atlantic Monthly*, for example, completely eclipse the role of settlers on other islands and the Ecuadorian context of their new world. But these relations are revealed in Strauch's book. She wrote of their preparations in Guayaquil, their journey aboard the "bedraggled" *Manuel Cobos* from the coast, and the aid of an "*indio* boy," Hugo, whom they hired to help them get settled on the island. That Ritter neglected these details served to construct Floreana as an island isolated at the end of the world. But foreigners were far from alone in the archipelago, often to their chagrin. When a couple from the United States, Frances and Ainslie Conway, arrived in 1937 seeking a deserted island, they were disappointed to find that they would live on Santiago with thirteen others: an Italian-Chilean man fulfilling his sentence as an Ecuadorian political prisoner, along with his wife and their eight children; a German who had fled military service and his Ecuadorian wife; and a Hungarian who was returning to the island after three years on Santa Cruz. When the ship that transported them from Guayaquil arrived on San Cristóbal, they were greeted by von Hagen's Darwin statue, in the port town near El Progreso, where the ship dropped off convicts and indigenous peons. The islands then, like now, were a place that brought together people from diverse cultural and socioeconomic backgrounds, all of whom had to find a way to live from the land and sea.

A dominant narrative of white settler memoirs is one common to settler colonial societies, that of "subdu[ing] the wilderness and mak[ing] a home for ourselves," as Ritter wrote.[133] In these stories the islands are presented as largely empty space where settlers' labor entitled them to claim ownership. Settlers

wrote extensively about the nature of the islands through description of the heroic travails that daily life necessitated. Strauch and Ritter found little time for
philosophizing after the work of surviving, from building a house and clearing
land to growing crops, hunting wild cattle and boar, collecting water, and pre-
paring food. Others told of clearing the then-tangled groves of lemon trees that
Villamil's colonists had planted and struggling to build fences sturdy enough to
keep wild cattle, burros, and hogs from devouring their plants.[134] Settlers closely
watched climate and weather patterns, waiting for months of gray *garúa* clouds
to lift from the highlands and worrying about gathering enough water during
the hot, sunny winter and spring. Indeed, a drought had precipitated much of
the drama surrounding the deaths of Wagner-Bousquet, Phillipson, and Ritter.
Island nature was for settlers a vibrant material force, not because of evolutionary adaptation, but because it shaped their conditions of life, particularly what
they ate. Settlers wrote extensively about food, from the process of farming and
how the plants they brought—which included "bananas, sugarcane, coffee,
yucca, sweet potato, otoi"—fared on the islands, to setting up kitchens and
learning to smoke meat.[135] They also wrote about the many visitors to the islands and the food and supplies they gave the settlers as gifts—a wheelbarrow,
dynamite, flour, sugar, cigarettes—helping to make island life possible.

Visitors, too, were intrigued by what the settlers ate. The botanist on the
Zaca expedition, J. T. Howell, visited with Strauch and Ritter, who shared his
plan for *"eine Garten-Symphonie"* that would provide for all their needs. But
Howell questioned their success in living apart from the modern world because the couple did not use any native plants. "I wonder how they can reconcile what they are doing with what they are trying to do, to live entirely apart
and away from civilization," he asked in a letter home. These vegetarians
raised chickens and ate a diet of "bananas, papayas, taro, sugar cane, young
coconut and date palms, and corn." But, he ventured, "How much more picturesque and impressive—if they were really staging a 'back to nature' stunt—
would be *Scalesia* cakes with wild honey for breakfast, native tomatoes for
luncheon, and galápagos' eggs for dinner!"[136]

The islands' animal life figured centrally in settlers' tales, as it had for most
visitors. But colonists had different relationships with island nature than did
naturalists. The giant tortoises so sought after by naturalists and earlier sailors
are scarcely present in these narratives; on Floreana the animals had been extinct for more than half a century, and on Santiago they lived too far inland to
be worth the trouble for the Conways. Instead, on Santiago, the native species
of greatest dietary importance was the Galápagos dove. Frances Conway

reported that the birds "were all too easy to kill and even easier to eat"—the Conways and the Chilean family would sometimes kill fifty in a day. The finches were notable for the havoc they caused; the Conways classed them with rats and cockroaches, calling them "'bastard birds,' partly because they looked like biological bastards, and partly because any other name seemed too good" for these birds who ate bean stems, pulled up corn sprouts, and cut through banana leaves.[137]

Settlers wrote far more about the introduced species they both depended on and struggled with. Strauch's closest friend on Floreana was not her prophet-partner, but a donkey. Bulls were both a useful source of meat and leather and a menace known to crash through garden fences at night. Settlers mainly ate goat meat, the men sharing the work of hunting and slaughtering these animals descended from those introduced by previous sailors. To endure the predations of wild "pests," the Conways relied on the other settlers for knowledge of how to keep rats away from food stores that did not come in cans or glass jars. They learned to protect crops from hungry goats by planting fences of cactus. But fences were not always effective. Ritter waged an all-out war on a seemingly invincible wild boar he thought might be the devil incarnate because it resisted multiple attempts on its life before eventually succumbing to four sticks of dynamite.[138] For settlers, the dynamic nature of the island was edible, and often predatory, rather than evolutionary. European and North American colonists had come to the islands seeking a quiet, isolated life where they could test their hands at living out the romantic dream of life on desert islands, even if this was not what they found.

Island Dreams

Whether settlers dreamed of finding prosperity or isolation in the islands, their diverse dreams speak to the imaginative possibility of islands. Geographical imaginations of the Galápagos as deserted islands, agricultural penal colonies, and an evolutionary laboratory all attempted to make the archipelago into a place where the order of the modern world was inverted.[139] A deserted island was an escape from the cold industrialization of northern cities; the penal colony a place to ostracize and control those who threatened idealized visions of the Ecuadorian citizenry and turn them into workers whose labor would profit *caudillos*; the natural laboratory an enclave for the study of prehistoric nature. Indeed, the limited size of islands and their common location at a distance from continental centers of modernity seem to invite grandiose dreams of reinvention. Yet as island studies scholar Lanny Thompson has written, the very notion of "islandness" is not a straightforward geographic reality but an understanding of space "constituted through the establishment of colonies, possessions, and territories . . . in terms of racialized and engendered projects of modernization, territorial expansion, and empire."[140] In other words, these different constructions of island space were each reflections of the social and cultural politics that shaped the modern world. The Galápagos Islands were not isolated from the flows of global history but were testing grounds for the creation of alternate versions of modernity. Competing visions for the Galápagos reflect the attempts to create a whiter, profitable Ecuadorian nation-state on the backs of penal laborers—as well as their empancipatory dreams. They reflect northern whites' desire to escape from the political turmoil of industrial life. And they reflect campaigns to save nature from natural resource–based development.

Over the past century, the dream of remaking the frontier world of the Galápagos into a sanctuary for nature and biologists has become hegemonic in a way that 1930s conservationists scarcely could have hoped. This geographical imagination painted the Galápagos Islands as sites of isolated, nearly untouched nature that, like the giant tortoises, are remnants of a prehistoric world and should be protected as such. This dream has had political effects beyond the creation of protectionist legislation. Seemingly straightforward assertions that the archipelago *is* a natural laboratory are claims to territory that directly challenge other understandings and uses of the islands. The idealized natural laboratory world eclipses, even delegitimizes, competing visions about what kind of modernity should emerge in the archipelago and precludes us from understanding the perspectives of Ecuadorian, European, and North American settlers who lived in the islands when conservation measures were first introduced. Yet despite the power of the natural laboratory imagination, these other worlds at world's end remain present in the archipelago and have shaped life there over the past century.

Like Latorre's all-seeing tortoise, conservationists routinely look back at the history of colonization on the Galápagos and lament its ecological consequences. Peon laborers had cleared thousands of acres of native forest for sugarcane and other crops. Settlers had introduced all manner of domesticated crops, some of which had already become invasive. Populations of giant tortoises had been further devastated—and replaced by cattle, goats, and wild boars as the dominant animal life on several islands. But we cannot adopt the perspective of the judging tortoise and condemn colonists who saw the world through different contexts than we now do, particularly not the people who had little control over their lives and did what they could to survive on the islands. Instead, we need to try to see the world as they saw it, for understanding various perspectives on island nature is no less crucial to the future of conservation today than it was in the 1930s. It is not only in terms of ecological impacts that the legacies of alternate imaginations are still powerful forces in the archipelago. The tensions that embroiled island populations about who should have access to the islands and the right to profit from their resources continue to shape the politics of life. During the decades that followed von Hagen's retreat from the Galápagos, conservation efforts would succeed only when their proponents found a way to align what had been competing goals for the archipelago—desire to escape from modernity, to secure Ecuadorian sovereignty, and to produce profit—with their vision of the archipelago as a natural laboratory.

But before that happened, the Galápagos became embroiled, like much of the world, in World War II. The bombing of Pearl Harbor in 1941 shattered the illusion of island isolation that had drawn foreign settlers during the 1930s. Days after the attack on US soil, the United States took possession of Baltra (Seymour) Island in the Galápagos, with the Ecuadorian government's forced compliance. On the low, flat shelf of the island, it set up an air force base from which to defend the Panama Canal.[141] For decades, US security hawks had questioned the appropriateness of leaving the archipelago under the management of one of the weakest states in South America. "Whose hands should hold the key to our Canal's western gateway and what are we going to do about it?" asked one columnist.[142] Such US news items were reprinted in the Ecuadorian press, reigniting long-standing nationalist concerns about state sovereignty and debate about whether or not the state should sell or rent the islands. In 1938, President Franklin Delano Roosevelt himself cruised to the Galápagos during a tour of the Pacific—he was ostensibly on vacation, although the trip was also well timed to substantiate US dominance in the region.

With the work of migrant laborers from Ecuador, the US military quickly established a military city in the middle of the Pacific that served as a way station for both the Navy and Air Force and housed twenty-five hundred troops awaiting deployment. The base included two runways (one of which stretched 8,000 feet) for two squadrons of heavy bombers; a 250-foot pier; storage containers for 1.39 million gallons of aviation gasoline, 150,000 gallons of motor fuel, 168,000 gallons of diesel oil, and 900,000 gallons of water (shipped in from Santiago and San Cristóbal); and laundry facilities that could handle the wash of forty-two hundred people. The base also had facilities to entertain the troops, including a bar, bowling alley, and shooting ranges.[143] Despite the amenities, bored soldiers saw the station as a kind of prison, calling the island "The Rock"—a reference to the island prison of Alcatraz in San Francisco Bay.

The base lasted only four years; it was dismantled after the war when the Ecuadorian government declined to allow the United States to lease the island for the next ninety-nine years.[144] Despite its short life span, the base was a turning point in the history of the archipelago. It was also a major source of employment for island colonists, particularly for men who worked on construction and helped to supply fresh fish and water. Although the US military stripped most of its wares when it left, the infrastructure left behind provided the home for an Ecuadorian military outpost that increased the state's ability to patrol its Pacific frontier. The airstrips on Baltra have since become the main tourist entry to the archipelago. And crucially, conservationists saw that if there was to be foreign control over the archipelago, it would not be achieved through formal colonial domination by the United States. To allay Ecuadorian resistance to loss of sovereignty, foreign influence would have to be more subvert, arranged diplomatically through transnational institutions.

The base also left behind an environmentally devastated island that so concerned visiting biologists in the 1950s that they revived campaigns to protect the Galápagos as a natural laboratory. But achieving their vision would require negotiating diverse visions for life on the islands.

5 • Making a Natural Laboratory

In which the tortoises become conservation icons, and the islands
become pristine

Twenty years after the US military left its base on Baltra at the request of the
Ecuadorian government, it returned to the Galápagos. On 20 January 1964, after a ten-day cruise from San Francisco, the California Maritime Academy's
Golden Bear, a 424-foot, 7,000-ton training vessel staffed with a crew of two
hundred cadets, sailed into Academy Bay at Santa Cruz.[1] The Galápagos
Islands were not part of the ship's regular Pacific itinerary. Instead, it was on a
special mission: to deliver more than fifty scientists, mostly North Americans
but also Misael Acosta Solís and a few others from Latin America, as well as one
hundred tons of their equipment for the two-month Galápagos International
Scientific Project (GISP).[2] It was an endeavor organized by two California scientists, Robert Usinger and Robert Bowman, to show that more than a century
after Darwin's visit, the archipelago remained an important site for biological
research. Aided by a US military underemployed in peacetime and eager to
make its presence known in the eastern Pacific at the height of the Cold War, the
scientists would spend six weeks conducting research in botany, ornithology,
entomology, geology, and marine biology, among other fields. A Quito newspaper deemed it the largest scientific expedition to the islands since the *Beagle*.[3]

The trip received much fanfare in the US press as well; reporters from the
San Francisco Chronicle and *National Geographic* tagged along, filing stories
that played up the natural drama of the islands and their promise for scientific
research. The *Chronicle* stories took readers along as two US Navy helicopters
shuttled scientists and local field assistants across the archipelago, where they
collected insects, plants, and songbirds; injected land iguanas with radioactive
isotopes to study the formation of goiters; and fed giant tortoises tiny radio
transmitters to study their digestion.[4]

But before the science got under way, there was a party. As soon as scientists finished unloading their equipment from the *Golden Bear,* they threw a grand fiesta in the tiny village of Puerto Ayora. Nearly five hundred guests attended—including nearly all the island's settlers and 107 dignitaries that the US and Ecuadorian air forces flew out to the island from Quito and Guayaquil. Members of the military junta that then governed Ecuador raised glasses with renowned international conservationists, including Harold Coolidge and Victor Van Straelen. The occasion was to toast, with ten cases of California bubbly, the inauguration of the newly constructed Charles Darwin Research Station (known in the Galápagos as *la estación*).

Five years before, in 1959—the centenary of the publication of Darwin's *Origin of Species*—the Ecuadorian government had declared the Galápagos a national park and allowed scientists working through UNESCO and the International Union for Conservation of Nature (IUCN) to build a complementary research station. In the intervening years, young European biologists sent by UNESCO had overseen the construction of station buildings, created a reserve to protect giant tortoises on Santa Cruz, and begun surveying wildlife. GISP was as much a celebration of these efforts as a scientific endeavor. After the party, one of the Ecuadorian generals was so enthralled with the pomp of big, postwar US science—and the opportunity to ride a giant tortoise—that he signed an agreement with the scientists that made the Darwin station the official scientific advisor to the state for the archipelago for the next twenty-five years.[5] Trip coordinators flew to Guayaquil, where they were awarded medals after a symposium with local scientists. It was a triumph for scientists and a turning point for the islands and their inhabitants, both human and nonhuman. GISP marked the beginning of a new era in which the islands would be governed as a national park for science where nature was to be preserved in accordance with its evolutionary history—that is, as a natural laboratory where pure nature was to be isolated from human disturbances.

As the name of the new research station makes clear, mid-twentieth-century scientists saw themselves following in Darwin's footsteps to work in a "natural laboratory of evolution," as had their peers during the 1930s.[6] The arid, volcanic islands of the Galápagos had provided key evidence for Darwin's theory of natural selection but, they argued, had remained relatively unstudied. At the expedition's send-off in San Francisco, Julian Huxley gave a public address in a crowded hall that dramatized the importance of the islands for Darwin. In an essay adapted from his remarks, he wrote, "It was on the Galápagos in the early autumn of 1835 that Darwin took the first step out of the fairyland of creationism into the coherent and comprehensible world of modern biology; for it was here that he became fully convinced that species are not immutable—in other words, that evolution is a fact." It was grand, inspiring rhetoric, if untrue. But Huxley's point was not to discuss the nuances of Darwin's process. It was to make sure that "the Galápagos becomes a living memorial of Darwin—not only a museum of evolution in action, but an important laboratory for the furtherance of . . . a truly Darwinian biology."[7]

Huxley's rhetoric was part of a powerful geographical imagination associated with the natural laboratory ideal. Public writing from those involved in 1950s conservation campaigns was full of allusions to the islands as a place "where time stood still," where, it seemed, they could witness firsthand the origins of life. "A visitor setting foot on these shores for the first time feels as though he has gone back to the secondary period, the age of the reptiles," one booster wrote. The islands were "truly a Noah's Ark for animals that have disappeared from the face of the earth."[8] The giant tortoises were both scientific objects at the heart of the natural laboratory and species to be saved in this Darwinian ark. Installed as the centerpiece of the logo for the Darwin station, a stylized Galápagos giant tortoise became one of the first endangered animals—like the World Wildlife Fund's panda and the Frankfurt Zoo's gorilla—branded to represent a new era of species-focused conservation. These animals who conservationists repeatedly framed as looking "prehistoric" were the keystone species not only for island ecology, but also for the vision of the islands as both a museum and a laboratory of evolution.[9]

Metaphors about pristine, Edenic nature, prehistoric species, and natural laboratories were common tropes of twentieth-century conservation campaigns around the world that dramatized the need to save endangered species in national parks. For midcentury biologists, the Galápagos were, or should be, a scientific Eden—a natural place isolated from the march of humanized time and industrialized modernity. The rhetorical flourish makes it sound as if conservationists needed only to protect what already existed. For example, what we commonly think of as constituting the Galápagos-as-natural-laboratory are the particular confluence of geological forces, ocean currents, and biological processes. Indeed, that is how biologists explained the evolutionary significance of the Galápagos in the 1960s, and it is how nature documentaries like the BBC's *Galápagos: The Islands That Changed the World* do so today. But these explanations miss half of the story. The Galápagos became a natural laboratory as much because of the interplay of geological and biological processes as because of the meeting-up of multiple social worlds in the mid-twentieth century.

Conservationists succeeded in establishing the Darwin station and their vision for island management only by forging an alliance that tied together the multiple competing understandings of the Galápagos Islands—as frontiers for development, island prisons, desert island escapes, and places for heroic adventure—that had shaped their prior history. This was no straightforward task and would require years of negotiation both before and after GISP.

Biological Reconnaissance

International efforts to conserve the Galápagos were rekindled after World War II when two young biologists each wrote to the new IUCN to express concerns about the fate of wildlife.[10] One was Irenäus Eibl-Eibesfeldt, an Austrian ethologist trained by Konrad Lorenz, the famous scientist of animal behavior. Eibl, as he was called, visited in 1954 during Hans Hass's marine biology expedition aboard the *Xarifa* sponsored by the International Institute for Submarine Research. The other was Canadian-American ornithologist Robert Bowman, who had followed in the footsteps of David Lack doing research for his dissertation at the University of California at Berkeley on the foraging patterns of Darwin's finches during 1952 and 1953. Both were alarmed about what they saw. Eibl reported that everywhere he was "met with traces of destruction": "Sea-lions with bashed in skulls lay rotting on the shores" and "birds with broken wings and smashed beaks lay about." "The sun-bleached armor of giant tortoises," he wrote, "was a mute testimony to the misdeeds of Man, the most predatory of all animals."[11] In the island settlements, colonists offered young tortoises and penguins for sale, as well as the furs of sea lions and fur seals—a continuation of a trade with visiting naturalists that at that point had more than a century-long history. The protectionist legislation passed two decades earlier was going unheeded.

The destruction left over from the US military base was equally alarming. On Baltra, where William Beebe had found a paradise for land iguanas, Eibl could find only one, its sun-shriveled body punctured by bullet holes. He gave a pass to the soldiers who, out of boredom, had evidently used the animals for target practice. At least that was Eibl's analysis—he was perhaps too quick to attribute the cause of the animals' disappearance to soldiers, for naturalists since Beebe had noted a lack of juveniles on the island.[12] But Eibl was less understanding of island settlers. He lamented their "most distressing barbarity" for it seemed to him they had killed sea lions and pelicans "without rhyme or reason and just for the sheer lust of slaughter."[13] What he did not recognize was that—right or not—these aggressive animals most likely had not been killed "senselessly" but by fishers protecting their catch.[14]

Slaughtered animals were not the only signs of a place in need of conservation. In the iguanas' place, introduced mice were now the dominant animal life on Baltra. They skittered along a cracked asphalt road and around what was left of The Rock: small huts with smashed windows and rusty shutters, as well as "empty munitions depots . . . and abandoned explosives dumps."[15] When the US Air Force left Baltra in 1946, it did so by "dynamiting, burning, or dumping into the Pacific Ocean whatever could not be pried up and hauled away. Jeeps, radio equipment, telephones, refrigerators, ovens, and other supplies and equipment were tossed into the sea." What the military could not move were "two hundred empty buildings and two runways."[16] The Ecuadorian government invited settlers to strip the buildings of whatever they found useful. Lumber from the base was reused in buildings across the archipelago, and several prefabricated buildings were moved entirely to San Cristóbal and Floreana, where Margret Wittmer picked the former base post office to serve as her new residence.[17]

But other abandoned pieces of the base were not useful. In 1960, the first biologist sent to build the Darwin station, Raymond Lévêque, complained that "American soldiers didn't hesitate to shoot at seals with machine guns, drop their bombs on the rocks where seabirds live, etc., etc. You find 'traces' of these soldiers everywhere: I found a piece of shrapnel atop the Barrington [Santa Fé] summit, and it probably wasn't the only one there. The northern coast of Santa Cruz hides many vats and various barrels, half-buried in the sand on the beaches, serving as a target for planes, etc." That there was no longer a working base, he thought, was "a very good thing for the 'protection of nature.'"[18]

Bowman's and Eibl's descriptions of devastation made an impression at the IUCN. The institution had formed in 1948, with the support of 107 nations, as an environmental complement to UNESCO to unite international networks to promote nature conservation. Its roots reached to the "world nature protection" movement of the early twentieth century that argued for the protection of natural treasures as global commons, not beholden to the rule of any single power.[19] Transnationalism was also at the heart of UNESCO, another postwar institution, created with the mission of preventing another global war by promoting international collaboration, education, and science. Julian Huxley, long a supporter of Galápagos conservation, was UNESCO's first director-general. As historian Edward Larson has shown, the Galápagos

project well fit Huxley's philosophy for UNESCO: "a scientific world humanism, global in extent and evolutionary in background."[20] An adherent of a
philosophy he called "evolutionary humanism," Huxley believed that UNESCO
should "relate its ethical values to the discernible direction of evolution, using
the fact of biological progress as their foundation."[21] A Galápagos research
station would enable the protection of wildlife, encourage the study of evolution, and educate the public about both, thus fulfilling scientific, conservationist, and cultural purposes. In turn, the educational opportunities, development
of science, and cultural exchange programs that the research station sponsored would encourage the evolution of society as a whole.[22]

Plans for a new global empire based on rational scientific management
were shot through with a paternalistic racial and cultural elitism.[23] UNESCO
and the IUCN were among the postwar institutions that comprised a "new
imperialism" premised on internationalism and faith in the expansion of capital, science, and technology to spread modern Western values and ensure
prosperity and peace around the world.[24] In the realm of nature conservation
especially, they replaced the previous efforts of colonial powers, particularly in
Africa. Although the IUCN was formed in an era of decolonization, many of
the same authorities who had worked for European colonial governments led
new postcolonial initiatives. Victor Van Straelen, who had been central to the
creation of the Albert National Park, moved from managing Belgian conservation in the Congo to become IUCN vice president—and later president of the
Charles Darwin Foundation for the Galápagos Islands, which administered
the Darwin station from Brussels. Yet aside from the career histories of such
individuals, the ideology and power structure of these institutions, based on
appeals to the supposedly universal value of nature and science, were firmly
embedded in histories of colonialism.[25] They tied a technocratic ideology that
had underwritten earlier British imperialism—that scientific rationality would
provide the most appropriate use of resources—to midcentury faith in science
as a route to progress.[26]

It was a powerful ideology that centrally shaped the role of foreign institutions in Galápagos conservation—and was welcomed by Ecuadorian officials.
In 1956, the IUCN—where Harold Coolidge then chaired a special commission to establish national parks—secured an invitation from the Ecuadorian
government to send its scientists on an exploratory mission to the archipelago.
Ecuadorian politicians were eager for international recognition of the state's
sovereignty over the islands, which remained on the fringe of national economic and cultural life. For the government, the mission was a well-timed

opportunity to back its claims to the islands with scientific authority. In 1956, a film about Norwegian explorer Thor Heyerdahl's recent archeological expedition to the Galápagos screened in European cities. The film buttressed Heyerdahl's theory that the Polynesian islands were originally populated by migrants from the South American continent. In the Galápagos, Heyerdahl's team found pottery shards that Peruvian scientists matched to similar artifacts of the Chimú people of coastal Peru. Ecuadorian diplomats saw the film—which made no reference to the islands as an Ecuadorian territory—as a Peruvian attempt to use modern archeology to make a historical claim to the archipelago.[27] Always a sensitive geopolitical issue, the Ecuadorian hold on the archipelago was at stake again in the mid-1950s because of disputed access to one of its most lucrative natural resources: tuna. The second "tuna war" began in 1952 (and would last until 1980) with the Declaration of Santiago in which Ecuador, Peru, and Chile claimed sovereignty over the sea two hundred miles from the coast—a claim contested by the United States, which recognized the states' territorial rights as extending only twelve miles offshore. Over the following years, the Ecuadorian navy began patrolling the Galápagos more regularly—and detained several US tuna clippers for illegal fishing. In light of these contestations, UNESCO's recognition would help anchor the state's claim to the islands.

With the government's endorsement, UNESCO arranged to send Eibl on a "reconnaissance" mission, echoing the language of military-backed scientific exploration. He was to survey the status of wildlife and report on the feasibility of building a research station. When US conservationists heard of the plan, they scrambled to find resources to send Bowman along too, not wanting this to become a purely European project.[28] Well aware of the value of publicity for their efforts, the Americans also sent along famed photographer Alfred Eisenstaedt and illustrator Rudolf Freund, who both worked for *Life* magazine. Such coverage was a core strategy that postwar conservationists used to enroll educated US publics in the cause of nature appreciation through popular science and nature writing, nature documentaries, and museum exhibits.[29]

In July 1957, the four men arrived in the Galápagos, where they spent three months "on the trail of Darwin."[30] An Ecuadorian navy ship chauffeured them around the archipelago; they visited sixteen islands—all the major ones and several smaller islands as well. Local residents—including Ecuadorians Gilberto Moncayo, the son of a "veteran tortoise hunter," and Miguel Castro and the German Angermeyer brothers—acted as their guides. It was a time, Bowman reported, filled with "some of the most other-worldly experiences of our lives," and it would inspire, for him, a lifetime engagement with conservation in the islands.[31]

An outing to Fernandina demonstrated Galápagos nature at its most wild, extreme, and unusual. Upon landing, the team was greeted by huge colonies of marine iguanas piled on top of each other that blended into the rocky shore. The editors at *Life* reused *Beagle* Captain Robert FitzRoy's description of the "crawling cliffs" of Fernandina as "a fit shore for pandemonium" to caption Freund's illustration.[32] Hiking inland, they climbed the cone of the island's central volcano, which rewarded them with a view "unsurpassed in all Galápagos."[33] "There is perhaps no more beautiful, or—depending on one's point of view—more repulsive spot in the Galápagos than Fernandina Island," Bowman wrote.[34] He described the island in exotic terms, saying it took days of "intimate contact with the weird world" to overcome his impression that "the physical and biotic elements were in temporal and geographic discord with the outside world":

> Where else on earth can one find enormous herds of ocean-venturing dragon-like iguanas, reminiscent of prehistoric times? Where else does one find a member of the typically Antarctic penguin tribe living in comfort directly on the Equator and feeding on the small fishes that teem in these "tropical" waters, which here are cooler than in any other equatorial region in the world? Where else can one find a giant of the cormorant family of birds with wings so reduced that, in mockery of the evolutionary process, they render the bird flightless? And all these biological productions are staged in a setting one might imagine to be more typical of the cooler parts of Hell or an adolescent planet.[35]

Bowman's rhetoric played up the evolutionary strangeness of the archipelago and its "other worldly" wildness.[36] Gone are exclamations about unusual human inhabitants that often appeared in naturalists' descriptions of the islands from the 1930s. Instead, settlers entered Bowman's account through his concerns about the extirpation of giant tortoises and other wildlife, something he contrasted with his own scientific appreciation of the species.

After a daylong hike on Santa Cruz, Bowman and Eibl spent four days observing tortoises. The journey from the coast had taken them through the islands' vegetative zones—from a cactus forest where prickly pear cacti as tall as trees lined the path, through tangles of shrubs and short trees infested with fire ants, along hibiscus-lined paths beside coffee and banana plantations, and into the "thick virgin forest" of leafy Scalesia trees and moss-covered Pisonia trees.[37]

When they finally reached tortoise country, they found a "graveyard" where tortoise shells were scattered across the landscape. Many had clearly been killed long ago, but other remains were fresh, a sign that settlers did not heed protection laws.[38] But soon they came across living tortoises, "magnificent reptilian monsters plodding through the dense forest underbrush with the ponderous gait of a tank."[39] They observed tortoises wallowing in shallow muddy pools, hiked down to drier nesting areas, and took measurements of the animals—one of whom was so heavy that four of the men could not lift it.[40] But they also observed the remains of tortoises killed by local settlers who relied on the animals for their rich fat.

The sight of these tortoises—living and dead—made a lasting impression. Bowman's report detailing the status of giant tortoises on each island concluded that they, along with land iguanas and fur seals, were most in need of protection from hunters. He thought the severity of settlers' poaching of tortoises, iguanas, flamingos, seals, and sea lions—all for food and pelts—warranted immediate action by the Ecuadorian government. Eibl agreed; this was not how wildlife should be treated. "Something must be done and done quickly," he later wrote, "to preserve these remarkable creatures for posterity."[41]

But it was not only for posterity that the giant tortoises should be saved. Eibl made the case for conserving these animals as evidence of evolution in a natural laboratory. He reminded readers that "it would not be enough just to preserve the tortoises on one island only." It might be thought, he continued, that as long as tortoise populations were thriving on one island, the extinction of those on other islands would not much matter. But this would not suffice, for it was the differences among island populations that made the species unique evidence of adaptive radiation: "It was the differentiation alone that prompted Darwin's teaching of evolution." Surely, he argued, it was "worth while preserving one of the most striking of all experiments in the history of species."[42]

Near the end of the trip, in November 1957, Bowman and Eibl completed their mission by selecting a site for a future research station. Dressed in worn field clothes, their beards grown out for weeks, they posed for a photo at the ideal spot: Tortuga Bay on Santa Cruz Island (fig. 5.1).[43] Compared with San Cristóbal and Floreana, both too much despoiled by colonists, and Isabela, then home to a penal colony, Santa Cruz was well situated in the center of the archipelago and had been colonized only some thirty years before, so was still relatively wild. They did not have a flag to stake in the ground on this uninhabited bay a few miles from the settlement; instead, they made a pile of lava rocks to mark their claim. Securing the site, though, would be a more complicated political affair.

Figure 5.1 Robert Bowman (left) and Irenäus Eibl-Eibesfeldt (right) look over the landscape they selected for a research station to be named after Charles Darwin at Tortuga Bay on Santa Cruz, November 1957, during a "biological reconnaissance" mission sponsored by UNESCO and US conservation organizations. (© California Academy of Sciences)

Negotiating a Natural Laboratory

When they returned from the islands, Bowman and Eibl each submitted reports to the IUCN outlining the results of their trip and their suggestions for a research station.[44] The bulk of both reports discussed evidence for concerns—biological intelligence Eibl and Bowman gathered about the status of native fauna and the threats that endangered them, from foreign fisheries to settler colonies and their introduced goats, pigs, and dogs. Eibl stressed the need for a permanent scientific presence in the islands, positioning a biological station as a center of conservation enforcement and research. As primary justifications for a research station, Bowman listed the importance of the archipelago to Darwin's work on evolution, the destruction of unique life-forms, the threats posed by settlers and tourists, and the "impediment to scientific study . . . due to lack of basic facilities on the islands."[45] In his view, a research station would serve four functions: assist in the protection of the Galápagos biota, serve as a center for biological and other scientific research, commemorate the visit of Charles Darwin, and serve as an example of international cooperation in the preservation of the natural resources. For these scientists, a research station would help ensure wildlife protection and facilitate their own access to do research in the Galápagos by making difficult field conditions on the islands more livable. Creating a station in Darwin's name would also firmly place modern research in his legacy.

Bowman's and Eibl's reports were published by UNESCO and formally presented to the International Congress of Zoology in 1958. Although it was an unorthodox situation, the Congress adopted a resolution on their behalf and created a Galápagos committee to see that the suggestions were carried through.[46] This international committee included several prominently placed scientific environmentalists, chief among them Huxley, who acted as chairman but primarily lent his name for stature and publicity. Luis Jaramillo, Ecuador's permanent delegate to UNESCO, represented the state and was a key figure in securing government support. Van Straelen served as president.[47]

In Ecuador, the national commission for the International Geophysical Year presented a project to reform the 1936 law to the Ecuadorian National Congress after a 1958 trip to the islands. This effort was led by Cristóbal Bonifaz Jijón, who would become Ecuadorian ambassador to France in 1961 and, with Jaramillo, the only other Ecuadorian on the original board of the Charles Darwin Foundation. In Paris, Bonifaz stressed the international scientific value of the islands during a UNESCO meeting, saying that their position at the intersection of multiple currents and periodic El Niños would allow scientists to better understand the history of the formation of the islands and their surrounding oceans.[48]

Even with credible backing of elite, transnational scientific institutions, the geopolitical issues that von Hagen had faced in the 1930s surfaced again as IUCN officials set to work to secure a foreign presence in the islands. They arranged for an official Ecuadorian request for technical assistance and applied for capital funds to begin construction. But the chief issue would be getting land rights.[49] As Bowman explained, it would be "a delicate diplomatic project which must not be forced, since in effect it means that an outside agency (albeit international in membership) wants to have authority to regulate some of the 'internal' affairs of Ecuador. It is as if the [committee] wanted the United States Government to turn over authority to prohibit dams, etc., in our National Parks."[50]

Europeans worried that Ecuadorian politicians were hesitant to commit to the project, lest they cede state authority over the islands, particularly to US interests.[51] To allay skepticism that the project was a "pseudo-international front" for US imperialism, a truly international effort to protect the Galápagos was necessary, Bowman explained to a supporter at the Philadelphia Herpetological Society: "if the Ecuadorians think for a moment that the Americans are monopolizing the [committee] or in any way trying to pressure the Government unduly, the whole affair will backfire."[52] Because of the "many foolish and ill-advised [US] schemes proposed for Galápagos," Bowman felt that the United States could not "take over the driver's seat."[53]

Yet while Bowman promoted the spirit of internationalism, privately he felt sidelined from the project by Europeans. He complained to Coolidge that "intrigue" in the IUCN "may ultimately lead to the downfall of the whole project if something isn't done immediately to get the situation cleared up." He felt that enthusiasm for the project in California was misread as "an attempt to monopolize the project" and cautioned that it would be "unwise . . . to ignore the American group," considering that the largest natural history collections from the Galápagos were in US museums, particularly the Cal Academy in San Francisco, and because of the presence of California tuna boats in the Galápagos.[54] While committee members agreed that American involvement in the administration of a research station would be essential for funding purposes—to make the station eligible for US government grants—the umbrella of the IUCN and UNESCO provided legitimacy to the project as an undertaking of transnational governance in service of the state rather than an institutionalization of US power over the eastern Pacific.

Eibl saw the situation differently. In a joust of bickering, he accused Bowman of having "a quite arrogant attitude toward the Ecuadorians" and said, "I think he did not realize yet that *Ecuador* is going to build a station on Galápagos with international help."[55] While the criticism of Bowman's arrogance may have been fair, the station was for decades much more an international undertaking than an Ecuadorian one—though state support was essential.[56] Under pressure from the director of the IUCN, Coolidge was forced to ask Bowman frankly whether he had ever published anything that might be read as disturbing to Ecuadorian authorities and jeopardize his ability to return to the islands.[57] Although Bowman denied publishing "un facheux article," he did acknowledge that his initial letter to the IUCN in 1955 "was much criticized for its frankness." But he stood by what he had written, saying "as it turned out, my comments about Ecuador (given in confidence) turned out to be true."[58] Bowman's condescension betrays an attitude of superiority that was all too common among scientists who disparaged local cultures and asserted that they knew best how to manage the islands.

Despite the trials of holding the network together, the committee succeeded in winning Ecuadorian support for a research station during a 1958 follow-up trip to the continent. But the situation was complicated when UNESCO's legal department stipulated that it would be inappropriate for either the organization or the lead envoy, Jean Dorst, an ornithologist who was then subdirector and later director of the Muséum National d'Histoire Naturelle in Paris, to sign agreements directly with the Ecuadorian government. To solve this bureaucratic problem, committee president Victor Van Straelen organized an official entity, the nongovernmental Charles Darwin Foundation, under Belgian law to act as signatory and collect funds for the station project. The Galápagos Committee established at the International Zoological Congress became the foundation's board of directors. Board members were a Who's Who of midcentury nature protection, including Van Straelen; Dorst as secretary-general; Harold Coolidge; Jean-G. Baer; Julian Huxley; S. Dillon Ripley, director of the Smithsonian; Sir Peter Scott; Jaramillo; Bonifaz; and Bowman.

On 4 July 1959, the Charles Darwin Research Station and the Galápagos National Park were officially created when the Ecuadorian government issued an executive decree establishing an emergency law that declared the islands to be "zones of reserve and National Parks." The decree officially recognized the Darwin station, to be administered by the Charles Darwin Foundation, and empowered it with the authority to determine which zones would be deemed reserves as well as which species needed protection. It also permitted the station to control or exterminate "menace" animals as necessary for conservation and prohibited "spontaneous colonization" for farming or timber extraction within protected areas to be designated by station officials.[59] With this decree, the committee secured scientific management of the islands with the cooperation of Ecuadorian military and civil authorities. Over the next few years, additional legislation negotiated with successive governments—including the decree signed following GISP—solidified foreign scientists' role in the archipelago and their vision for island management. Scientists had secured the political authority to manage the archipelago as a natural laboratory where nature was to be protected according to scientific guidance provided by experts—a technocratic vision for management on which postwar institutions were premised and that powerfully shaped Latin America in the twentieth century.[60] But making a natural laboratory on paper was quite a different undertaking from making one in the islands, as the early directors of the Darwin station soon found.

Life in a Laboratory

The geographical imagination of a natural laboratory suggested a place of pristine nature that could be protected and studied by scientists. But geographical imaginations, as geographer Jessey Gilley has argued, often "simplify the knowledge of places by making complex and confusing realities seem simple, with sharp lines of differentiation."[61] Conservationists entered the islands with ideologies that oversimplified a complex place with multiple intertwined socio-natural histories. As scientists arrived to make a natural laboratory in the Galápagos, they did not enter the stereotypical space of a hermetically sealed lab, isolated from social life.[62] Instead they found a place where the competing understandings and uses of the islands that had evolved over more than a century of colonization were very much still alive.

In 1960, about two thousand people lived in the archipelago. The vast majority were Ecuadorians descended from those who had come, willingly or not, at the turn of the twentieth century. Many others had migrated to the islands more recently with government aid after a 1949 earthquake destroyed much of the Andean city of Ambato and leveled several surrounding villages. While the state encouraged colonization of its Pacific frontier, the Galápagos Islands were at the time governed as part of Guayas province and, aside from the port captain and territorial chiefs, had little official government representation. Franciscan missionaries more than the state served the needs of island populations. Several Europeans and North Americans had also settled in the islands during the years since the ill-fated Floreana settlements, most choosing to live on Santa Cruz. Many of these colonists came independently in search of a life away from modernity and established lasting roots in the archipelago. Among them were the German Angermeyer brothers, shipbuilders who often acted as guides for visiting scientists; the Belgian DeRoy family, parents of now-renowned nature photographer Tui DeRoy; Forrest Nelson, a US draft dodger who built the first hotel in "Eden," on Santa Cruz; and the Norwegian Kastdalen family, which still runs a dairy in the islands. Many of these settlers set up farms in the highlands or houses on the western edge of Academy Bay, on a point still referred to as El Otro Lado, the other side, accessible only by boat and removed from the center of town and Ecuadorian settlers.

Nineteenth-century visions of the islands as prisons also remained. Shortly after the US base closed in 1946, the Ecuadorian government established a new penal colony on Isabela to the west of Puerto Villamil at an old outpost of the US military base. The intention, initially, was to redeem criminals through agricultural labor and give them an opportunity to start over with an island plot after their sentence had been completed. But after sending three hundred *penados* and thirty police, the continental government largely ignored the colony. With a mix of forced labor, guards who were themselves stationed in the islands as a form of punishment, and a harsh landscape, the colony soon devolved into a place of brutal oppression. Peons were forced to build a wall of lava blocks more than thirty feet high and almost four hundred feet long, transporting boulders from the highlands and even the far side of Sierra Negra. When one Ecuadorian official visited the construction site, he reported finding human skeletons scattered at its base—some curled into the fetal position, others with bullet holes between the eyes. The lieutenant in charge—known for his brutality—explained that the deaths were the consequence of failures to complete a day's labor. Today, the wall these prisoners built, now called Muro de Lágrimas, Wall of Tears, is a regular stop on tours of Isabela, where under the hot sun visitors marvel at a wall that contains nothing except the tragic memory of its construction.

In 1958, the prisoners of this penal colony revolted, capturing police during a three-day standoff in which the other residents of Puerto Villamil took refuge inside the Franciscan mission. The chaplain, Jacinto Gordillo, negotiated with the prisoners and managed to keep peace until the rebels escaped on a fishing boat. (They later commandeered a North American yacht that took them to the coast, where they were eventually caught and sent to the Panóptico jail in Quito.)[63] Conservationists, who were then in the midst of negotiations to protect the Galápagos as a national park, feared the government would send more police to reestablish the penal colony; but instead, the island prison was formally closed, and a new era was ushered in as conservationists began to arrive during the new decade.

The continued use of the Galápagos as a penal colony was not the only challenge to conservationists' desire to preserve the archipelago as a natural laboratory. In 1960, a new contingent of settlers arrived from the United States with utopic plans to build a new society on scientific, rather than political or religious, principles. In 1959, Don Harrsch, an unemployed former tugboat skipper whose formal education ended before high school, advertised in a Seattle newspaper: "WANTED: Swiss Family Robinson. Is your family one of the 50 adventurous families with the spirit of America's early pioneers needed to establish a model community on a beautiful Pacific island?"[64] He had never been to the Galápagos but saw the islands as a laboratory for a new kind of colony, which he called Filiate Science Antrorse, a mix of Latin and English meaning "Together with science we move forward"—a motto strikingly similar to UNESCO's mission. Harrsch sought "stable" families and a "balance of the sexes" for a colony "dedicated to furthering scientific research": "The biological laboratory work will give us an insight into the chemical and biological makeup of animals, and pursuing this work will give us much insight into the makeup of a human organization. One cannot disconnect sociology, psychology, and biology and expect to come up with the correct answers."[65] The payoff for colonists, for their investment of $2,500, would be the "making of everyone a millionaire blessed with a wealth of knowledge, health, happiness, and, most of all enable its members to create and leave for their children a better world to live in."[66]

One hundred and six people signed on—twenty-two families and fourteen single men, including "aircraft workers, farmers, truck drivers, firemen, salesmen, a janitor, a plumber, and some school teachers."[67] With their funding, Harrsch flew to Quito and met with Teodoro Crespo, an attorney who would soon be named the head of the National Colonization Institute, which had been established in 1957 more in response to calls for agrarian reform on the continent than to deal with foreign settlers in the Galápagos. Harrsch put down a $30,000 deposit on an abandoned coffee plantation at El Progreso and a refrigeration plant on San Cristóbal and officially founded the Island Development Company.

But Crespo was skeptical about the project: "They proposed to cure everyone in the islands of whatever disease ailed them, to give everyone free dental care, to reeducate the islanders to their way of thinking, with belief in God relegated to a minor role and great emphasis placed on scientific achievement. . . . I do not think there is any future for a colony with its head up in the clouds instead of having sensible, down-to-earth ideas."[68] A professor of sociology at the University of Washington who helped Harrsch write a constitution for the colony also thought it would fail—though he and his colleagues planned to study the colony from afar.[69]

The colony's reception in Ecuador and in the Galápagos was less than smooth. When the colonists arrived in March 1960, a sign reading "Bienvenidos Millonarios" greeted them at the dock at Puerto Baquerizo Moreno. It must have been a surprise, for Harrsch had led the new settlers to believe the island was unpopulated. He also had led Ecuadorians to believe the colony would open a $2 million hotel on the island, promising riches for all in this new Hawai'i. Yet while some islanders welcomed the newcomers, others were skeptical, believing the government had sold the entire island to these foreigners.[70] On the continent, the colony became a political issue in an election year. Leftist candidate Antonio Parra Velasco, rector of the University of Guayaquil, riled his supporters with charges that the Americans were invading the islands. Dozens of young supporters bombarded the US consulate in Guayaquil with rocks, chanting for the *estadounidenses* to leave the islands. Parra Velasco lost the election to Velasco Ibarra, but rumors that the colony would soon send five hundred people to San Cristóbal led the government to send several hundred Ecuadorian settlers, mostly Indians, to maintain a population balance in national favor.[71] These concerns that the colony was a US imperialist takeover threatened the success of concurrent negotiations for the Darwin station; Harrsch's colony was one of the "foolish" US schemes that Bowman worried would undercut conservationist plans.[72]

The colony did not last long. The *Los Angeles Times* sent a reporter and photographer to document the colony's progress, but their dispatches told of utopic plans gone awry: crops failing because of heavy rains and feral animals, an outbreak of hepatitis, colonists' inability to fish, a broken refrigeration plant, and unclear land tenure that meant the colony could not buy the plantation outright.[73] Social life was also a challenge, marked by infighting among settler families as well as cross-cultural issues of class and race. Although he espoused some egalitarian ideas, Harrsch expected to himself be a new kind of *patrón*—demanding that, as president of the colony, he have a nearly incontestable voice, as well as a larger salary, house, and Ecuadorian servants. Harrsch, who was openly racist in an interview with the sociologists, claimed that this class structure was suggested by Ecuadorians he consulted as necessary to gain the respect of island populations.[74] While some of the US colonists befriended Ecuadorian counterparts, others drew clear lines between themselves and Ecuadorian settlers, also betraying racist attitudes: "We Americans cannot live" as the Ecuadorians lived, one told the *Times* reporter; "We cannot eat the same diet these people eat. We cannot live in a house built of poles with mud plastered on the windward wall . . . with a cane-thatched roof."[75] By December, the colony had all but dissolved. Harrsch was voted out as president, and most of the settlers had returned to the United States, complaining that the island was far from paradise. One told the *Times* reporter: "There is nothing . . . absolutely nothing in the Galápagos in the way of making a living . . . in achieving a better life. A totally different life, yes. But a better life, never."[76]

The same year that these misguided US colonists arrived on San Cristóbal, a twenty-eight-year-old Swiss ornithology student named Raymond Lévêque arrived on Santa Cruz. He had been hired by UNESCO as the Darwin station's first director and charged with building the new station and implementing conservation plans. It is remarkable that his efforts fared better than contemporary colonization schemes. At what seemed the edge of the world, he entered into a "contact zone"—a place of encounter where actors from differing social, political, and geographic backgrounds meet.[77] Building the research station would require negotiating divergent understandings of the islands among the scientists Lévêque represented and island residents whose disparate past histories shaped their respective geographical imaginations.

When Lévêque arrived he knew little about the Galápagos, having learned of the archipelago only a year before through a nature documentary made by French filmmaker Christian Zuber. Images of island wildlife little prepared him for his arrival at the tiny village on Santa Cruz. His first steps there did not remind him of stepping into a prehistoric age. After a weeklong journey on a "stinking" ship, he was left on the beach with a pile of luggage and the name of a local resident, Miguel Castro, who had helped scientists before him and would serve as his host for the first weeks. The village of Puerto Ayora consisted of a few dozen houses, a Franciscan mission, a couple bodegas stocked every three months when transport ships came, and an office of the port authority. Communication was trying: Lévêque could hardly speak Spanish and depended on Castro's ability to speak English, as well as on other multilingual colonists.[78] He complained that nobody had told him that the island's only connection to the mainland was a monthly boat. After a few months he wrote to the project coordinator in Paris, Jean Dorst: "It makes me mad to be trapped on this archipelago of misery."[79] He described what he saw as "terrible living conditions, terrible . . . they're little barracks, you know, little shacks. No proper furniture, nothing. No running water. Electricity, they had about two hours or three hours maybe in the evening 'cause there was a generator for the community on the beach."[80] Worse he thought was settlers' ignorance of the value of the archipelago. He wrote to Dorst that they "absolutely must" screen nature documentaries like Zuber's in the islands, "because the large majority of Ecuadorian settlers don't know anything about anything, and like I already said, ignorance destroys many things."[81] His supremacist views were a typical denial of the knowledge of Ecuadorian residents on whom foreign conservationists would heavily depend in the coming years.

Lévêque enjoyed his fieldwork on bird populations in the archipelago, but building a research station was quite different from surveying a natural laboratory. As he attempted to begin construction—buying supplies and coordinating labor and permissions—he encountered practical and political problems that revealed "all difficulties of doing anything in the Galápagos." Lévêque's letters home reflected frayed emotions and a disdain for Ecuadorians with undertones of racism: "I must say that these Ecuadorians are low. Liars and thieves, all of them, from the first to the last, starting with the president of the republic, the man called Velasco. I grin and bear it, but there are moments when I've had enough, and I may well burst one day." Lévêque himself was difficult to get along with—by this point he had made enemies with a settler from the United States. Two months later he was equally fed up with "'Monsieurs' in offices," complaining to Dorst: "It seems that there's an intricate conspiracy to place as many obstacles as possible in front of me." The project took its toll: "I had a real nervous breakdown" at the end of his two-year stay, he told an interviewer in 2000.[82] The isolation and rugged conditions of desert island life that charmed other European settlers were maddening for Lévêque. Once he returned home, it took him two years to recover from the exhaustion and depression caused by the strain of isolation, a demanding job, and a failed love affair.[83]

But before he left the islands, Lévêque managed to begin construction of the research station—though not at the site at Tortuga Bay that Bowman and Eibl had selected. Castro took Lévêque to see their proposed site by boat, but the opening to the bay was guarded by a "bar" of waves several feet high that made entry dangerous. The land route was little better. Colonists had made two narrow paths through dense, dry vegetation from town to the bay, where they collected lobster and sea turtles. But Lévêque found that the land in this arid zone was "truly Galápagos, 100% . . . completely un-traversable for all kinds of vehicles in all directions . . . the lava fields are riddled with cracks in all directions, and there are many instances of land caving in—not to speak of vegetation." This landscape might be relatively pristine, but more important for Lévêque was that it ruled out vehicle transport (not that there were any on the island at the time—Bowman would bring the first Jeep to the island on the *Golden Bear*). It made him think twice about even going out for a walk: "It would take *tremendous* work just to walk" to the bay from town; even without a load to carry, one could cover only a little more than one mile in an hour. Tortuga Bay was, Lévêque surmised, "as isolated from the village as an island. This is good for a laboratory, since we won't have too many neighbors, but it's a bad thing from all other perspectives."[84] So Lévêque rejected the Tortuga Bay site for the

Darwin station headquarters. Today the site remains undeveloped and one of the prized beaches on Santa Cruz. Now accessible by a paved mile-and-a-half path from Puerto Ayora, the site is popular with tourists and island residents— one of the few easily accessible places to see the more remote Galápagos.

Instead of Tortuga Bay, Lévêque chose a location just to the east of the village at Academy Bay where the station remains today. Even though it was not as picturesque as the half-mile-long white sand beach at Tortuga Bay, Lévêque thought the site was ideal because it was accessible by road and was bordered by a beach and two "practically impenetrable *barrancos*" (steep lava-rock cliffs formed by faulting as the land surrounding Academy Bay had dropped down). This left only a half mile bordering the village that needed to be blocked from entry by settlers. The station's isolation, Lévêque thought, would be bolstered by a cultural buffer between the station site and Ecuadorian colonists. Half a dozen white families lived in an area called Lundt at the eastern edge of the village, where a Norwegian family had lived years before. What's more, the town cemetery was nearby, and "no true Ecuadorian would ever want to pass by at night, for any reason. They are very superstitious!"[85] If the station could not be completely remote, he would use the island's physical and cultural geography (or at least his stereotypical assessment of it) to achieve privacy. But securing the physical isolation of the research station was not Lévêque's only concern.

DEVELOPING THE LAB

One of scientists' greatest challenges in making the Galápagos into a natural laboratory was confronting the multiple visions for island development that had shaped the history of the archipelago over the previous century. Foreign scientists' goals of protecting the archipelago as a prehistoric evolutionary Eden were in tension with settlers' goals of a better, if not necessarily prosperous, life in the islands. Eibl had questioned human settlement there altogether in his report to UNESCO following the reconnaissance mission: "On the whole colonization conflicts with nature protection."[86] Lévêque, like Bowman and Eibl before him, saw island colonists as the main threat to conservation efforts. Indeed, he recommended a "complete halt" to colonization efforts, lest fishers and hunters destroy the nature he was working to protect. This was only the first time foreign scientists would propose such a move in order to achieve their dream of a biological sanctuary.

Lévêque faced this conflict head-on. He fretted about island colonists' "misunderstanding" of the Galápagos and the UNESCO project, complaining to

Bowman that they seemed to imagine UNESCO as a commercial enterprise like the United Fruit Company. At the time, the US-based banana and sugar monopoly held plantations through much of the Caribbean and Central and South America and was commonly referred to as *el pulpo* (the octopus) for its extensive reach into foreign territories and governments.[87] United Fruit had come to Ecuador in the mid-1930s in the wake of the cacao crash and played an instrumental role in making the nation the world's leading banana producer in the 1950s. Although the company's tentacles never had quite as strong a hold in Ecuador as they did in other countries, it was well known as a foreign *patrón* that controlled not only local economies, but workers' social lives as well.[88] Lévêque's comparison was a telling illustration of the consequences of international scientists' attention to the islands and how Ecuadorian colonists made sense of the presence of white northerners making territorial claims. It suggested that Ecuadorians in the Galápagos saw the Darwin station in a similar light—as a powerful foreign entity come to control use of natural resources and shape their lives and livelihoods with its employment opportunities and rules. It was not the "misperception" that Lévêque believed, given the extent to which conservation would shape people's livelihoods on the islands in the coming decades.

Lévêque blamed the "awfulness" of such competing understandings of the Galápagos on government officials who had never been to the islands and "who don't even know apparently what the Galápagos are!"[89] Bowman argued that the Darwin Foundation needed to persuade the government that colonization would not succeed financially: "Every new scheme results in more and more destruction of the very things that make Galápagos an attractive site for tourists. Galápagos *means* marine iguanas, giant tortoises, tree-like cacti, flightless birds, lava and fiery volcanoes, among other things. To destroy these is to destroy Galápagos, for then there is no reason for anyone to go to Galápagos, except to witness what ill-advised projects can do to destroy a jewel once world renowned."[90]

But the clash that foreign scientists saw was not inevitable. Acosta Solís addressed concerns about economic development and nature protection explicitly in his contribution to the GISP proceedings (one of only three of the forty preliminary studies to focus on "applied science"). He framed his essay in terms of the debate between preservation and conservation that animated US wilderness politics.[91] Protection, Acosta Solís wrote, was about preserving endangered native species for aesthetic and scientific reasons, whereas conservation was focused on maintaining natural resources for continued economic exploitation in the future, regardless of their status as native or introduced species. Both approaches, he thought, were necessary in the Galá-pagos. He recommended establishing strict reserves on islands "worthy of protection as a living museum of nature."[92] But on islands where people al-ready lived—San Cristóbal, Floreana, Isabela—there could be no such strict reserves. Santa Cruz was the exception where a reserve for tortoises and yet-undeveloped agricultural land should be delimited into specific zones. Indeed, when the first management plan for the national park was issued in 1973, it largely followed Acosta Solís's recommendations. Island territory on the four inhabited islands was divided into multiple zones that ranged from urban zones to agricultural buffer zones to park reserves, with other uninhabited and "most pristine" islands, like Fernandina, established as the zones of strictest protection. He also recommended that the Ecuadorian government "prohibit further colonization . . . except under regulations for the sound scientific use of lands," to be based on detailed studies of soil quality and oceanographic studies to allow for rational land use and a "scientific basis for a prosperous fishing industry." Acosta Solís warned, "Destroy the renewable resources of [the] sea, and overexploit the natural richness of the soils, and Galápagos will become a wasteland, a monument to man's greed."[93]

Another GISP scientist, Carl Koford, who worked for the US Public Health Service in San Juan, Puerto Rico, also wrote on economic development. He made several suggestions for improving agriculture, including cattle breed-ing, as the Norwegian Kastdalen family had already done, hybridizing an im-ported bull with local feral cows to "produce fine dairy cattle"; applying scientifically informed agronomy, including crop rotation and fallowing to increase land productivity; managing game hunting and fishing; and improv-ing the supply of potable water.[94] As Darwin Foundation board member Cristóbal Bonifaz wrote in 1963, "Economically, we have the right to think that today's studies, in many cases, will be sources of wealth in the future." He suspected this wealth would come in the form of development of the archi-pelago's fisheries, where "scientific work will have immediate industrial use."[95] For these scientists, nature protection should coexist with the develop-ment of island communities, albeit limited in scope.

But the foreign directors of the research station did not want to encourage local agricultural development. Instead they focused on wildlife surveys, pro-tection of endangered species, and control of introduced species. Station direc-tors after Lévêque actively tried to curtail extractive industries in the islands. One recalled proposals from various entrepreneurs to farm sheep, mine sulfur, raise deer, and harvest marine turtles—all of which he deterred, often in the face of major complaints.⁹⁶ These proposals did not all come from capitalists ignorant of conservation concerns. Sheep ranching was suggested by Bonifaz, who helped Lévêque negotiate a 1961 decree for island protection.⁹⁷ But for Lévêque and subsequent station directors, Bonifaz's recommendation revealed the failures of a developing state. Mid-twentieth-century developmentalism judged states on their ability to manage and successfully exploit their natural resources for economic development. Northern conservationists thus critiqued extractive colonization in the Galápagos on two grounds. First, the failures of most colonies to them meant that Ecuador had not efficiently developed island resources, thus justifying foreign management. And second, because of the environmental consequences of previous colonization, they argued that their conservationist expertise was necessary to protect island nature. From their perspective, both reasons justified foreign intervention. Yet foreign scientists also knew that the Ecuadorian government would not support their vision for a natural laboratory without some kind of strategy for development. Instead of strategies they saw as conflicting with conservation, they advocated a vision for development *based on* nature protection: tourism.

Tourism was a common strategy of mid-twentieth-century wilderness preservation; it had long been a central feature of US national parks and was also advocated by the IUCN in Africa, where elite game hunting evolved into a profitable tourism safari business. As Eibl had written in his report to UNESCO, tourism would fit both scientists' desire to protect pristine nature and the state's desire for development. Galápagos conservationists had looked to tourism in the 1930s, and by the late 1940s, the growing industry was a focus of Ecuadorian politicians. Nature tourism would provide a source of revenue, reinforce an imagination of the islands as pristine, and fit well with Huxley's mission for the national park as a place to educate world publics about evolution.

The Galápagos project was a clear early example of the central role that nongovernmental conservation organizations have played in the process of value production through tourism. Darwin Foundation scientists built on the geographical imagination of the archipelago established through Darwin's and Beebe's writing to create the value of Galápagos wildlife as commodities that could be encountered—for a price—by elite tourists.⁹⁸ The scientific value of the archipelago became fused with its economic value. At the time, it seemed an essential strategy.

In their campaign to negotiate the research station, foreign conservationists focused on state interest in tourism to persuade the government of the profitability of their vision. Bowman explained to a supporter that the "greatest hopes" of the Ecuadorian minister in charge of island colonization were

> in tourism, once a first-class hotel is built. Unfortunately, these people do not realize that the main reason people come to Galapagos is to see the giant cacti, the unusual reptiles, tropical penguins, etc., not because of the cultural attributes of the people, institutions, etc. It (the islands) are classical ground because of Darwin's visit. Ruin the biota, and there is nothing left. Iguanas, tortoises, etc. *are* Galapagos. We need to make this point, I think, in the resolution. Present policy on Galapagos is to pay little attention to the natural elements which will in future attract tourists.[99]

Tourism would depend on the quality of wildlife encounters, thus making nature protection central to, rather than a challenge to, state interest in development. Bowman translated the archipelago's scientific value based on a Darwinian imaginary of Galápagos nature into an economic value through tourism. An editorial in *El Comercio,* one of the state's leading newspapers, supported this vision for the research station because of the recognition it would bring to the state: "The Galápagos marine biology station will elevate the name of our country in all latitudes and environments. In the shadow of this scientific work can be developed a form of organized and controlled tourism that looks and collaborates, but does not destroy, the rare examples of marine and terrestrial life here safeguarded by an exceptional conjuncture of circumstances not found anywhere else on the planet."[100] This potential for Galápagos tourism was first officially acknowledged as guiding policy in the 1964 decree signed with the dedication of the Darwin station.

Over the coming years, studies by international development consultants elaborated the vision for national park tourism. In 1966, Ian Grimwood, a national parks expert, and David Snow, an English ornithologist on the Darwin Foundation board, made recommendations about how the state should organize its new national park to have control over tourism and find "the best means of exploiting its economic potential."[101] British ornithologist and advertising executive Guy Mountfort, who helped found the World Wildlife Fund, cautioned that tourism could mean "irreparable" disaster for island ecosystems but thought the industry would bring in foreign currency desired by the government. What's more, organized tours would "indoctrinate tourists with the principles and needs of conservation," and their donations could help fund the research station.[102] It was a plan endorsed by Darwin station biologists who trained and issued permits to naturalist guides and worked with cruise companies to set visitor sites around the archipelago, dictating that tourist ships adhere to trip itineraries instead of allowing them to cruise the islands at will.

The following year, the international consultancy firm Arthur Little, Inc., reported on the results of a tourism market analysis and economic feasibility study, making recommendations to the government about how to structure the tourism industry, how to price tours, and how to best serve the demands of the elite foreign tourists who were likely to visit. The typical tourist would be an "able-bodied grandmother" who would expect a cruise ship that offered "creature comforts comparable to those in [her] own home." She would be "vigorous and adventuresome" and a "good sport" but should "not be expected to run, jump, or get wet."[103] The consultants outlined the kind of boat needed to best serve her—with one hundred first-class rooms with private bathrooms and plenty of hot water—as well as the need for expert guides and regular itineraries to established visitor sites. Although many of their suggestions never came to fruition—including picnic tables and bathrooms at visitor sites on remote islands—the report laid out the basic structure of cruise tourism that still exists today. Of principal concern to both scientists and government officials at the time was not the welfare of island populations, but rather how best to serve elite tourist visitors.

LAND FOR THE LAB

The tourism strategy did not end Ecuadorian colonization, despite conservationists' suggestions. But it did reorient a sense of who the islands were for. No longer would they be an escape for European migrants, a jail for society's unwanted, or an open frontier of possibility for colonists looking to live off the land. Instead the islands would be a sanctuary for nature, preserved in the service of an elite global tourist who could afford the opportunity to visit. This singular vision for the future, however, was challenged by the continued arrival of new colonists and their desires for development. In response, conservationists struggled to secure the boundaries of their natural laboratory. The understanding of island space that had underlain penal colonies was now inverted: instead of a place to banish prisoners, the islands became a fortress of conservation where nature would be enclosed to protect it from human use.[104] The laboratory metaphor justified a series of social and physical inclusions and exclusions, including opening the islands to scientists, conservationists, and tourists and closing park territory to local residents, who were restricted to the 3 percent of the terrestrial area of the archipelago designated for human habitation and agriculture.[105]

This new vision of land use for nature preservation in the service of global science and foreign tourists became entangled in debates about agrarian reform that dominated Ecuadorian politics during the 1960s. The late 1950s and 1960s was a period of state-promoted frontier colonization in Ecuador, largely motivated by the desire to both secure sovereignty in frontier areas (challenged principally by Peru) and bring uncultivated land into production. Many of the Ecuadorian colonists who went to the islands at the time did so seeking land and opportunity. According to one station director, they arrived "with an erroneous conception of the economic possibilities in Galápagos."[106] But for decades, the Ecuadorian government had encouraged island colonization by making it relatively simple to acquire land. Amid mounting calls for continental agrarian reform to distribute land held overwhelmingly by elite *hacendados,* the state had encouraged island colonization by offering homesteads. This was facilitated first by the Colonization Institute that Crespo led and later by IERAC, the Ecuadorian Institute of Agrarian Reform and Colonization, which had been founded as part of agrarian reform efforts in 1964 led by the military government of Ramón Castro Jijón. Land redistribution reflected a competing vision for economic development in the Galápagos; for the junta, it was to be "the cornerstone on which to build a new, harmonious, just, and dynamic Ecuador."[107] Reforms were aimed in part at assuaging the threat of a communist uprising among landless peasants but were largely worked out between a repressive military and landowners.[108] Claiming land in the islands was easy before the conservationists came; as one early *colono* recalled of the nearly three-hundred-acre claim she and her husband made, one just would go and say, "OK, this is where I'm going to be. You went, cleared it, enclosed it, and that was it. You were a property owner."[109] Darwin station scientists wanted to end such claims, but their own land claims for conservation were interpreted by colonists as an indication that the value of land in the archipelago was rising—which spurred more of the land claims that conservationists sought to discourage.

Lévêque and subsequent directors in the 1960s tried to make the station project fit their preexisting ideas about pristine environments, where the nature scientists wanted to study would be protected from colonists. But his experience echoes what numerous histories of conservation have shown: that Westerners with dreams of preserving "pristine" nature never enter into empty spaces, even in remote islands like the Galápagos. Political ecologists and historians have detailed how, to create the pure natural space of an ideal national park, conservationists have attempted to clear landscapes of their human residents and limit access to natural resources, often through violent processes of appropriation underwritten by colonial and national governments.[110] This work demonstrates that conservation territories are not spaces of apolitical nature as the natural laboratory metaphor suggests, but rather contact zones of political contestation in which environmental governance regimes rework the subjectivities of local residents by erecting boundary fences, limiting hunting and forest use, and introducing new livelihood strategies.[111]

The story of instituting conservation in the Galápagos was neither as bloody nor as fiery as were some of the worst experiences in the United States, Latin America, and Africa.[112] But, as in many places, the making of a natural laboratory did involve the curtailing of livelihood strategies and the forcible appropriation of land. One of the often-repeated narratives about conservation history in the Galápagos is that the 97 percent of island territory that constitutes the national park was carved out around settlements that existed when the park was founded. This is partially true, but it is too simplistic an explanation. This narrative implies that conservationists took only empty space, that creating the natural laboratory was not a politically contentious project. Marking the boundary between protected parkland and zones of human habitation actually took more than a decade of negotiation that included violent confrontations and legal battles.

When scientists first planned the station, Dorst thought they might succeed in protecting western Santa Cruz (where the island's tortoise population was concentrated) and perhaps the whole islands of Española (home to migratory seabird colonies) and Fernandina (the archipelago's "most pristine" large island). But protecting this much territory would be "undoubtedly difficult."[113] Lévêque, though, suggested claiming all of Santa Cruz that was not yet settled: "I'm not exaggerating," he wrote. "This would give us the right, at least theoretically, to surveil the island, permitting us to prevent people from wandering everywhere, destroying as they please."[114] Lévêque was particularly concerned not only about poaching of individual animals, but about land clearance, grazing, and the introduction of foreign plants and parasites that had a "direct effect on the natural equilibrium" and could indirectly cause the extermination of key species like Darwin's finches because of habitat loss. "Everything fits together in Nature, and keeping a biotope intact is more important than the individual protection of its inhabitants," he reminded Dorst.[115]

Governmental decrees issued in 1961 and 1965 gave Darwin Foundation scientists the authority to include in the national park land that was not settled. But the difficult, and unpopular, job of actually delineating the boundary foreign scientists gave to their Ecuadorian counterparts. Visiting scientists had long hired local men to assist them on research trips—when conservation efforts started, they also employed them to survey native species, hunt feral animals, and keep an eye on other residents' use of natural resources. At times this work put—and continues to put—local conservation workers in difficult positions because they had to police their neighbors, friends, and family. The station's first conservation worker (a position financed by the New York Zoological Society), Miguel Castro, was tasked with persuading colonists not to eat giant tortoises and was often put in the middle of controversies between the station and settlers.[116] Nature protection, particularly rights to timber and sand for construction, was often a source of tension with a population accustomed to living largely under their own authority in a place with little state presence.

In 1968, the Galápagos National Park began in earnest when the state sent two park guards to work on conservation in the islands. Juan Black and José "Pepe" Villa were young graduates recruited for the post from a forestry school north of Quito run by the United Nations Food and Agricultural Organization. In interviews, Villa, now in his seventies, told me that a government minister had personally requested he take the position. His country needed him, he was told, but these remote islands known on the continent as a place of banishment for convicts had not been where he was planning on working after graduation. Nonetheless, months later he found himself on a ship with Black bound for a place he had never been before. After a crash course in island natural history from Darwin station scientists, Villa was asked to delineate the boundary between the new national park and the agricultural area on Santa Cruz. Staking the boundary line was supposed to prevent further land claims. But it was a difficult process because attempts to draw a clear line confronted both fuzzy ecological edges and contested notions of the boundaries of locals' property, complicated by their customary use of forests as commons for hunting and timber collection.[117]

Villa spent months camping along the boundary with Castro. On Sundays they would have meetings with settlers to explain their work. The best way to do this, Villa told me, was to become friends with the farmers and drink with them in their homes. Indeed, neither marking the boundary nor the larger project of making a natural laboratory was purely a conflictive relationship with local residents. As geographer Juanita Sundberg writes, zones of encounter are places of "contestation and conflict, but also connection, empathy and contract."[118] Although Lévêque's letters to Dorst reflect a European culturally isolating himself, he also developed close friendships with Ecuadorian settlers, as did many other Darwin station scientists. When trying to understand the cultural shift that the scientists' arrival effected, I heard stories of Franciscan priests preaching conservation values from the pulpit—one Monsignor even declaring during the GISP visit that the Catholic Church supported Darwinian evolution.

Villa would invite farmers to walk the boundary line with him as they discussed conservation. But, he said, sometimes it was a bad situation that involved negotiating the placement of farmers' fences that he saw as encroaching on park land; he told me, "I had to say, no, that's not yours, but they had been living here four or five years."[119] Villa told me he moved the boundary line on one farm to try to protect a patch of leafy Miconia shrubs, but the farmer was furious. Most people carried arms (to hunt feral animals), and a couple of times they shot toward him, although not right at him. The son of one US colonist who had been involved in a property dispute told me that these early settlers were people who sought an independent life and did not appreciate the new imposition of authority.

Despite the months Villa and Castro spent in the *campo*, conflicts between conservationists and settlers persisted. In 1971, Darwin station director Peter Kramer reported to UNESCO that the national park's borders were in danger because territories to be included on Santa Cruz were being invaded by IERAC.[120] In 1974, in the midst of a second, stronger push for agrarian reform and frontier colonization fueled by new Amazonian oil profits, IERAC sent representatives to the island to remap the boundary with park guards in response to colonists' complaints. The previous year, the state government had made the Galápagos a province (no longer under the jurisdiction of the Guayas province)—largely in response to residents' calls for representation.[121] But the institution did not uniformly defend settlers' land. Several families took their cases to the continent for legal review. Most lost their appeals, and several were relocated from areas on Santa Cruz and Santiago deemed to be within the park—some choosing to return to the mainland rather than to stay in the Galápagos.[122] They were compensated for land lost, although not for their labor or crops.[123] The boundary of the national park was not legally settled until 1978, when UNESCO intervened by making it a condition for naming the islands a World Heritage Site. Yet the border has remained a site of contestation that park guards regularly patrol.[124] Making the Galápagos into a natural laboratory has never been as complete a project as nature documentaries suggest.

When the *Golden Bear* sailed into Academy Bay, the scientists who coordinated the trip saw it as promising a future of scientific research and nature protection in the islands. The founding of the Darwin station and national park marked the beginning of a period in which the islands would be governed as a natural laboratory—a place where nature should be prehistoric, and biologists could study the origins of life. The transnational environmentalist institutions, tourism companies, and Ecuadorian leaders behind this vision elevated the Galápagos onto a global pedestal as an evolutionary Eden where Darwinian biology was enshrined for the world's appreciation. This geographical imagination succeeded because of the way it drew together what had been conflicting ideals about the islands as frontiers for development, penal colonies, island escapes from modernity, and biological field sites. In this new vision, Darwin replaced Crusoe as the heroic adventurer, rare animals instead of peons were enclosed in island reserves, and the development of nature tourism sold the opportunity to escape to desert islands to mostly foreign tourists.

Tourism has been a central strategy of conservation since the founding of the Galápagos National Park. This point is often lost in concerns about overdevelopment on small islands that are so prevalent in conservationist discourses today. Such debates tend to pit conservation and development as competing forces. Yet while conservationists opposed some forms of development on the islands, they knew their visions for island management would be impossible without a thriving tourism economy. The founding of the national park and Darwin station was not so much an enclosure of the Galápagos from global economies, but was rather, as the geographer Christophe Grenier has argued, a key moment of the "geographical opening" of the Galápagos to the outside world.[125] Yet the archipelago had long been embedded in circuits of global capitalism. More important was how the vision of a natural laboratory was premised on a development strategy that would fundamentally change the islands over the coming decades. The creation of this purported space of pure nature outside of modernity was a project so steeped in the logics and economies of modern life that even as it protected the islands' prized species, it would paradoxically bring the archipelago ever more fully into the throes of modern development. Indeed, the oversize *Golden Bear* was a harbinger of the scale of a future that no one in the mid-twentieth century imagined. When the islands had just one hotel and no paved roads, it would have been nearly impossible to predict how successful this alliance between conservation, science, and tourism would become.

Yet mass nature tourism was just one of the changes the creation of the Darwin station and national park brought to the Galápagos. These institutions also began experiments in governing island life—both nonhuman and human—that attempted to fit both into the imaginative confines of a laboratory.

6 • Restoring Evolution

In which the sex lives of tortoises, and goats, become a problem . . .

and a strategy

If you travel to the Galápagos today, you will likely have the opportunity to meet Diego. He is often surrounded by a crowd of camera-wielding tourists hovering over a low, lava-rock wall toward the back of the breeding center at the Galápagos National Park headquarters on Santa Cruz. They are trying to get a good shot of this celebrity tortoise who is a star of the giant tortoise breeding program. This three-foot-long animal is, as tour guides tell it, a *"supermacho"*— a stud who over the past forty years has sired nearly a thousand offspring. When I volunteered at the center, Diego had lived in a corral with one other *macho* and six *hembras,* or female tortoises, for decades. They are all from Española, a small, dry, easily accessible island where hunting by nineteenth-century sailors devastated the tortoise population. In the 1960s and 1970s, as conservationists searched the island to conduct population surveys, they found only fourteen tortoises—so few that they feared the animals would not find each other to mate. So they brought them all back to the new Darwin station with the hope of breeding the tortoises in captivity to ensure the survival of the species. Diego, one of few named tortoises at the center, joined these animals in 1977 from San Diego—his namesake city where he had lived for forty years as part of the zoo colony Charles Townsend established in 1928.[1] These fifteen tortoises were all that remained of the *C. [Testudo] hoodensis* species that John Van Denburgh named in 1907; saving the species from extinction depended on their ability to reproduce.[2] Thanks in no small part to Diego's virility—and that of another, unnamed, male who, unbeknownst to keepers until recent genetic studies, has sired even more offspring—the number of individuals of the species has rebounded: since 1971, park guards have "repatriated" more than eighteen hundred *C. hoodensis* juveniles to their Española homeland.

Like Lonesome George, Diego is an icon who can help us understand the history of Galápagos conservation. His life story illustrates how the goals and methods of saving species from extinction have changed since the days when Rollo Beck and other collectors preserved taxidermied specimens for museums.[3] Diego's return to the Galápagos marked a shift toward in situ protection and captive breeding of charismatic animals, made possible through the creation of the national park and Darwin station.[4] With this change in *where* saving species took place came a concomitant change in just *what* was being saved: no longer was the focus on collecting and preserving type specimens to stand eternally in museums as proxies for species; nor was it on saving individual animals alive in zoos as exotic exemplars of dying species. Instead, the focus moved to saving *living* species as the central objects of a natural laboratory of evolution. What is at stake in Diego's reproductive prowess, and the actions of multiple other individuals, is not only the flourishing of his species, but also the future of much broader processes of evolution.

But just how does one save evolution? One conservation biologist I asked said the goal of conserving the Galápagos was "to maintain evolutionary processes in an unaltered state."[5] She wanted to "conserve the real Galápagos . . . an evolutionary laboratory," which involved "maintaining evolutionary processes, pristine ecosystems, [and] species."[6] For another scientist, this translated into work that aimed to protect the unique creatures that have evolved in the Galápagos so that they might continue to do so in a changing world—an interpretation I think the majority of his colleagues, and the public, would support.[7] But that was not how a group of fifty-eight Galápagos-based and international conservation biologists summarized their "Biodiversity Vision" for the archipelago at the turn of the twentieth century. For them, the goal was to go "back to Eden" by restoring the islands to their condition in 1534, the year before Tomás de Berlanga's ships drifted into the archipelago and the islands entered the annals of Western history.[8] Is it possible to *restore* evolution?

As a goal for conservation, restoring an evolutionary Eden seems a paradoxical objective considering that Darwin taught us that life is always moving forward, never backward. He made it clear, as philosopher Elizabeth Grosz reminds us, that "time, along with life itself, always moves forward, becomes more rather than less complex."[9] Why has such an un-Darwinian goal so captivated conservationists? Is it possible to conserve the *process* of evolution in a world where isolation is a thing of the past? What happens when conservationists try?

If the species of natural history, although preserved as independent objects on museum shelves, were constituted through animals' entanglements in nineteenth-century expeditions, fieldwork, specimen preparation, and museum studies, then what kinds of multispecies relationships go into restoring evolution? Galápagos conservation is a place-based endeavor in which evolutionary understandings of life—what species evolved over hundreds of thousands of years on which islands—inform a biopower that values and aims to "make live" certain endemic species, such as the giant tortoises.[10] "Biopower" was philosopher Michel Foucault's term for forms of productive power concerned with the management of populations—a focus he traced to Darwin's understanding of life.[11] In practice, wildlife biopower in the Galápagos is translated into conservationists' fixation on the sex lives of tortoises. From a Darwinian perspective, the fetish is understandable; reproduction, coupled with individual variation and natural selection, was for Darwin the heart of the process of evolution.[12] Restoring evolution as a process required conservation efforts focused on reproducing the tortoise species—feeding and cleaning up after Diego and other animals in captivity, collecting and incubating their eggs, and caring for tiny hatchlings until they are large enough to be released in the wild. But the fixation on sex extends beyond intervention in the reproduction of desired populations.

The chief threat to the natural laboratory are the many introduced, or "alien," species that did not evolve in the archipelago over the *longue durée* of evolutionary history but have managed to land there more recently, with a bit of help from human travelers, and to reproduce. Many of these introduced species—such as goats, rats, and blackberry—have had such reproductive success that conservationists see them as invasive threats to native species—an ecological problem framed in the language of a nationalistic battle. According to the logic of restoring an evolutionary laboratory, the proliferation of these invasive species is undesirable. It is *unnatural* selection. Their ability to adapt to island ecosystems is portrayed as the ecological ruin wrought by the imposition of human history in this would-be Eden.[13] Alien species disrupt the isolation of island ecosystems that was central to the evolution of unique flora and fauna. Restoring an evolutionary Eden means eliminating them. These species have been "made die" as the targets of control and eradication campaigns that have slaughtered hundreds of thousands of animals over the past fifty years.[14] If conservation is centrally about sex and the reproduction of life, then it is equally about death and the extinguishing of life.

Attempts to save evolution in the Galápagos have involved what environmental philosopher Thom van Dooren sees as the "violent-care" of conservation work, acknowledging that caring for endangered species often involves violence toward them and other species.[15] The survival of some species has come to depend on the elimination of others. How do conservationists, island residents, and we—as members of the global public who are ostensibly, along with native species, the beneficiaries of this work—sit with this? What does it mean to restore evolution? Should this be the goal of conservation? These questions do not have easy answers, but understanding the work of conservation is essential for beginning to address them. Let's begin as Galápagos conservationists did in the 1960s, by surveying the islands.

Surveying Evolution

On a sunny day in the mid-1960s, Carlos Ochoa sat eating lunch on a small hill overlooking a parched landscape sprinkled with thickets of brush and short, spindly trees.[16] Four days earlier, he and a *compañero* had hiked up here following a wide burro trail from Bahia Sardina on the eastern coast of San Cristóbal looking for giant tortoises. Below him in the distance was a small cluster of shade trees that seemed promising, so he walked over for a look. Ochoa told me this story one afternoon some forty years later as we sat in the courtyard of his house in Puerto Ayora. He was born there—his father had been in the Ecuadorian military and stationed at Baltra in 1942. When Ochoa was nineteen, he started working for the new Darwin station, helping to construct the first laboratory building and later serving as a conservation worker, one of several young Ecuadorian men who traveled the archipelago surveying wildlife. On this trip, Ochoa's boss, Miguel Castro, had sent him and another man in search of tortoises on the dry northern stretch of this island, far from the settlement of Puerto Baquerizo Moreno, which lay at its southern tip. The hunt would be a challenge because the tortoise population had been considered nearly extinct since the Cal Academy expedition in 1905–1906. But as he neared the trees, Ochoa found a large scat; descending farther toward the plain, he discovered a giant tortoise lounging in the shade of leafy green trees. "*¡Venga! ¡Aqui hay una tortuga!*" he called back to his companion, who was still eating lunch. They were supposed to leave for the five-hour hike back to the coast, but Ochoa's discovery prevailed, and the two men continued exploring the area. They found nine tortoises that day, members of the species Van Denburgh had classified as *C. [Testudo] chathamensis* and warned was "nearly extinct."[17]

Scientists and conservationists who worked in the *campo* during the 1960s and 1970s tell many stories about discovering populations of tortoises in places where they thought the animals had disappeared. Field surveys were

the basis for a new era of studying the animals that focused not on morphological comparison, as systematic classification had, but on the animals' behavior and ecological relationships. The island laboratories gave modern
naturalists a "far better ringside view of the process of the differentiation of
species . . . than anywhere else in the world," one biologist wrote.[18] The species were not only exemplars of evolutionary processes of diversification, as
they were in museums, but were the *living* subjects of nature's ongoing evolutionary experiments. The foreign scientists UNESCO sent to the islands went
there with training in Darwinian natural history, but they knew little about the
status of key species they were charged with protecting. Surveying the islands
was among their first priorities, and they relied on local men to do much of
this work.

The months Castro and his team spent in the *campo* from the mid-1960s
through the early 1970s amounted to more time in the field in the Galápagos
than anyone had previously spent. As these men lived on deserted islands
and searched for tortoises—as well as albatross, boobies, rails, penguins, and
iguanas—they made the most complete survey of the islands that had
been conducted since turn-of-the century collecting trips. The work was
often harrowingly difficult. As José Villa, one of the first park guards, told me,
"the conditions were very bad, we didn't have any equipment, we didn't have
any communication."[19] With no radio to call back to the Darwin station, they
were not always even sure when the station's research boat, the *Beagle II*, or
another ship would be back to retrieve them, leaving them unsure about how
to ration water.

But these men's knowledge and labor made—and continues to make—
field science and conservation possible. They accompanied Darwin station
scientists and visiting biologists, acting as field guides and often porters as
well. One of the first station directors, Roger Perry, wrote that after one trip
with Camilo Calapucha, a member of the Yumbo tribe who had worked on
Santiago as caretaker of a salt mine until it folded in the mid-1960s, he would
not go on a field expedition without him. Such was the value of Calapucha's
knowledge and skills for the safety and success of field ventures. These workers made crucial contributions to science and conservation, and they were
gratefully acknowledged by several foreign scientists who credited their work,
dedicating theses to friends and co-authoring publications. But too often these
men were eclipsed in official histories of the station or mentioned only as
shadowy presences in published papers of visiting naturalists, their names
forgotten.[20]

These conservation workers discovered hundreds, if not thousands, of tortoises living on Santa Cruz, Santiago, and Volcán Alcedo. Following the 1957 reconnaissance mission and the 1964 Galápagos International Scientific Project, foreign scientists had feared the imminent extinction of tortoise populations. But, as Rollo Beck and Arthur Slevin had found some seventy years before, early field surveyors revealed that the tortoises were not in quite the dire straits that naturalists feared. Although they were not university-trained ecologists—few had formal education past grade school, as there was no high school on the islands in the 1960s—they gathered extensive knowledge about tortoise population numbers, sex ratios, feeding practices, and reproductive behavior that was essential for saving these species in situ and became the basis for conservation. These surveys brought to life as ecological beings the species Van Denburgh had described on the basis of taxidermied specimens.

A few weeks after Perry arrived in the islands in 1965, Castro took him up to "tortoise country" on Santa Cruz northwest of the research station at Academy Bay. After a day's trip on horseback and a night at La Caseta, a small shack where another islander, Laurtaro Andrade, spent weeks at a time protecting the tortoises, the men went to El Chato, a relatively flat area in the humid zone just below the highland hills. There, Perry wrote, they found "a green hollow where a dozen tortoises were grazing. Their smooth shells, darkened by the rain, looked like polished boulders in the mist. There was something infinitely archaic about the sight of those animals quietly feeding in the pale light and stillness of morning." As Perry watched the scene in awe, imagining the tortoises in a kind of timeless existence, Castro "was already busy upending the tortoises, peering beneath, measuring and making his notes. It was from these data that a picture was emerging of the numbers, sexes, and movements of the tortoises."[21] Castro developed a system for marking tortoises by making notches in their carapaces, allowing him to estimate population size (fig. 6.1). Looking underneath the tortoises told him whether they were male or female, as male tortoises' plastrons, or under-shells, are slightly concave, which facilitates their ability to mount females during copulation.

In 1968, Castro was joined by a young ecologist from the United States, Craig MacFarland, who was there to do a preliminary study for his zoology dissertation at the University of Wisconsin–Madison. MacFarland would go on to direct the Darwin station and then serve on the Darwin Foundation board for decades, but he got his start in Galápagos conservation by spending two years in the field with Ochoa, Villa, and other local workers. He also worked alongside his wife, Jan, and eventually their baby daughter, who spent

Figure 6.1 José Villa, one of the first two guards of the Galápagos
National Park Service, marks a tortoise on Pinzón, late 1960s. Photo
taken by Miguel Castro. (Used with permission from the Charles Darwin
Foundation Archives)

much of her first year living in tents and having her diapers changed on tortoises' backs. When he first arrived, MacFarland planned to do "pure ecological" work on tortoises: he wanted to measure ecosystem function in a rising
field that was dominated by quantitative methods of analysis. But as he struggled to compile data—finding, for example, that many of the tortoises he
wanted to weigh exceeded the five-hundred-pound limit of his hanging scale—
he realized that "applied" knowledge for conservation was more practical. So
he focused on basic population ecology, tortoise behavior studies, and natural
history.

With funding from the National Geographic Society, the MacFarlands
spent months camping among the giants, observing their behavior. The

animals would arise at dawn, spend most of the day grazing, and settle in for the night about 4 p.m., as the sun began to wane (on the equator, sunrise and sunset vary little from 6 a.m. to 6 p.m. throughout the year). They often rested in muddy pools that helped regulate their temperature and protect them from mosquitoes that lined up to feed on the tissue between the scutes of their carapaces. Yellow warblers, vermilion flycatchers, and lava lizards would also gather around the tortoises, feeding on flies that constantly buzzed in their eyes. But most impressive was watching ground finches approach a tortoise, "hopping and chirping" until the tortoise stretched out her legs and neck. She stood still while the finches hopped on her legs, neck, and head, pecking at her skin. They were removing ticks, the MacFarlands realized. "The tortoise didn't even blink when the finches pulled ticks from the corners of her eyes and from her nostrils. Throughout the 5½ minute cleaning session, the animal stayed stock still" until the birds had flown away.[22] The MacFarlands had observed one of the wonders of cross-species mutualism in the islands.

Both the MacFarlands and Castro's team spent considerable time observing tortoise reproductive behavior, which MacFarland described for *National Geographic* readers: "The rutting male tortoise stalks about looking for mates and sniffing the air for their scent. Spotting a female, he chases her down and usually begins courtship with intimidation, ramming her with the front of his shell and nipping at her exposed legs until she draws them in, immobilizing her."[23] During nesting season, early in the year, they followed female tortoises as they migrated to dry lowlands near the coast. MacFarland spent ten days following one tortoise as she attempted to dig a nest but was repeatedly thwarted by the rocky ground. Finally, about four o'clock one afternoon Jan called Craig over for what seemed like a promising attempt. Using her "tremendous strength and agility," the tortoise "excavated [a nest hole], using only her hind legs—probing, shaping, measuring. Her forelegs held the front half of her 150-pound body raised above the ground in an unchanging position."[24] She dug for five hours; when her labored breathing increased pace, she wriggled and waved her tail and then dropped her eggs ten inches to the bottom of the nest. Using her foot, delicately not to crack the eggs, she spread the pile into a single layer and then urinated on the surrounding dirt to make a thick mud cap for the nest. By the time she finished, it was 3:30 a.m. For anywhere from three to eight months, the eggs would incubate in the nest, warmed by sunlight, until hatchlings would dig their way out.

By observing this last part of the reproductive process, the researchers saw what was endangering the tortoises' survival. It was not so much the human hunting, which had worried George Baur and Rollo Beck, that now threatened the tortoises, but introduced mammals—particularly rats; wild hogs, or *chanchos;* and feral dogs—which would dig up nests or eat hatchlings as they emerged, leaving nothing but ruts in the ground, broken eggshells, and tiny chewed tortoise shells.[25] Finding ways to protect young tortoises thus became a top priority for conservation work.

During Roger Perry's tour of "tortoise country," Castro and Andrade took him south from La Caseta to La Reserva, the one-hundred-square-mile tortoise reserve where Andrade's primary job was to hunt animals who preyed on tortoises. That morning, he had been out hunting pigs with the aid of his dogs. Perry recounted how "he drew from his pocket a string of smoke-blackened snouts, his tally, which Castro silently counted and entered in his notebook" before throwing the snouts to the dogs. *Chanchos* were a plague for the tortoises because they had an uncanny ability to sniff out nests. "One pig can destroy dozens of nests in a night and never miss an egg," MacFarland wrote. They are intelligent, one park guard told me; "they know when nesting season is and live close to the area where females lay eggs; that is why people have to hunt nearby."[26]

Knowing what the hogs knew, conservationists could protect the nests. Castro built small lava-rock walls around each nest so that pigs could not get to them. Months later, he would return to "liberate" newly hatched baby tortoises. Park guards still do this work, although they now protect nests with wire mesh rather than rock walls (which lets more sunlight fall on the nest—the warmth it provides is essential for successful development). While volunteering with the park, I accompanied a liberation mission to the nesting zone below La Caseta. The trip down was an arduous trek—at least for a hiker accustomed to clear paths over solid ground. We marched through grasses taller than me— the park guards always in front, hacking away with their machetes—mucked through mud that held onto our rubber boots (one volunteer stepped right out of his when the mud refused to let go), scrambled up a lava cliff, and then followed a mostly dry streambed full of loose lava boulders washed down from the highlands during the last El Niño rains. But the hours-long journey was well worth it as we helped new tortoise hatchlings no bigger than my palm emerge from their nests and quickly scurry into the brush to hide.

But protecting nests was not enough on all islands, nor was it feasible when long trips were required. On Pinzón, surveyors found about a hundred adult tortoises, but no young—only rats and evidence of their predation. MacFarland published photographs of baby tortoises eaten by rats, only their carapaces remaining. The images sent a poignant message about the destruction rats caused. In 1965, Perry and Castro decided that they would need to take eggs from nesting sites back to the station to incubate and hatch while they found something to do about the rats. This was the beginning of the breeding center.

On other islands, goats were the problem. In the early 1960s, on Diego's home island Española, surveyors found a lone tortoise feeding on a fallen prickly pear cactus, surrounded by fifteen goats sharing this prize feast. Finding the tortoise was a promising sign, but herds of a hundred goats—thought to be descended from just a few introduced in the early twentieth century—had degraded vegetation. Yet goats, Perry wrote, "prevent[ed] regrowth by consuming anything within reach and churning up the brittle crust of soil" on this eight-and-a-half-mile-long island, where a volcanic peak had long since eroded into a "rolling plateau of rocks and brown volcanic soil, [now] covered with mesquite, acacia, and scattered groups of prickly pears."[27] The wild goat, another biologist wrote, was a "destructive, hungry, thirsty beast during the dry season. He not only eats fallen [cactus] pads but is observed sometimes to gnaw through the trunk of a giant cactus, cut it down and then eat its fleshy parts."[28]

From such field observations, new conservation priorities emerged in the 1960s. A century before, Darwin had anticipated the havoc that introduced species could wreak on islands where native species had not evolved ways of protecting themselves.[29] His observation was born of the kind of ecological understanding of species interrelations that he saw as the motor of life, and which motivated mid-twentieth-century research on the relationships between evolution and ecology. But the ecological story that early conservation workers described was not one of the evolutionary triumph of the survival of the fittest. Instead, rats, pigs, goats, and even feral dogs and cats were proving to be more "fit" to island ecosystems than the tortoises and other native species that had evolved there over hundreds of thousands of years. For conservationists, this was *unnatural* selection: the result of an ecological invasion caused by early sailors and island colonists who had brought these animals along with them.[30]

Together, introduced human and nonhuman species were disturbing, even destroying, what ecologists then understood as the balance of nature that was supposed to have been worked out through millennia of evolutionary processes. Saving this natural laboratory would mean saving its evolutionary history and restoring the islands' past isolation from human worlds. To achieve these ends, scientists used early ecological surveys to devise a two-pronged plan of captive breeding and rearing of endangered species, and hunting to purify the natural laboratory of introduced species. Through these methods, Galápagos conservationists strove to save the environmental conditions that had produced prized endemic biodiversity. By doing so, they inserted themselves as agents of *conservationist* selection.

The Work of Reproduction

As part of my research, I volunteered at the breeding center, then run by the national park under the direction of Fausto Llerena, the park guard for whom the center is now named (Centro de Crianza de Tortugas Terrestres Fausto Llerena). Don Fausto, as everyone calls him, was then in his early seventies and had spent more than thirty years caring for giant tortoises.[31] Following his lead, I and the other volunteers—naturalist guides completing annual service hours and the teenage children of visiting biologists—spent mornings feeding and cleaning up after tortoises. On Mondays, Wednesdays, and Fridays we delivered to the tortoises wheelbarrows full of leafy *porotillo* branches (an introduced tree that farmers in the highlands use as living fence posts to stake their land) and prepared "salads" for the young using machetes to julienne giant leaves and dice stalks of *otoy*, another plant introduced by early settlers primarily to feed livestock. Feeding time often drew a crowd of tourists who watched the tortoises rise from their usual lounging repose and lumber toward food, sometimes even clambering right over each other, as though their corral-mates were boulders in their path. On Tuesdays and Thursdays we cleaned, picking leftover salad out of the pens that house juvenile tortoises and sweeping scat and thoroughly gummed *porotillo* branches from feeding platforms in the nine corrals that house adult tortoises. We also scrubbed gunk from their cement wading pools with wire-bristle brushes while the tortoises lazed in the sun nearby, uninterested in our actions if they did not involve food. Lonesome George, however, would often amble over to challenge us, stretching his neck high in an attempt to prove he was taller than his human keepers and thus show his dominance.

This daily maintenance was part of the routine of care that park guards and station scientists had worked out over the decades since Miguel Castro had brought the first Española tortoises and eggs from Pinzón to the station in the 1960s. Since then, facilitating giant tortoise reproduction has become a standardized routine that also includes monitoring nesting, incubating eggs, and taking measurements and recording data as tortoise hatchlings grow.[32] Examining how tortoise life is produced in the breeding center highlights the care work, intimate relationships, and assemblages of human and technical labor that are central to the continuation of the species. It also demonstrates aspects of captive breeding that tour guides do not highlight—the years of experimentation aimed at improving low early rates of survival. The work of saving species is both more technically difficult and ethically complex than triumphant headlines or quick tours suggest.[33]

I confronted this ethical murkiness firsthand at the end of my time as a volunteer, when Don Fausto and the other guard who worked with him, Daniel, taught me how to excavate nests the *hembras* had recently laid. It was August, in the midst of nesting season, and for several days the other volunteers and I had been "helping" Don Fausto and Daniel excavate nests, mostly by taking pictures of this work, which we found far more exciting than scrubbing pools. In the small nesting zones within the corrals, where guards had cleared lava rocks and brought in soil from the nesting area in La Reserva, they identified nests that to me were almost indistinguishable from the rest of the ground, tapping them with the handles of their ever-ready machetes to confirm a hollow hiding under the hard-packed cap. Using the knives, they would wrench off the dirt caps, somehow never puncturing an egg, then carefully retrieve the billiard-ball-sized eggs and put them in a plastic tub lined with vermiculite, being careful not to rotate them and dislodge the membrane from the shell, which would affect the embryo's development. They penciled an X on top of the egg as well as the number of the nest and the *hembra*, before we walked them back to the office to put in an incubator (fig. 6.2).

One morning, standing over a new nest, they said it was my turn and handed over a machete. With more effort than either man had shown, I eventually removed the cap, nearly a foot wide and six inches deep, revealing thirteen white eggs, one of which I seemed to have broken. I worried about the consequences of my inexperience, but they said I shouldn't—indeed, they told the other volunteer and me to throw the eggs from this nest into the forest behind the corral, even challenging us to hit a cactus. I hesitated, sure that I had misunderstood their Spanish or that they were pranking me. Why were we not treating these eggs—new potential life of a prized endangered species— with the same care as the others? But goaded on, I pitched an egg toward the cactus, missing. We took turns disposing of the eggs from this nest; I hit the cactus on the third try and watched the egg's yellow yolk drip down the trunk, feeling unsettled.

Afterward, as we walked back toward the office, Don Fausto explained that these tortoises, which lived in a corral at the back of the center out of public

Figure 6.2 Park guards Don Fausto Llerena and Daniel excavate a tortoise nest at the breeding center on Santa Cruz, August 2011. (Photo by the author)

view, were a new breeding group from Floreana. The population on that island went extinct shortly after Darwin's visit, but geneticists had recently found survivors, or their descendants. Some of these were tortoises living on Volcán Wolf at the northern end of Isabela; DNA analysis revealed dozens of hybrid tortoises, the product of interbreeding between species from different islands. Some of these tortoises had genes from the Floreana lineage. But the genetic analyses also revealed other Floreana tortoises already at the breeding center who had been living there for decades, unbeknownst to keepers. They had been collected from island settlers in the 1960s, so conservationists did not know what islands they were originally from and had kept them in *mixtos* pens separated by sex so they would not breed. Now in a new corral of their own, male and female Floreana tortoises were spending their first mating season together. It would be an experimental year. The incubators were already full of eggs from these tortoises, which, sexless for decades, had produced a lot of eggs. Besides, most of the eggs in the nest had been misshapen, oblong instead of round, or encased in shells that were too thin and would not produce successful offspring. By having us dispose of these eggs, guards were reducing the number of potentially unhealthy hatchlings and saving room in the incubators for other eggs.[34]

We did not bring any eggs in that day, but usually the routine of care work continued back in the office, behind a gate that separates the juveniles' pens from the part of the breeding center open to tourists. Here, Don Fausto and Daniel weighed each egg and recorded it in their logbook before packing it in another tub to be incubated. The incubators were not the modern ovens used in many zoos, but two tall wooden cabinets, each with another cabinet inside with shelves full of plastic tubs covered in dark plastic. The cabinet was heated by a hair-dryer, rigged to cycle on and off to maintain a standard temperature— one cabinet set to 29.5 degrees Celsius, the other to 28 degrees—a temperature differential that would produce female and male hatchlings. "Hot chicks and cool dudes" is how the tour guides put it. Like many reptiles, giant tortoises are temperature-sex dependent, meaning that the embryonic development of their reproductive organs is influenced by the temperature of incubation. In the wild, varied amounts of sunshine falling on each nest, or parts of nests if shaded by vegetation, ensure that populations are sex-diverse. In the breeding center, caretakers incubate two-thirds of each clutch to be females and one-third to be males, a ratio designed to facilitate quick future population growth.

Though they have now been replaced with modern incubators, the hair-dryer-heated cupboards were decidedly low-tech, Rube-Goldberg kinds of machines that Howard Snell, a conservation biologist who built them in the mid-1980s, told me were the most appropriate devices for the islands. The station hired Snell, who had first come as a Peace Corps volunteer, as well as other Ecuadorian and US biologists to improve the success of the breeding program. In the early years, only about 26 percent of each clutch of eggs hatched into healthy juvenile tortoises, in part because of poor incubation. The first incubators used in the 1960s were made of lava rock with an oven compartment warmed by the sun; as they were located just off the beach, however, they seemed too close to cool sea air. A second model, farther inland, was heated with tungsten bulbs and worked better, but Snell wanted to test sex dependence on temperature, which required a technology that offered more control. He had commercial incubators shipped to the island, but erratic power surges fried them. Finally, he built the cabinets, a solution that offered both control and easily replaceable heating components. They dramatically improved breeding success—hatching rates went up to about 70 percent.

Snell's team also revamped the rearing pens. The original rearing house, built in 1970 with funds from the San Diego Zoological Society, was a round cement building with small pens, concrete floors, and one-thousand-watt tungsten bulbs to provide light and heat. But the hatchlings did not grow well in this space. Scientists thought that the cement floors were too cool and the tortoises suffered from lack of direct sunlight. Snell added outdoor runs to the building and built additional outdoor pens, comparing the health of tortoises kept in each. The outdoor pens decreased mortality from 14 percent to 1.6 percent and conservationists soon moved all the tortoises outside.[35] The design has since been replicated in two new breeding centers on Isabela and San Cristóbal.

These experiments, coupled with decades of care work, made the success of tortoise breeding possible. Reproducing tortoise populations is a socio-natural process that enfolds human labor, technologies, and individual tortoises in the genealogy of these species. Only through these joined efforts could Diego, and the unnamed Española male, become *supermachos*. Celebrations of the virility of these tortoises deflect attention from the routine work of continuing the lives of their species. It also deflects attention from the fact that their virility may not be such a good thing for the species: genetic analyses have shown that because the third Española male has hardly sired any offspring, and males have been mating with the same *hembras* for decades, the effective population size in terms of genetic diversity was not that of 15 individuals, but only 5.7, making a population bottleneck due to little individual variation even worse.[36] In response, keepers tried to move Diego and the other male into different corrals so that they would mate with different females. But the experiment did not take because Diego, distressed, tried to return to his previous corral, knocking himself against a lava-rock wall until he had bloodied himself, so keepers put him back.

Diego is not the only tortoise who has resisted plans to control his reproductive activity. The sex life of Lonesome George was for decades a core concern for conservationists. George, keepers told me, was shy and antisocial after being brought into captivity in 1972. It must have been quite an adjustment. As one biologist pointed out, it is remarkable that *he* was the last tortoise found on Pinta—his proclivity for hiding out may well have saved him from the fate of other Pinta tortoises. But this was not what he was expected to do at the breeding center. The great lengths conservationists went to in their attempts to get him to breed are popular fodder for tourists, a comic tragedy of human-tortoise relationships. In the early 1990s, the Darwin station's director of herpetology,

Linda Cayot, a conservation biologist who had done her dissertation work following giant tortoises on several islands, put two females from Volcán Wolf (*C. becki*) in George's corral. These Volcán Wolf tortoises were the population scientists thought were most closely related to the Pinta animals. She hoped to socialize George in case a Pinta female was ever found and to try to get a hybrid offspring so that the Pinta lineage would continue in some form. But George showed little interest in the females, prompting speculation about his health and sexuality. Perhaps George was gay, the guides used to joke; or maybe he was deeply shy around lady tortoises; or perhaps his weight affected his performance. When I knew him, what turned out to be a little less than a year before his death in 2012, folds of flab rolled out of his rear shell, although keepers told me this was a healthier state than he had once been in, when a veterinarian was called in to devise a special diet for him and eliminated his regular papaya treat. These stories further anthropomorphized this tortoise with a human name, projecting onto him common cultural anxieties about sexuality and obesity that made him, and thus conservation work, more relatable.

Scientists also speculated about possible causes for George's weight problem, adding biological and ethological concerns to psychological hypotheses. Perhaps his weight diminished his testosterone levels. Or perhaps, after years of living alone on Pinta, he needed to learn reproductive behavior. Several advocated putting another male tortoise with George to demonstrate and help socialize him, but others worried about the potential psychological damage of introducing an aggressive male into the corral of an already-shy tortoise.[37]

Instead, Cayot recruited human help. In 1993, a Swiss biology student, Sveva Grigioni, spent three months trying to get a sperm sample from George. She proceeded slowly, first practicing with other males, letting them become accustomed to her presence and then covering her hand with cloacal fluid from female tortoises before manually stimulating the males. After some practice, Cayot told me, Grigioni could get a sperm sample from some of the males in fewer than fifteen minutes. She moved on to working with George, trying thirty-two times to make him ejaculate. She never succeeded. She did report, however, that during her stay George became more active and interested in the female tortoises—she observed several unsuccessful mating attempts.[38] Cayot told me George also seemed to become attached to Grigioni; before she would begin her sessions, she would post a sign explaining what she was doing on the tourist observation deck and often ended up talking to visitors about her work. As she did this, George would often amble over, seemingly eager to see her.[39]

Finally, in 2008, one of the *hembras* with George laid a clutch of eggs. Would they hatch, and would George, at last, reproduce the Pinta lineage? Keepers and tourists waited with bated breath. Souvenir sellers printed T-shirts—cartoon versions of George with the words "Lonesome No More" and his "arms" wrapped around two female tortoises, each wearing a pink bow and heels. But, to much disappointment, the eggs did not hatch. A year later another clutch of eggs had onlookers holding their breath, but again the eggs were not fertilized. In 2011, park guards switched the two *hembras* in his corral to adult Española females who had been bred in captivity—genetic analyses showed they were more closely related than the Volcán Wolf tortoises. But George showed little interest in the new females and died the following year. A necropsy revealed that an anatomical deformity had prevented him from reproducing. But conservationists' hopes for the species live on. After George's death, tissue and blood samples were frozen and stored north of Quito; Ecuadorian policymakers had not wanted these pieces of national patrimony to be taken out of the country to the cryogenic "Frozen Zoo" in San Diego, where biological samples of many of the world's endangered species are kept. The sequencing of Lonesome George's genome, and future de-extinction technologies, may provide unknown possibilities for his species.

Conservationists are motivated by an intense desire to see tortoises thrive, reflecting the cross-species empathy and hope at the heart of conservation. Yet their fixation also reflects the deep-seated anthropomorphism at the core of caring for charismatic animals like George. He was a symbol of desire to save nature, but also of human social anxieties about love and belonging. His story is gripping, an affective entanglement between people and animals cast in human terms that are easy to identify with. Nonetheless, Lonesome George serves as a frustrating reminder that human desires for animal reproduction are not always fully realized. Animals never completely conform to our plans for them. This lesson arises repeatedly in attempts to control populations of invasive species as well.

Killing for Conservation

In 1957, fishers left three goats on Pinta—something they did periodically as a means of sustaining themselves far from home, much as sailors of previous centuries had. A little more than a decade and a half later, the goat population on this twenty-three-square-mile island had irrupted. In 1974, Darwin station director Peter Kramer reported to the American Association for the

Advancement of Science that in the previous three years park guards had shot at least thirty thousand goats on the island. The exponential growth of a population on an island with plenty of vegetation and no natural predators was a problem of "alarming proportions." The island looked as if it had been swept by a fire, with all the vegetation growing below three feet gone. For Kramer, this dramatic case made it clear that "measures to control introduced organisms and to prevent further introduction must be given the highest and most immediate priorities."[40] Hunting was to be a central line of conservation work.

"That was our profession, to kill animals," José Villa told me of his work as one of the first park guards. For both conservation workers and scientists, hunting introduced animals was a regular part of fieldwork that accomplished two goals: adding to their food supply and reducing the presence of threatening species. It has remained a core part of conservation for the past fifty years. But over the past twenty years, emphasis on this more violent side of the "violent-care" of conservation has grown. The management of invasive species, and ideally their eradication, is now the target of multimillion-dollar campaigns to protect island environments and is big business in environmental contract work. In the Galápagos, recent campaigns to ensure the survival of tortoises and other endemic animals have involved killing tens of thousands of hogs and donkeys and hundreds of thousands of goats and rats. Like producing life, this side of conservation is also an affective entanglement. The multispecies relationships of killing take an emotional toll. They also raise ethical questions.

When confronted with the violence of crane conservation, Thom van Dooren asked, "On what grounds do we kill some animals to save others?"[41] Species eradication, he argues, is driven by particular ecological imaginaries that justify killing plants and animals that do not "belong." These imaginations provide moral comfort even as they condone, even demand, killing in the name of conservation.[42] In the Galápagos, the ethical rationale that informs conservationist selection, making invasive species categorically "killable," and even *demanding* their death, relies on both ecological and evolutionary logics.[43] The evolutionary history of these common species and the historical entanglement with humans that introduced them to the islands makes them aliens, out of place in a natural laboratory. Their ability to replicate rapidly makes them ecological threats. Although conservationist logics condone killing these animals, however, the actual work of killing has not always been so ethically straightforward for the hunters.

Decades after his time as a park guard, José Villa remembered hunting with disgust. He told me of killing thousands of animals on Pinta during the early 1970s, when conservation hunters would make three visits a year of about twenty days to each of the central islands—Pinta, Rábida, Marchena, Española, and others. In places, they would kill hundreds of goats in a day, sometimes with guns, sometimes with machetes or even rocks. It was "terrible," he said—difficult to see the blood and to take life; "probably I couldn't do it now." Yet then it had been his job to train others to do the work, something he found difficult because many of these men were from rural areas and "love[d] animals." "People from rural areas don't like to kill animals," he told me. They did this work only because botanists said the animals destroyed native vegetation. "But for us it's not the same," he said. "We raise these animals. We didn't want to kill them." He did not like talking about the killing; only when he would drink with friends would they discuss it.

Killing work was not the kind of natural resource management for which the Food and Agriculture Organization forestry school had prepared Villa. By describing his personal discomfort with the slaughter, he pointed to different understandings and engagements with nature between the rural people he identified with and the scientists who were directing conservation efforts. What distinguished foreign scientists from Ecuadorian workers, according to Villa, was a willingness to frame species commonly kept by rural households as killable en masse. Burros, for example, were incredibly useful animals who people relied on for transportation; park guards did not want to kill them. But the animals' stomping hooves, foraging, and propensity for rolling on their backs in the dusty earth were trouble for tortoise nests and native vegetation.

Other park guards, though, were less anguished by the work. Daniel, who spent much of the 1990s hunting boars on Santiago before working at the breeding center, enjoyed the twenty-day campaigns, during which guards camped at eight different *casetas* across the island. He loved hunting, loved being out in the field. The men hunted at night, when the boars were out; they used dogs who would sniff out the boars and give chase. It was difficult work in the beginning, without much gear besides boots and knives and dogs who were scared of the boars' tusks. But on a good night, the three men in each camp might kill six or seven boars. Daniel enjoyed the camaraderie and simplicity of field life and told me of eating *seco de chivo* (goat stew) and *fritada de chancho* (fried pork)—delicious cooked over a fire.

These campaigns went on for decades, with varying foci and levels of intensity. On small islands, guards eradicated goats completely in the early years. But on larger islands, like Santiago, introduced populations were much more difficult to eliminate, so conservation efforts focused instead on control. It was an unending task; the production of "natural" island ecosystems necessitated thousands of hours of continual human labor, all of which was to be kept away from visitor sites and out of sight of tourists. By the late 1990s, after three decades of hunting, Washington Tapia, who was one of the guards who had hunted on Santiago, told me that it was clear that "by land we were never going to succeed—it was impossible. Every month we killed three thousand or four thousand, and when we came back it was like nothing had happened."[44] The goats reproduced too quickly and were spreading. On Isabela, they had crossed the Perry Isthmus, a field of sharp 'a'ā lava thought to be impassable that separates the northern and southern halves of the island. In the mid-1990s, naturalist guides who took tourists up Volcán Alcedo reported fields once dominated by grasses now denuded by goats. Photographs from the time show tortoises and goats in a barren, brown landscape. In one photograph a goat stretches high above a tortoise to reach green leaves from a tree branch— a poignant image that encapsulates fears of introduced species out-competing native herbivores.[45]

Conservationists needed a new strategy. They looked to New Zealand, where similar concerns about protecting native island fauna from invasion had led to the development of specialist conservation hunting. There, sharp-shooters hunted from helicopters, using "Judas" animals, collared with radio transmitters, to track their prey. An international workshop and an exploratory trip persuaded conservationists that this was the model they should follow if they ever wanted to clean the Galápagos Islands of unwanted animals.[46] As Cayot explained, island eradication campaigns require the death of every last animal; failing to find even one pregnant female means the population could regenerate. And finding every last animal required a level of precision, and speed, for which guards previously had not had the resources. They had been hunting in a "folkloric way," one local conservationist told me: "They didn't have proper equipment, they didn't have any strategy. They were just going and shooting whatever moved, without doing anything in a systematic way."[47]

More systematic, modern hunting would require considerable financing, which came from the Global Environment Facility (GEF), an environmental financial organization backed by the United Nations and several multilateral development banks. The Darwin station and national park received more than $18 million for the "Control of Invasive Species in the Galápagos Archipelago" project—the largest grant GEF had given. They raised another $32.5 million of cash and in-kind cofinancing from local and international institutions to implement a "Total Control Plan" for permanent control of invasive species, including a big-push eradication campaign, institutional capacity building, and quarantine systems to prevent future introductions; they also established a $15 million trust fund for future eradication work.[48] The core of the project ran from 1998 to 2006, focusing primarily on eradicating the goats on Santiago and Isabela. One project leader framed this conservationist selection in biblical terms: "We are rather like a collective Noah, deciding with a biblical coldness which life forms will be able to accompany us on our new journey in the Ark."[49]

This flood was to wash the islands in a deluge of bloodshed and goat carcasses. It was a big-budget, high-modern hunt of epic proportion. Sixty local hunters were trained in precision hunting and used automatic rifles for ground and air campaigns with helicopters to speed the process and allow access to difficult terrain. The arsenal for this assault included "0.223-caliber rifles, exploding bullets, telescopic sights, global positioning systems, two-way radios, nylon nets," and a million rounds of ammunition imported from the United States—an order that prompted a quick phone call from the US State Department, apparently concerned about the prospect of an armed insurgency in this Latin American state.[50] The hunting crew also included more than seventy dogs, many of them especially bred and trained in New Zealand, who sniffed out and herded goats, helping their human companions during ground campaigns. The eradication campaign was not survival of the fittest that pitted one species against another, but a tangled web of multispecies killing.

Companion hunting animals have long been used in the Galápagos and around the world, but eradication hunting also involved nonhuman aid from target species. To locate difficult-to-reach populations, the Galápagos team used Judas goats tagged with radio telemetry devices to lead hunters to a group of target animals. The use of Judas animals comes from nineteenth-century cattle drives in the American West, but it has taken on new life in eradication hunting, when, as Judas Iscariot betrayed Jesus, tracker animals betray their kin by helping hunters locate herds to kill. The method, as one conservationist wrote, "exploits the gregarious nature of goats," leveraging their sociality as a lure.[51] Hunters argue that the use of Judas animals allows for more humane killing because of the speed it facilitates. But even if Judas hunting might be more humane for target goats, it certainly is not for the Judas animals themselves who repeatedly see their packs killed.[52]

Judas work not only exploits animals' sociality but their reproductive biology as well. In addition to collaring goats with telemetry trackers and tagging them in red paint, conservationists also sterilized Judas goats. Pregnancy made female Judas goats "ineffective" work animals; instead of seeking out new reproductive partners, expectant mothers added to the populations they were supposed to help eliminate and spent time away from herds to give birth and care for their young.[53] Traditionally, hunters shot newborn kids to encourage the Judas mothers to seek out a new pack. But one of the Galápagos project leaders, Karl Campbell, an Australian conservation biologist, did his dissertation research on improving the efficiency of Judas goats. The ideal Judas goats would be sterilized, but in a way that did not negatively affect their sociability, as castration did for males. Campbell tested methods—such as tubal occlusion for females and epididymectomy for males—that would retain the animals' "sexual motivation" while preventing reproduction. The more sex, the better. "Nymphomaniac behaviour would be desirable in JG [Judas goat] operations," Campbell wrote, but "it is unknown how to induce this condition."[54] Through his experiments on female goats, he developed a new model of Judas goats put into prolonged estrus with hormone implants. These "Mata Hari" goats, named after the Dutch courtesan who spied for Germany during World War I, would continue to attract males without becoming impregnated. Campbell's solution to this conservation problem folded the biological cause of goats' invasive force to work in hunters' favor.

This project aiming for total control over nature has been construed as a major success in the world of island conservation, demonstrating the feasibility of large-scale eradication programs. Hunters used 800 Judas goats to kill 152,292 of their kin on Isabela and Santiago. On Isabela, with a large number of goats, the cost per kill was $47.91; on much smaller Pinta, with fewer goats, it was $1,290.63 per goat.[55] All the kills were geo-located with GPS units and mapped.

Yet as anthropologist Paolo Bocci reports, not even all the hunters who worked on the project could find the will to totally eradicate goats. "We were OK with killing goats, but eradication seemed like too much!" one hunter told him, so sometimes they spared juveniles and pregnant females. But, "it is something that we kept to ourselves." To ensure kills, however, project leaders began demanding a piece of ear from each goat kill marked with a GPS point as evidence that the killing work was done.[56]

The technocratic, economized view of death presented in project reports puts the focus of killing on efficiency and deflects it from ethical and political issues. In my interviews, particularly with residents on Isabela, where community relationships with the National Park Service have long been contentious, residents challenged the project from multiple perspectives. Some took issue with the idea of eradication because goat meat is a dietary staple in Ecuador and local people depended on the animals. Conservationists did not target domestically kept goats, but locals thought feral populations should be *usefully* rather than categorically killed. They found goat eradication wasteful because of wasted meat (all the carcasses were left in place), wasted opportunity to employ more local hunters, and wasted resources that they thought should have been spent on other causes, such as clean water.[57]

But local resentment of the project went beyond utility. In a place where residents have long defended their right to live in the islands, some charged that conservationists wanted to eradicate human residents as well as the goats—that the helicopters buzzing over town were actually threatening the residents themselves. The conservationists I interviewed assured me that this claim was false. They understood arguments about animal waste but used ecological and logistical arguments to defend their actions. Many local hunters wanted to work on the project, Cayot told me, but they did not understand why eradication required quick, precise methods: "Nobody got the difference between killing the first 90 percent and the final 10 percent. Everybody could kill the first 90 percent, that's easy. It's the final 10 percent that's really hard. And you can only get there if you kill the first 90 percent really, really fast." The urgency of eradication also explained why the goat carcasses had been left to rot. Cayot said it would have been logistically impractical to remove tens of thousands of goat carcasses from remote, equatorial volcanoes before they rotted: "If your goal is eradication, you don't have time for anything. If you were going to kill one hundred thousand goats and take the meat off and have a refrigerator ship down at the coast, your helicopters would be [constantly running goat carcasses]." She also argued that the nutrients from the goats' bodies should be left in the ecosystem. While the goats were alive and grazing, conservationists considered them a dangerous, unnatural threat to island ecosystems. But dead and no longer able to reproduce, their unnaturalness changed as the nutrients in their decomposing bodies became a resource that should remain in the ecosystem. But what of the effect of leaving a million rounds of ammunition in the ecosystem as well?

The aftereffects of goat eradication call into question the political and ecological feasibility of restoring an evolutionary Eden. Project reports show triumphant photographs of vegetation rebounding on Volcán Alcedo, and studies have shown that Galápagos rail and hawk populations are also rebounding. But also necessary, it became clear, was better understanding of the role goats had come to play in island ecosystems. Conservation biologists knew they denuded landscapes, but what became apparent after the eradication was just how much they had also held in check introduced plant species—such as *mora* (blackberry), dense thickets of which are an especially pernicious problem in the agricultural highlands of Isabela and Santa Cruz—as well as other woody vegetation, which now covers much of Pinta.

Political responses also threaten the long-term success of eradication. Shortly after Project Isabela ended, someone introduced six goats to Santiago. Another goat soon appeared on Darwin Island, far in the northwest corner of the archipelago. Park wardens found and killed these goats—at a cost of $5,000 each—but the reintroductions demonstrate that eradication is a contingent achievement that is always susceptible to acts of defiance.[58]

Galápagos residents have multiple grounds for criticizing goat eradication, but their experience nonetheless invokes a shared history with goats and other island species. That history is discounted in conservationist logics that judge what belongs in the islands according to what evolved there. Eradication campaigns depend on a reductive understanding of ecology that focuses on a few key species and an idealized view of a past state of nature. The presumption is that either tortoises or goats, but not both, can, and *should*, survive. But as van Dooren argues, "no matter how much and how many we kill, our goal cannot be to put things back to how they once were. Rather, we are killing to produce *valued natures*."[59] Yet such values are not universally held. Goals of total eradication eclipse thorny political questions about how different groups value nature. As goat reintroductions warn, such elisions have consequences. Conservation needs a broader frame for understanding the stakes, and entanglements, of eradication work that acknowledge multiple, and conflicting, ways of valuing nature.

Caring for tortoises and killing goats are entangled practices of conservationist selection. Thinking through the ethics of saving nature means, in the words of philosopher Donna Haraway, "staying with the trouble" of the violence of killing and the power-laden question of whose values matter.[60] Saving species is not a question of sorting the natural from the unnatural. Such binary frameworks posit some clear division between these two realms—like the year 1535, or the boundary line that divides island territory between national park and human-use zones—that miss the dense entanglements that comprise life. The lives of giant tortoises are embedded in evolutionary entanglements with finches and the islands' native Opuntia cacti—and so too goats and rats and park guards and conservation volunteers. Saving giant tortoises inextricably knits the species together with the lives of these other beings in ecosystems

that are always in flux. These power-laden multispecies entanglements become part of what the species are. There is no return to Eden. Conservationist
selection is a potent biopower, but that goal is a utopic dream that is just as
much ethical and moral as it is scientific and technical. And besides, as much
ecological research shows, it is simply not possible.

Re-tortoising

Eradication work has cleared space for restoring tortoise populations—or
"re-tortoising," as Galápagos conservationists call the work of restocking islands with tortoises raised in captivity. In 2015, park guards and scientists who
were surveying Pinzón after the most recent attempt at rat eradication found
ten tiny tortoises, the first known to have hatched successfully on the island in
more than 150 years. The discovery was reported in *Nature* and made international headlines.[61] Pinzón, though, is not the only eradication and reintroduction success story. On Diego's home island, reintroduced tortoises have been
breeding in situ for years. By re-tortoising the islands, conservationists have
achieved one of the core goals of restoration: they have secured the ability of
tortoises to reproduce on their native islands and to continue their species'
evolutionary lineages, opening up possibilities of future life that seemed impossible a half century ago. But ecological studies on the effects of eradication
and re-tortoising have revealed a Darwinian truth—that life is a complicated,
interdependent web that does not move backwards, but in surprising ways.

Española, for example, "is a beautiful island, it's a pristine island, but
maybe it's completely in a novel ecological state," conservation biologist James
Gibbs told me. A professor at the State University of New York College of
Environmental Science and Forestry, Gibbs has worked in the Galápagos
since the 1970s, when he got his start as a field hand with Peter and Rosemary
Grant's finch research. For more than a decade, he has studied tortoises reintroduced to Española, examining their effects as "ecosystem engineers" that,
much like dam-building beavers, actively modify their environments.[62] Gibbs
is cautiously optimistic about the success of conservation. About half of the
eighteen hundred juvenile tortoises "repatriated" to the island have survived
their transition to life in the wild. It is a statistic that sounds dismal, he acknowledged, but compared with breeding-and-release programs with other
species around the world, it is actually quite impressive. Tortoises are hardy
animals, Gibbs pointed out, and can tolerate treatment that more delicate animals may not withstand. Still, how much death is an acceptable by-product of

the quest to restore wildlife? And is this death natural? Most of the fatalities on Española are, Gibbs thinks, attributable to desiccation because tortoises overheated or failed to find enough water. This is a normal part of tortoise life; tiny tortoises face lots of challenges, and low survivorship rates are likely the reason why female tortoises produce several clutches of multiple eggs every year. Those that have survived on Española are doing well, Gibbs thinks, and the offspring produced in situ look healthy, with a different hue and slope to their shells than those bred in captivity. The reintroduction program has been successful enough that he thinks the park could now stop sending new babies back. With time, the current population should grow to replace the two thousand tortoises he estimates once lived on the island.

But Gibbs's research also shows cause for concern. The future population growth of the eight hundred or so tortoises on Española is limited by available vegetation, particularly the island's native Opuntia cacti.[63] On dry islands like Española, tortoises rely on these treelike cacti for food, shade, and moisture; the tortoises evolved their saddle-shaped back so that they could stretch their heads up to reach cactus pads. Tortoises also disperse and fertilize the cactus seeds they eat as they defecate while ambling across the island.[64] But Opuntia are not recovering well following the goat eradication in 1978, despite the reintroduction of tortoises.

The removal of goats seems to have changed the relationship between grasses and woody vegetation that now dominates parts of the island— including *palo santo* (Bursera), cordia, Jerusalem thorns (Parkinsonia), mesquite, and croton, a scourge for fieldworkers who return from hikes covered in stains from its sticky sap. These species are all native, but by analyzing soil cores, Gibbs has found that their current density amounts to a thousand-year anomaly. The land cleared after heavy goat presence during much of the twentieth century has facilitated a rapid expansion of woody plants rather than the grasses and Opuntia on which tortoises normally feed. Additionally, although tortoises are known for trampling down roads, they avoid this woody vegetation—even for them, it is almost impossible to pass through. It appears to be worse still for the islands' endemic waved albatross, which need long runways of cleared land in order to take flight and land.[65] Even without a dominant invasive species, the ecosystem has entered a novel state—and the resurgence of one native species is not necessarily good for others.[66]

The Española case highlights the limits of ecological control in restoration work and the impossibility of returning to a past state of nature. On other islands, the challenges differ, but the difficulty remains. As everyone in the

Galápagos will tell you, no two islands are alike—a fundamentally Darwinian point, and one that underscores the difficulty of restoring evolutionary processes. On Santa Cruz, tortoises are not counted upon to drive back invasive plant species, but they appear to enjoy eating them. For the past decade, wildlife biologist Stephen Blake has been researching tortoise migration and ecology, tagging tortoises with GPS radio telemetry units to track their movements up and down island volcanoes. He also studies their role as seed dispersers. I spent an afternoon in his office lab picking seeds out of tortoise scat—the average scat contains 464 of them, he has found.[67] Many of the plants they ate and dispersed were native, but their favorite apparently—as judged by the prevalence of seeds—was guava (*Psidium guajava*), an "escaped" agricultural plant that has become highly invasive on Santa Cruz and Isabela. The tortoises could well have aided this invasion—they often wander out of protected park zones and onto agricultural land to eat grasses on cleared land.

The spread of these plants is a continual challenge for conservationists. As part of the GEF program, botanists spent $1 million attempting to eliminate guava and thirty-five other invasive plants, but succeeded in eradicating only four.[68] The worst of these invasive plants, in the opinion of botanist Mark Gardener, is *mora*. Its dense brambles are nearly impassable for anyone without a machete, and its seeds scatter, making it practically impossible to eradicate. It covers at least seventy-four thousand acres of park and farmland in the islands. Indeed, Gardener says, "as far as I am concerned, it's now a Galápagos native, and it's time we accepted it as such."[69]

In 2011, Gardener, then director of restoration at the Darwin station, was one of nineteen scientists who authored an essay in *Nature* that sparked intense debate in the conservation world. He and his colleagues admonished conservationists not to "judge species on their origins" and even suggested that they should accept invasive species as integral parts of modern ecosystems.[70] But in a field structured around the desire to return to a more pristine baseline, accepting "novel ecosystems" was anathema.[71] When the field of conservation biology came of age in the 1970s and 1980s, it was fundamentally shaped by belief in balanced ecosystems—as ecologist Daniel Botkin explained, the idea that "nature, left alone, will achieve a single state," a timeless "constancy of form and structure."[72]

Over the past decades, ecologists have shown that balanced ecosystems exist only in idealized models. As Botkin put it, "The one thing we can be sure about nature in the future is that it will be different from today, because nature is always changing."[73] But it has been hard to square this recognition with the mission of Galápagos conservation. What would conservation be if not the work of *re*storing, *re*-tortoising the islands? Gardener's views cost him his position at the station. Does accepting invasive species mean throwing in the towel on conservation? Must ecologists accept, as one put it, a new "homogenocene" world of diminished biodiversity and species richness?[74] Is the future a world covered in kudzu? (That, at least, is one plant Galápagos conservationists *were* able to eradicate before it became a plague.) Steeped in language about species "colonization," "invasion," and "repatriation" that betrays the discipline's colonial roots, the very purpose of conservation biology has been to defend the home territories of native species. Would conservationists no longer do this? Geographers Paul Robbins and Sarah Moore argue that the novel ecosystems debate reflects an existential crisis for the "Edenic sciences" of conservation and restoration biology. Conservationists are gripped by an "ecological anxiety disorder" because of the dissolution of the very idea of pure nature in an Anthropocene world of climate change and global human influence.[75]

In the Galápagos, such anxieties shaped decades of debate about what to do with Lonesome George's home island, Pinta. If George would not, or could not, reproduce, then how should conservationists approach restoring this ecosystem? Should they introduce tortoises from another island? Leave the island alone? The debate split conservationists into two camps: "species purists," who argued that it would be unnatural and thus wrong to re-tortoise the island with any animals other than those from the *C. abingdonii* species; and "island ecologists," who wanted to put nonnative tortoises on Pinta because they had historically been part of the ecosystem and so were essential to restoring ecosystem function.[76]

After park guards killed the goats who had taken over the island in the early 1970s, native vegetation—Scalesia, Opuntia, and Bursera—rebounded, but ecologists feared that the absence of an herbivore would decrease their success.[77] For decades, however, concern for those species was trumped by others more concerned about adhering to the purity of the island's evolutionary history. Howard Snell, who in the early 2000s was splitting his time between teaching at the University of New Mexico and serving as director of science at the Darwin station, told me that putting nonnative tortoises on Pinta would be a mistake: "[The idea that] a tortoise is a tortoise, so I'll take this one from over here and plop it over there—it's the designer ecosystem model, it's the Disneyland model that I personally feel is a convenient model, an easy model, but it is a fundamentally flawed model if you really value natural systems, because it's not a natural system." Instead, he thought the island should be left without tortoises as a reminder of past human destruction; these sins should not be papered over by using nonnative tortoises to restore the ecosystem.

This debate went unsettled until 2005, when after a winter of heavy rains, Washington Tapia, who was then head of protection for the park, decided that something needed to be done. At a workshop that year, conservationists and visiting scientists took up the Pinta question again. Tapia had been to the island while goats were living there and returned about six months after the eradication campaign. "I saw that there had been a change that was not normal . . . much more green than before, species that need open areas that were not recuperating, and other plants creeping into areas where they should not be." This sparked attention to ecological restoration work that went beyond goat eradication: "We had two options: wait and only monitor the changes and document what happened, or really do a process of active restoration and put back on the island its herbivore." A shakeup in station leadership in 2000 meant that much of the old guard who had opposed returning tortoises to the island was no longer in charge, opening space for intervention.

One suggestion was to re-tortoise Pinta with Española tortoises, which genetic analyses showed to be the most closely related species. But with long-shot hopes that George might still reproduce, conservationists were hesitant to start introducing animals from another species. Moreover, scientists were not all reconciled to the idea of putting nonnative tortoises on Pinta and were also concerned that juvenile tortoises would have little effect for decades in the densely vegetated island until they were grown. The project reached a standoff when foreign scientists started calculating what size of tortoise to restore based on how much vegetation they could eat. It was "rubbish," one local conservationist told me—scientific minutiae that hampered conservationists' ability to do anything productive on the island. Instead, he suggested what he considered a "brilliant" solution: return adult tortoises from the *mixto* corrals at the station, sterilized so they could not reproduce and thus would not interbreed with each other or future repatriates and contaminate the evolutionary purity of the lineage.[78] By tagging the animals with radio telemetry trackers, scientists could also monitor their behavior and impact on vegetation.

It was a popular solution. In 2009, a Houston veterinarian led a team that sterilized forty tortoises—experimental operations, he noted, which vets are not regularly called upon to perform. They removed the ovaries of fifteen female tortoises (one of whom died from complications) as well as the male's phalluses.[79] Intervening in the reproductive biology of these animals changed their relationship to conservation. At the breeding center, these tortoises had been considered extraneous, not useful for conservation because they were hybrids. But sterilized, they would become experimental objects of restoration, freed to live in the wild on Pinta.

The reintroduction of these *mixtos* in May 2010 drew international news crews to the remote island. As a news anchor from New York recorded sound bites about his age in "tortoise years," park guards with tortoises strung up on poles—like the old tortoise hunters used to do—carried them from the park's work ship halfway up the island (fig. 6.3). Cayot, who had wanted to see tortoises on the island for decades, was thrilled: "The moment they hit the ground, they were ready for action. They immediately began moving off through the vegetation, knocking down whatever stood in their path, finding juicy plants to forage on, and exploring their new world."[80] It was hard not to identify with the tortoises and anthropomorphize them, another conservation biologist told me: "the way they were exploring the habitat, it's hard to not think, oh they've been in captivity forever and now they want to explore. It's just a great feeling, both putting tortoises back on Pinta and also releasing these captive organisms into the wild again. It feels good."[81] The project was as much about playing out conservationist desires to return to wild nature as about restoring ecosystem function.

Yet while Cayot and the global media were enthused about returning tortoises to Pinta, not all scientists agreed. Peter Pritchard, a turtle and tortoise conservation expert who has worked in the Galápagos, told me the project was "ridiculous": "Getting a number of tortoises of unknown origin, cutting off their male parts, and putting them on the island—one wonders what on earth is going on? What is the point? They can't breed so they are either going to live there for a few years or maybe quite a lot of years and then they start over. So I thought it was one of the silliest things I've heard in a long time."[82]

Is it silly? A great feeling? Both? There is no one clear perspective by which to judge the project. As far as rewilding experiments go, though, returning sterilized tortoises as ecological analogs for the Pinta species seems less preposterous than recent proposals to rewild the American West with African elephants, lions, camels, and cheetahs or to genetically re-engineer wooly mammoths.[83] The experiment with the sterilized tortoises was also less extreme than one Gibbs told me he would love to see as a scientist: releasing a "hybrid swarm" of tortoises, tagged, tracked, and genotyped, and then "letting evolution go to work" to see "what traits get selected for and against, how they use the islands. It'd be consistent with Galápagos as a laboratory of evolution. But I think it's a bit too esoteric for most people."[84]

Figure 6.3 Park guards of the Galápagos National Park Service "re-tortoise" a sterilized giant tortoise strapped to a pole on Pinta, May 2011. (Used with permission from the Galápagos Conservancy)

Instead, Elizabeth Hunter, who was then doing her master's research in Gibbs's lab, lived on Pinta for two summers monitoring how domed and saddle-backed tortoises acted as ecosystem engineers. She found that larger domed tortoises did not migrate to areas with cacti but instead preferred higher, moister elevations and thus would not help disperse cacti seeds. The animals had beaten down trails and in doing so helped to condition the island for additional releases. Hunter also found that the tortoises gained, on average, about twenty-five pounds—a clear sign they were thriving in their new surroundings.[85] Through Hunter's work, the Pinta project has served as a pilot for reintroduction of nonnative tortoises to other islands.

In July 2015, for the first time, park guards released juvenile tortoises to an island that was not native to their species. Tapia, who now works with the Galápagos Conservancy, and park guards took 201 of the Española babies to Santa Fé, a small, low island just to the southeast of Santa Cruz where the native population has been thought extinct since at least the Cal Academy expedition. The shift in strategy was run by the Giant Tortoise Restoration Initiative, led by Tapia, Cayot, and Gibbs. They are also conducting research on ecosystem dynamics and the introduced analog tortoises as engineers.[86]

The emphasis of restoration projects has shifted from strict fidelity to recreating a past pristine world to a focus on ecosystem function and replacing key species. But the experience of this work poses questions about to what degree it is possible to restore the history of evolution. How many species can be managed, and how much effort should be spent to do so? If there is no clear baseline to return to, and if nature always responds in unpredictable ways, then how do conservationists decide what to do? In the midst of global climate change, how can they ensure a resilient future for valued creatures and ecosystems?

Ecological surveys of the islands continue to reveal surprising findings. In February 2019, during an expedition on Fernandina, biologists and park guards found a single adult tortoise far in the remote interior of the island. She was proof that the *C. phantasticus* species, which had not been seen since the Cal Academy expedition, was not in fact extinct. Nearby were trails and the occasional scat—suggesting that this tortoise was not alone. Park guards took her back to the breeding center on Santa Cruz with hopes of beginning a breeding program for the species.[87] This lone female tortoise, sure to become an icon like Lonesome George and Diego, is a reminder that even in the face of something as final as extinction, our knowledge is not always as certain as it seems.

She is also a reminder that the possibilities for restoration continue to shift. Conservationists have worked intensely over the past half century to protect evolution in the natural laboratory of the Galápagos. In many ways, their work has been remarkably successful—today some twenty thousand giant tortoises are thought to live in the wild and reproduce in situ, and the eradication of goats and rats set a precedent of possibility for island conservation around the world. Yet this does not amount to restoring an evolutionary Eden. The language of restoration, so prominent in conservation biology today, is misleading. The choice is neither a return to an Edenic past nor acceptance of a novel ecological state. Evolution, as conservationists well know, is far more complicated than that. Nature is always evolving, always novel. The idea of restoring evolution is born more of powerful conservationist imaginaries than of biological theory. It is an inherent contradiction that plays out as conservationists debate how to best protect endangered species and their island habitats. Acknowledging that evolution is always novel does not mean abandoning conservation. It does not mean abandoning quarantine systems that attempt to protect the archipelago from new pests, like the parasitic fly *Philornis downsi* that threatens Darwin's finches and other avian species on almost every island.[88] But acknowledging that evolution is always novel does mean grappling with difficult questions instead of relying on a presumed Edenic baseline as the foundation on which conservation stands. It also means acknowledging that the closer one gets to nature, whether chasing goats, hacking away at *mora,* or trying to sexually stimulate a sterile tortoise, the better one sees how nature shifts and moves—and exceeds human control.

Attention to the work of saving species, in all their entanglements, shows that nature can never be something completely removed from human influence, even if it escapes our grasp. The species themselves are produced through multiple, and often contradictory, relationships that entwine natures and cultures—they are themselves multispecies entanglements. The work of breeding giant tortoises and eradicating invasive species involves practices that are both caring and violent that complicate the simple ethical message conveyed in fund-raising campaigns that ask wildlife lovers to donate to save giant tortoises. Conservation biopower is based on normative ideas—informed by biblical references, romantic nostalgia, and scientific experimentalism—as much as evolutionary history. These normative ideas motivate people to care about the natural world. And they work—for some people. But we cannot forget that the straightforward plea to save charismatic endangered species is not a true picture of the moral stakes of conservation. A responsible ethics of conservation must recognize that the will to save nature is a matter of contested politics.

The simplified rhetoric of restoration is a political tool—a way of imagining conservation as if it were not active selection weighted with ethical valuations and assertions about who and what belong in the archipelago, but rather the straightforward act of returning order to a messy world. But there is not one *real* Galápagos to restore. There are vast stretches of the archipelago where nature is relatively untouched, where giant tortoises live who likely have never seen a human being. There is value, and wonder, and awe in that worth protecting. But there are also thousands of human beings living in the Galápagos who likely have never seen a giant tortoise in the wild. Conservation must also find a way to make space for their lives in this entangled place—a task at which it has been far less successful.

7 • Laboratory Life

In which the tortoises are part of experiments in ordering island life

In October 2011, I took a taxi from the sandy streets of Puerto Villamil on the southern coast of Isabela into the highlands. As the road crept out of town, it cut through sheets of lava and dense, dry scrub before ascending into the humid, but cooler, green flanks of Volcán Sierra Negra. After a fifteen-minute ride we neared two enormous Ceiba trees that are registered as some of the notable gigantic things of Ecuador. The driver, trying to sell me a tour of the island the following day, told me they had been planted centuries ago by pirates. But it was not these giants I had come to see. Just past the trees was my destination: Campo Duro, one of several private farms in the islands where tourists can see giant tortoises outside the confines of the breeding centers. I had been here a few years before while touring the island's sights with friends, but this time I returned to camp for a night among the tortoises at this eco-lodge.

Wild tortoises are rare in this part of Sierra Negra. The area was once native habitat for the *C. guntheri* tortoise species, but this was where Rollo Beck photographed machete-hacked casualties of oil collectors at the turn of the twentieth century. The tortoises at Campo Duro were captive—transplants from the breeding center in Puerto Villamil who had been brought to the farm through a partnership between the national park and the landowner, Wilfrido Michuy. For Don Michuy, the tortoises anchored his small eco-tourism business, providing a draw for tourists. For the national park, Campo Duro provided a safe space for returning these tortoises to a more wild environment—although they still needed watchful care.

Isabela is sometimes called the Wild West of the archipelago, a fringe where local norms can have more standing than national law and where tortoise poaching is not uncommon. Island park guards were afraid to repatriate captive-bred animals to the wild on Sierra Negra because they feared local residents might kill them. At first, Don Michuy was concerned too. For three years while he was building his house at Campo Duro, he slept on a wooden platform above the tortoise corral. His neighbors did not all care for the tortoises as he did. Jokes they made about the *siete sabores,* seven flavors, of tortoise hit too close to home. But nearly two decades later, the tortoises at Campo Duro have been safe and the farm has become a model of alternative Galápagos tourism—small-scale, locally owned, and land-based.

Camping among the tortoises at Campo Duro is not tourism in Darwin's footsteps. This tourism is not oriented around cruising through a bubble of what looks to be pristine nature, imagining oneself as a heroic explorer in an evolutionary Eden. There are no statues of Darwin in the highlands, and the landscape here bears many traces of the island's social history. If tourism at Campo Duro follows in anyone's footsteps, it is that of *los colonos,* the colonists who have wrought their lives in the islands by working with nature over the past two centuries. This kind of tourism is about human entanglements with the natural world. It can teach us much about what it is like to live in a laboratory. Indeed, there are many places like Campo Duro in the archipelago that can show us how island residents, and nonhumans too, have negotiated ongoing experiments in how best to order relationships between nature and society. Let me take you on an alternative tour. But before we begin, we need to take a detour to understand how visions of the Galápagos as a laboratory have changed since the 1960s.

A New Laboratory

Campo Duro is an experiment in a new kind of laboratory—not a living museum of pristine nature, but what conservation biologists call a "socio-ecological laboratory." Over the past two decades, this vision has superseded the laboratory-as-fortress-of-nature ideal that was at the heart of conservation efforts in the mid-twentieth century. The ideal of pristine island labs where nature was isolated from social life has fallen out of favor as the basis for conservation. Policymakers in the Galápagos have turned away from fortress conservation and toward a recognition that, as one national park publication put it, "there is a clear need in Galápagos to abandon the historical perspective of the separation of humans from nature, which only exacerbates conflicts between conservation and development." To do this, they proposed a socio-ecological systems (SES) model that would break the "nature vs. society dichotomy and build . . . bridges between the two artificially separated worlds."[1] In this new scientific geographical imagination of the archipelago, the goal of conservation is no longer "to conserve in perpetuity ecological and evolutionary processes."[2] Instead, it is to promote sustainability by better understanding relationships between social and ecological systems. It is a paradigm shift in ordering island life.

Around the world, not just in the Galápagos, SES models have captured the imaginations of ecologists and conservation biologists. They have become nearly as powerful a way of understanding relationships between nature and culture as were visions of preserving pristine environments that influenced the form of many national parks in the early and mid-twentieth century. It is worth pausing, then, to dive into the world of the Galápagos as imagined in models of a socio-ecological laboratory to understand how this framework orders life.

The SES model is a powerful way of narrating history, but one that is foreign to most historians, anthropologists, and others who study social life. The SES model presents the history of the archipelago as a series of feedback loops, shown in figure 7.1 as intertwined, sideways figure eights. Each loop represents a different era of Galápagos history, shown as four periods in the downward arrow at the left: extractive exploitation (1535–1832), colonization (1832–1959), wilderness conservation (WC in the figure; 1959–1998), and conservation-development balance (CDB in the figure; 1998–present). As is the norm in SES models, each loop is further broken down into four phases: growth, consolidation, release, and reorganization. The language of the four periods is familiar for environmental historians for whom natural resource extraction, colonization, conservation, and sustainable development are foundational concepts. But the terms used to name the four subphases of these periods are not. Historians are not wont to see the world in neatly ordered loops of growth, consolidation, release, and reorganization. These concepts instead come from ecological theory developed to describe the flows of energy and nutrients through the natural world understood as a cyclical system.[3] The ecologists and economists who developed the SES approach broadened this framework to include social life, understood as discrete, but interacting cultural, sociopolitical, and economic systems. The components of these systems—power relations, values, legal and political economic institutions—do come from social theory; but fitting their interactions into neat, looping flows, even reducing these constructs to component pieces of social life, is an unfamiliar, and worrying, way of using this work. For me and many others

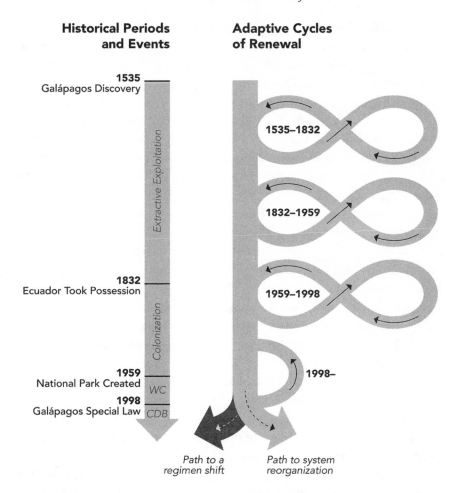

Historical Periods and Events

1535
Galápagos Discovery

Extractive Exploitation

1832
Ecuador Took Possession

Colonization

1959
National Park Created
WC
1998
Galápagos Special Law
CDB

Adaptive Cycles of Renewal

1535–1832

1832–1959

1959–1998

1998–

Path to a regimen shift

Path to system reorganization

Figure 7.1 Galápagos as a socio-ecological system: historical periods (left) and adaptive cycles (right) (Design by Meghan Kelly; adapted from González, J. A., C. Montes, J. Rodríguez, and W. Tapia. 2008. Rethinking the Galapagos Islands as a complex social-ecological system: Implications for conservation and management. *Ecology and Society* 13 (2): 13. [online] http://www.ecologyandsociety.org/vol13/iss2/art13/). (Used with permission)

trained in social theory, it raises substantial concerns about the utility of such models for understanding the world.[4]

This is not to say the diagram is without merit. As frameworks for understanding the organization of life in the Galápagos, SES models are an improvement over old-fashioned understandings of the islands as a natural system devoid of human shaping. Any understanding of the Galápagos that leaves people out is so fundamentally incomplete as to be incorrect. Policies based on it are doomed to failure. That the SES authors recognize this is the model's greatest strength. They believe, as I do, that integrative ways of understanding the myriad processes that intertwine natures and cultures are essential for building a future for the Galápagos that is not principally marked by extinction and ecological crises. We also share a focus on the complexity of modern life and how processes operating at multiple scales—local, national, global—all come together in this place that has too often been cast as isolated at the world's end.

The argument on which they base the model is deeper yet. The authors acknowledge that the natural laboratory model, which positioned local residents as antagonistic to the environment, exacerbated socio-environmental conflicts in the archipelago. That model was at the root of the crisis declarations that swirled in the air when I first went to the Galápagos. They argue that most of the contemporary problems in the islands began during the period of wilderness conservation. This argument draws from the analysis that geographer Christophe Grenier first made in the late 1990s when he published his book *Conservación Contra Natura* (Conservation Against Nature; originally published in French). The creation of the research station and national park in 1959, he argued, accelerated the "geographical opening" of the islands to networks of global tourism, creating a fundamental contradiction at the heart of wilderness conservation: the desire to protect the archipelago's geographical isolation while simultaneously positioning it within world capitalist networks.[5] When Grenier first presented his ideas in the islands, they were not well received. What he had to say was not what conservationists wanted to hear. Ten years later, though, his insights were at the core of a new approach to island management, and Grenier was invited back to serve as the Darwin station's first resident social scientist.

Despite these changes, two significant problems with the SES model jump out. The first is the strangely abstract way of narrating history that is out of line with the stories I have told thus far in this book. Models are necessarily reductive frameworks for understanding the world, but the ways in which they represent life have political stakes. System-based frameworks can be useful tools for simplifying complicated worlds, but they suggest that history moves in a looping structure of regular patterns, guided by a functional rationality. Yet, much as Darwin understood the evolution of biological life, social history does not move along a rational path. It is contingent, happenstance. While it is not wholly unpredictable, it certainly is not fated. Asserting otherwise eclipses the power of individual actors and collective actions to shape the world. It also leaves us incapable of seeing how they have shaped the nature of the Galápagos.

The second problem becomes clear when we consider the downward arrows of history on each side of the diagram. They are not terribly subtle signals of an entrenched normative conservation claim: that since the beginning of a human presence in the archipelago, the course of history has been a story of decline. Ecuador claimed the archipelago after centuries of exploitation, establishing agricultural colonies that grew until the "back-loop of collapse and reorganization" when international scientists became aware of the "threat to native ecosystems" and "overexploitation" of endemic species led the system into the next phase of conservation.[6] Such declensionist narratives are common in the world of conservation and used to be in Latin American environmental histories as well.[7] They present a story of human destruction of nature until intervention by heroic, and often foreign, conservationists, with racist and colonialist undertones suggesting that whites knew best how to manage nature. Yet the SES model continues with a new chapter despite wilderness conservation efforts: a plunge into crisis that will continue if the model's lessons are not heeded. This dystopian future is represented by an arrow shown in red in the original version of the figure that skews in the opposite direction of the other loops, a trumpet sounding an alarm. The Galápagos was poised at a precipice caused by the contradictions of tourism-reliant wilderness conservation. The SES system teetered between continuing in an exogenous model or finding a way to reorganize into an "endogenous" model that would revolve around "safekeeping of natural capital." Only proper management, and a shift in social values, could lead into the desired, but elusive, state of "conservation-development balance."[8]

Such political messages are not an unusual feature of SES models. Like all models, they are not a description of the world as it is, but rather normative visions for how the world should be.[9] Consider the implications the authors drew. The SES plan envisioned a comprehensive reworking of local culture from a continental or "insular" lifestyle to an *isleño* one that would be more appropriate for island life. Conservationists presented a comparison of cultural traits: the insular lifestyle was one where people had a colonist mentality, strong cultural links to mainland communities, and sought the same level of services and consumption patterns as on the continent. In contrast, an *isleño* lifestyle would be based on recognition that being an island *resident* (rather than a *colonist*) meant developing an identity based on a strong sense of place that recognized the special nature of the archipelago and meant accepting limitations to consumption.[10] The social-ecological laboratory would change the relationship between people and the environment, making the Galápagos an experimental space for working out new environmental and economic subjectivities.[11] Instead of protecting natural resources from exploitative colonists, this new model laboratory would involve an experiment with a much further-reaching prescription for creating the kind of island resident who, like Don Michuy, would care for nature and profit from doing so.

The SES model presents two possible futures: a crisis of colonist lifestyles or a sustainable future based on an island culture. But this is a false choice. Reducing politics to two predetermined outcomes is a danger of relying on reductive models to order life. The model's vision for the desired future is no less utopic than the misguided dreams of the Europeans who came to the Galápagos in the 1930s seeking to play out the life of Robinson Crusoe on deserted islands. The SES model makes the same mistake that midcentury conservationists made as they fit competing geographical imaginations of the islands into their vision of a natural laboratory: it fails to respect the fact that people are living in the islands who do not share conservationists' way of understanding the world. It also fails to account for how people and nature together make history, and in doing so remake each other as well.

To understand how life has evolved in this laboratory over the past fifty years, we need to step outside the conceptual space of the model. We need to visit places that can give us a window into what life has been like for the people, and plants and animals, who have been caught up in ongoing experiments of arranging, and rearranging, relationships between nature and society to meet shifting conservation goals.[12] This trip does not follow a cruise itinerary but instead a series of multispecies transects, a method I borrow from ecologists who use transect walks to measure species abundance. I adapt them as a method for tracing boundary lines that structure relationships between nature and culture, and for noticing and narrating the entanglements of humans and nonhumans that resist and redraw these lines.[13]

The Path into Town

This tour begins on Baltra, where the airstrip first paved by the US military in the early 1940s has been converted into the Galápagos Ecological Airport, the world's first "green airport" where the majority of the more than two hundred thousand tourists who visit the archipelago begin their journeys after a two-hour flight from Guayaquil.[14] After paying your national park entrance fee—$100 for foreigners, $50 for foreign kids and those with most South American passports, and $6 for Ecuadorian nationals—you collect your luggage, which has been screened by specially trained dogs to prevent the entry of drugs, seeds, and nonnative organisms. Instead of heading to a cruise ship, you climb onto a bus for a short ride across the island. Along the way, you can still see the flat, crumbling foundations of military buildings. When you reach the Itabaca Channel that separates Baltra from Santa Cruz, you load your luggage on the roof of a ferry for the three-minute crossing. Across the channel, you then load your things onto a van sent by your tour company, into the bed of one of the dozens of waiting white double-cab *camionetas* that serve as taxis, or—much cheaper—onto the municipal bus that also goes to Puerto Ayora, on the southern side of the island. Regardless of the method of transportation, you take the same route, for it is the only one: a twenty-six-mile, two-lane road that bisects the island, heading straight southwest, up the slow rise of the parched, uninhabited northern side of Santa Cruz that is in a perpetual rain shadow. Along the way, you see many signs of the entanglement of human and natural history, if you know what to look for and what you are looking at, for few tour guides will point them out.

Signs on the side of the road read *"Despacio: Cruce de Aves"*—Slow, Bird Crossing. They are markers of the environmental costs of a flourishing tourism economy. Often, their point is punctuated by the carcass of a recently killed bird lying nearby. These birds, conservationists say, do not know to stay out of the way of vehicles that fly down the long hill because they did not evolve in a modern landscape and have not adapted to highways and speeding trucks. While I was living in the islands, speed limits to protect birds were an issue of contention, for these were not just any birds killed by speeding drivers. Many were Darwin's finches, which were made iconic through scientific study and are potent symbols of Galápagos nature. But these exemplars of evolution had collided with circuits of global tourist capitalism as drivers transported passengers from the airport to their hotels and cruise ships and back. The birds' tiny carcasses were fissures in the alliance between science, conservation, and tourism. Farther along the road, tortoise-crossing signs are also markers of the uneasy intersection of a fast-paced economy and these slow-moving creatures.

Conservationists had not wanted this road to be built. In the 1970s, a trip from Puerto Ayora to Baltra had to be taken by boat, a journey of several hours.

A network of donkey trails connected Puerto Ayora to the highland settlements of Bellavista, Cascajo, and Santa Rosa. The road would potentially connect the villages with the Itabaca Channel, providing much quicker access to
the Ecuadorian military planes that brought people and goods to the islands.
But the road would have to cut across park territory, something the Darwin
station director at the time, Peter Kramer, argued against. He countered local
residents' calls for the road contending that it was an "anachronistic attitude
and dangerous error to believe that highways attract tourism."[15] A highway
across the island was precisely the opposite of what the Galápagos needed, he
thought. Although he supported paving a road to connect Puerto Ayora to agricultural centers in the highlands, he did not think it needed to cross the
north side of the island. The park territory should be kept without such infrastructure development, he maintained. Instead, he thought a fast boat to
transport the slowly growing number of tourists would be more appropriate
for the islands. Kramer lost that debate. The road became a marker of development, infrastructure that reflected, and facilitated, what Grenier saw as the
geographical opening of the Galápagos to networks of global tourism. Yet it
was not the road that attracted tourism and development so much as it was the
efforts of Darwin station conservationists and the national government to sell
the Galápagos as a destination for nature tourism.

Also along the road on the north side of the island are two rock quarries and
the island dump. They are not visible from the road. Instead, they are marked
by entrance signs that remind all who pass by that Santa Cruz is home to a
prosperous economy, from construction to waste. Santa Cruz is the center of
Galápagos tourism, home to the headquarters of the national park and Darwin
station. The island was the last of the inhabited islands to be colonized but is
now home to the largest population—about eighteen thousand legal residents,
according to the 2010 census. Much of the stone and gravel from the mines
recently have gone to El Mirador, a new neighborhood on a hill with a view to
the sea along the southern end of the road, just as you arrive in Puerto Ayora.
It has been built over the past five years to accommodate the growing urban
population. The taxi driver I have relied on for years in Santa Cruz is building
a house there for his retirement, after migrating to the islands to work as a
park guard when he was young and then changing careers as the physical labor and extended time in the field became too much.

El Mirador was controversial. It was pushed by the longtime mayor of
Puerto Ayora, but conservationists opposed the development because it expanded the city onto parkland. To acquire land rights, the city swapped land

with the national park, exchanging the 175-acre El Mirador parcel for 250 acres of former agricultural land in the highlands. In the new development, the city promised better provision of services than in the other, unpaved streets toward the back of Puerto Ayora where residents have built homes over the past twenty years as the tiny village that existed in the mid-1960s expanded back from the waterfront toward the *barranco* wall that serves as a natural boundary to development.

The growing economy is evident too in the number of vehicles you pass on the roads. The vast majority are white *camioneta* taxis, which circle the main streets of Puerto Ayora. Few residents own private vehicles. The expense is prohibitive, and it is difficult to get a permit, making trucks and *motos* significant markers of status. Most residents rely on communal taxis or buses fitted with open-air rows of wooden benches that are a common form of transportation in rural Ecuador and Colombia (called *las chivas;* literally, goats). For a quarter, the *chiva* will pick you up on the side of the road in the highlands and take you into town. Bicycles too are a popular form of transportation for both locals and tourists. Often adorned with placards reading *Un auto menos,* one less car, they are a reminder of a perennial lamentation among the environmentally minded who see motorized transport as a marker of "continentalization"—the importation of lifestyles common in continental Ecuador that environmentalists see as ill-suited to, and inappropriate for, island life because of their reliance on imported, and heavily subsidized, gasoline.[16]

This continentalization, what the SES model identifies as an insular, *colono* mentality, has been a long-running concern of conservationists. A 1989 essay by the park director argued that even though economic planners saw a confrontation between conservation and development in the islands, he disagreed: everyone worked with the well-being of Galápagos residents in mind. The trouble, from his perspective, was that planners introduced development schemes from the continent as models of how island residents should live without recognizing that the islands present natural limits different from those on the continent.[17] During my first trip to the islands in 2007, the Darwin station director told us that these were the very same issues at the heart of Ecuadorian president Rafael Correa's and UNESCO's crisis declarations thirty years later. They had been a problem, he said, since state-led development began to take off with the designation of the archipelago as a province with its own government in 1973, a change demanded by citizens largely in response to the national park's territorial claims in the 1960s. In 1980 the federal government created the Galápagos National Institute (INGALA, by its Spanish acronym) to support development through infrastructural improvements. The state development bank opened an office in the islands in 1983, providing a mechanism for distributing oil revenues from the Amazon basin of Ecuador (the Oriente) to stimulate economic growth in the archipelago. The developments were crucial for the islands' nature-based tourism economy. They also marked the beginning of the long-brewing crisis in the islands.

In the 1980s, community leaders largely saw their roles as developing the islands' small settlements, particularly Puerto Baquerizo Moreno on San Cristóbal, the government center, and Puerto Ayora, the center of tourism and conservation. From 1950 to 1990, island population grew by a factor of ten, from just over a thousand to nearly ten thousand. Community leaders took the growth as a sign of progress, causing conservationists to fret over the pressure such growth put on what they saw as the islands' fragile and scarce resources.

During my first visit in the midst of President Correa's and UNESCO's declarations of a crisis in the archipelago, we heard repeatedly that the problems in the Galapagos stem from the 3 percent of the islands where people live—not the 97 percent that is national park. The editors of the semiannual INGALA *Galápagos Report* issued after the crisis declarations summarized it thus: "Analysis of the principal catalysts of change, those that influence the dynamics of the system and represent the root cause of the present crisis, clearly shows that current problems are not rooted in the natural system but rather in the socioeconomic and cultural systems."[18] Nature could not be at fault. The reason this evolutionary laboratory did not resemble an Edenic paradise must be social. Society and nature remained reduced to two distinct realms that did not so much interact with each other as impact each other, either by setting natural limits or, conversely, by degrading the environment.

The authors of the SES model also echoed this refrain. Their concerns about scarce island resources replayed those of eighteenth-century British political economist Thomas Robert Malthus, who argued against welfare support to the poor on the basis that the island of Great Britain could not produce enough food to feed its growing population.[19] Although Malthus's dire predictions have been proved wrong time and again, his anxieties have long underwritten environmentalists' fears about global overpopulation. Similar worries about the ability of the Galápagos Islands, with their limited patches of arable land, to support a growing population fueled conservationist policies that for decades sought to minimize migration and development. Malthusian framings ask us to weigh human population against natural resources, as if they sit on either side of a teetering scale. But this distinction does not hold. Although

marked boundaries divide the park from human-use zones on inhabited islands, the idea that the Galápagos are 97 percent nature and 3 percent culture
is an imagination produced by territorial maps and governing institutions.
The teetering scale also entirely misses the political history that produced
such contested divisions.

Looking back on this history, the local residents I spoke with more often
pointed to conservationists, rather than to island nature, as what limited
them—in particular charging foreign Darwin station scientists with a history
of opposing development. In the early years of Galápagos conservation, foreign scientists who led the Darwin station recognized that conservation
success was directly linked to local education and "appropriate" economic development. The road was just one example of tensions about what kind of
development was appropriate. In 1972, Kramer acknowledged a history of
conservation efforts that had focused exclusively on the preservation of wildlife and had neglected the development of community needs. The first park
management plan, adopted in 1974, envisioned most island residents as destructive resource users. Their access to water, sand, lumber, and feral game
needed to be controlled by park guards who established special use zones for
permissible resource extraction and patrolled the boundaries between settled
areas and areas of restricted protection.[20] UNESCO had sent only wildlife
specialists—not agricultural scientists, fisheries experts, or sociologists.[21] Proposals to work on sustainable development had not turned into concrete projects, although Kramer reported to UNESCO that he hoped the government
would pursue these plans.

His position was common among foreign conservationists in the islands.
Although they recognized the importance of development both for the community in which they lived and for the success of conservation, they saw this
as outside the purview of the Darwin station and the national park. Those
institutions focused on *natural* science and conservation; social development was the responsibility of state institutions. The conceptual and territorial
division of the islands that was fundamental to early conservation was also
reflected in governing institutions. The national park and Darwin station managed 97 percent of the archipelago, with funding and technical support from
the IUCN, the World Wildlife Fund, and other international environmental
NGOs. Provincial, municipal, and parochial governments managed urban
and agricultural life, with scarce funding available from a state with a long
history of indebtedness to foreign governments and international development banks. Conservationists often saw conflicts with the local community

over development and resource use as an inevitable consequence of their work. It was a pattern that would continue over the coming decades despite policy attempts to integrate conservation and development: conservationists decried population growth and poorly managed tourism while poorly funded munici-palities struggled to provide services to increasing numbers of residents.[22]

The road that bisects Santa Cruz is an important reminder of this political history. Traveling it shows us that the conceptual division of the islands into 3 percent human use and 97 percent park does not hold. The road cuts across these zones, a line that reminds us that attempts to bound nature and culture are also attempts to draw lines around local economies and development. It reminds us that the tourism economy links the park, the city, and the agricul-tural highlands.[23] But if we turn off the road, we can see that economic devel-opment and native species are not necessarily on a collision course.

Rancho Mariposa

After you crest the top of the island, driving through low-slung clouds, the road descends in a winding route through farm and ranch land in the humid highlands that are the most biodiverse area in the archipelago. Just after the small village of Santa Rosa is Rancho Mariposa, a place I visited several times in multiple guises. I went as a tourist with my mom as the last stop on the tour we took of the archipelago so that we could see giant tortoises in the wild. I visited as an interviewer to chat with the farm's owner, Steve Divine, and as a friend for an evening picnic in the clouds. And a few times I was there as a not-terribly-useful field hand accompanying ecologists studying giant tortoise migration up and down the sides of this dormant volcano. This is where I first walked an ecological transect early one misty September morning in 2011 with Fredy Cabrera. Cabrera is a local ecologist who works on the Giant Tortoise Movement Ecology Programme run by conservation biologist Stephen Blake. Rancho Mariposa is one of several farms his team uses as entrance points for transects in the highlands of Santa Cruz.

Rancho Mariposa is named after a horse ranch Steve's father Bud Divine had in Arizona in the 1930s. Steve's parents were among the first settlers on Santa Cruz; when they arrived in 1949 only about a hundred people lived on the island. They had set off after the war to sail the world but never made it past the Galápagos. Steve's mother, Doris, worked for the US military in the Panama Canal Zone, and Bud worked at the decommissioned base on Baltra for a while; but eventually they bought a house from a Norwegian settler on a point on El Otro Lado. Steve and his wife, Jenny, still own it, as well as a tour company based in town. Bud and Doris got the land in the highlands in 1960 when they joined the Colonia Velasco Ibarra in the midst of the US utopian settlement on San Cristóbal when the government was offering large home-steads. Rancho Mariposa was nearly five hundred acres initially, though it is smaller now, some of its territory claimed decades ago by the national park.

Cabrera's job was to collect data each week as he walked the transect. Like the road, the transect crosses agricultural land and parkland. It is a path that can show us how social life, economies, nature, and science are entangled in ways that overflow the boundary between the 97 percent and the 3 percent. The transect stretches from the road down the rocky drive into the ranch, across the boundary into the national park, into the tortoise reserve that was the first protected space of the national park, and terminates in the dry low-lands of western Santa Cruz, where female tortoises lay their nests. To get started on this journey, Cabrera and I walk down the lane toward the Divines' house, passing cattle in fields to our left and right. Steve had warned us about the *vacas bravas*; be sure the aggressive cows were enclosed, he told us, or we should not try to pass. When Steve was growing up in the 1970s, the farm was primarily a cattle ranch, as many of the highland farms of his neighbors were and still are. Santa Cruz provides most of the beef consumed in the Galápa-gos; the industry has been an economic mainstay since the islands were first colonized.

Now Divine focuses primarily on tourism. In 1975, he took the first guiding course offered by Darwin station scientists and has worked as a naturalist guide since. Even before the training, he took small groups of tourists around the archipelago on converted fishing boats that had none of the luxuries that cruise ships now offer. Passengers and crew alike bathed in the sea, built campfires on deserted beaches, and plucked lobsters from the water for din-ner. After decades guiding on cruises, he turned his attention back to the ranch in 1988. He built a restaurant with a wide veranda overlooking a cleared, hilly field below where giant tortoises can often be found munching on the pangola grass his father planted years ago for the cattle. For many years, tour-ists could wander around the fields and take pictures of the tortoises. Divine recently closed to tourism because increased governmental oversight and per-mits made the business financially unviable, but Rancho Mariposa used to be a common stop for cruise passengers who otherwise might not have the op-portunity to see a giant tortoise. On the veranda, Divine would serve you tea made from homegrown lemongrass and perhaps some local *aguardiente* as well. On a clear day, the veranda has a stunning view west to the calderas of Isabela.

The tortoises did not use to come up the mountain this far. Steve's dad took tourists to see them in the reserve, following a trail from the farm to La Caseta—the derelict shack that park guards and scientists used to use as a base camp for extended stays in the *campo* before it was possible to drive a truck here from town in half an hour. Seeing tortoises then meant spending ten to twelve hours on a mule and camping in tar paper–covered shacks. It was quite a different experience from the thirty-minute stop tour groups commonly make now, made possible, in part by changing tortoise migrations.

Understanding these migration patterns is the focus of Cabrera's work with Stephen Blake's team. Blake previously worked in the Congo, using radio telemetry to track forest elephants. When his wife's job as a veterinarian took them to the Galápagos, Blake translated his skills to another megafaunal herbivore. For more than a decade now, he has been tracking tortoises as they amble up and down the slopes of Santa Cruz and Volcán Alcedo on Isabela and across the much flatter Española.[24] Giant tortoises are creatures of habit who follow well-tread paths. On Santa Cruz they migrate seasonally, from the green highlands where food is ample throughout the drier *garúa* season to the sunnier lowlands in March for the rainy season where sprouting greenery makes the landscape a "veritable salad bowl." By June, as the lower island dries out, it becomes an ideal place for female tortoises to dig nests and lay clutches of eggs. But as the easy food supply disappears, most adult tortoises make their way back more than four miles to the highlands, walking an average of just eight hundred feet a day for more than two weeks so that they can spend the fall and winter in the highlands, resuming the much more sedentary lifestyle for which they are known.[25]

To gather this data, Blake has outfitted a few dozen tortoises with GPS units custom made in Germany. Each contains an accelerometer, similar to what is in a smartphone, a memory chip, and a radio transmitter. The units take readings of locational data every hour, and Cabrera hikes various *picas* (trails), on the island every week to download the data, using a three-foot radio telemetry antenna to find the tortoises. This is high-tech natural history exploration.

On one of my visits to Rancho Mariposa, I spent a morning with Cabrera and another volunteer replacing one of the transmitters. We removed one that had been placed too high on the front of the tortoise's carapace where it snagged on barbed wire fences. The process started off easily enough—we found the tortoise lounging in a small muddy pool near the cattle corral and managed to coax it out of the water. But as soon as Cabrera started chiseling away at the epoxy-glued transmitter, the tortoise took off, doing his best to escape us. We finally got him to hold steady in a muddy rut next to the corral— a struggle with this big old animal who weighed considerably more than I did and whose legs were considerably stronger than mine. Cabrera worked quickly, but the chiseling was a loud, slow process. My job was to gently kick the tortoise's hind legs so that he would keep them withdrawn in his shell instead of powering away from us.

I do not think the chiseling *hurt* the tortoise, but he clearly did not like the process. I imagine that the loud clanks of the hammer on the chisel and repeated reverberations were unsettling. But not much is known about how tortoises sense the world. Once during a tour of the specimen archive at the Darwin station, I asked the scientist leading our group whether the tortoises could feel through their carapaces. He had been explaining how carapaces are akin to human ribcages, only an external rather than internal shelter for tortoises' lungs. He did not understand my question and looked at me with what I took as a mix of perplexity and derision before moving on to another question. What it might have felt like to a tortoise to have someone sit on its back was not something that had interested biologists. But Blake and Cabrera were concerned that they not harm the tortoises they track. After the old GPS unit was off, the process of applying the new one was much quicker and less disturbing. As the tortoise calmly hunkered in the grass, I scrubbed off the remaining residue and cleaned the spot for the new transmitter on the rear of the carapace before Cabrera secured it in a bed of epoxy (fig. 7.2). The tortoise seemed to barely notice at all.

Continuing our transect walk, Cabrera and I crossed the open fields below Divine's veranda, walked between towering, dense copses of bamboo, and continued down the road that reaches the bottom half of the property. As we went, he checked the rain gauges he had set up along the way and periodically pulled out the antennae to search for tagged animals. There were a lot of tortoises here, few of them tagged with GPS units. I counted a hundred before losing track. It seemed much easier for them to move along this road that was at once a tortoise highway, *pica*, and access drive than it was for me, my *botas de caucho*—rubber boots that are de rigueur in the *campo*—twisting and sticking in the deep mud.

Carlos Ochoa, the park guard who traveled the archipelago tracking tortoises in the 1960s and 1970s, told me that these tortoises migrated so far up the western slopes of Santa Cruz now because of introduced species. He called

Figure 7.2 I scrub the carapace of a giant tortoise to be fitted with a
GPS tracker for a migration ecology study at Rancho Mariposa,
Santa Cruz, October 2011.

El Chato, the tortoise area a bit farther north, a "disaster." There are two *pozas* (pools) there where the tortoises used to water. But they go farther up to cleared farms now because El Chato was *"llena de plagas"*—full of plagues—a phrase I heard many times uttered by park guards, scientists, and highland residents alike. This area was once native Scalesia forest, but most of the trees now are Cestrum, Cedrela, and Cinchona—all introduced during the twentieth century. They have spread through the forest to such an extent that the national park cannot hope to eradicate them. The tall Cedrela trees, Spanish cedar, are resistant to termites and rot, their reddish wood a useful source of lumber. There is a steady market for Cedrela in town where it is used to make furniture, and selling lumber is a good source of income for landowners and laborers. But Ochoa lamented that this market is what keeps the national park from doing anything about the proliferation of the trees.

Mora (blackberry) is also a plague here. As we got closer to the boundary between the Divines' farm and the park, we picked through thin patches of the plant; where it is thick, its thorny brambles are more effective than barbed wire. Divine told me that *mora* did not become a plague until the mid-1990s. He used to have it under control, until he went to the United States for a while and was not present to be sure his laborers kept it clear. Now he has a lot of it because he does not like to use pesticides. He keeps the tourist area of the farm clear, but he cannot afford to clear all his land—to clean ten hectares three times would mean an outlay of $15,000. The $3 a head he charged tourists was not sufficient. Scientists, Divine told me, need to do more work on *mora* so that it does not invade and further harm endemic species; but they are too focused on getting their doctorate degrees and doing their own projects.

At the base of the property, Cabrera and I crouched down to slink under the barbed wire fence that marks the property line. The lowest wire is about two and a half feet off the ground, strung high so that giant tortoises are able to pass underneath. This is common on Santa Cruz and encouraged by the national park so that tortoises are able to migrate freely. But not all landowners make their boundaries tortoise-friendly. Some stack logs along the base of their fences so that tortoises cannot pass, particularly to protect agricultural fields, where the animals are known to feast on young pineapples and can cause considerable crop loss as they trample over fields. Farmers sometimes will overturn an intruding tortoise in an effort to teach him not to return. The animals can right themselves, but it costs them considerable effort.[26]

Just across the road that borders the park, we pass La Caseta. As we walk down the path, rockier and narrower here, Cabrera tells me about growing up in Bellavista and how he came to know this *monte*, this unmanaged forest, as he learned to hunt feral goats and wild hogs, the *chanchos salvajes*. He tells me stories of getting lost with his cousins as a teenager. Miles from the coast, where the land levels out and the trees are shorter, but still dense, it can be nearly impossible to tell which way is up the mountain and which way is down. We talk about the tourists who every now and then set out for a hike unaccompanied and go missing in the highlands. We stay close to the *pica*, which is marked by park guards who put a log or a rock in forking tree branches—very easy signs to miss. I ask, with a bit of trepidation, whether we are likely to see any of the aggressive, tusked *chanchos*, knowing this is the area where Laurtaro Andrade was stationed in the 1960s to hunt them and that park guards still protect tortoise nests. But Cabrera has seen more of them recently on the other side of the island. I ask him whether people on Santa

Cruz hunt tortoises. He shrugs—not really. He does not have much to say about that, although he knows it used to be common, before he was born. Not such a problem here anymore—not like it has been on Isabela.

Eventually Cabrera has collected enough data for the day and we turn around to hike back to Rancho Mariposa. For Cabrera, the hike was a routine part of the process of scientific knowledge production and an opportunity for him to spend time in a landscape he has known since childhood. For me, it was a journey into the thick knots of nature and culture in the Galápagos that bind together practices of science, conservation, tourism, and local development. The tortoises at Rancho Mariposa exemplify these entanglements. They are not icons of a prehistoric past. They are wild, yet they inhabit a landscape transformed by agriculture and introduced species. They are endangered species as well as objects of tourism and conservation science. Some, with their GPS units, are even cyborgs, fully hooked up to the technoscientific apparatus of modern life.[27] The tortoises are boundary objects. This is not so much because they anchor the translation of scientific knowledge into conservationist practice and ecotouristic ethics; such a linear, one-way process is rarely a reflection of real relationships among science, conservation, and tourism.[28] Rather the tortoises are boundary objects because they live at the intersections between agriculture and conservation, tourism and science, native and introduced species. They inhabit, even embody, these boundaries. Although life at Rancho Mariposa seemed quite good for the tortoises, boundaries can be a troubling place to live. Let's go back now to Campo Duro to understand some of these troubles.

Campo Duro

After a turn off the main road, Campo Duro's drive leads to a green lawn encircled with palm trees, an introduced species. Although palms and lawns are ubiquitous tourist landscapes on tropical islands, they are an unusual sight in the Galápagos. The arduous work of clearing the land is what earned Campo Duro its name, Don Michuy told me. We spent an afternoon exploring the property while he told me about its history. He was not the first to work the land; the farm had once belonged to the Gil family, the first colonists on Isabela, who came from the mainland and established a penal colony, coffee plantation, and tortoise processing business at the end of the nineteenth century. During the 1940s and 1950s, when the island was in its second phase of use as a penal colony, the farm was one of the places prisoners stopped as they moved between labor camps on the shore and in the highlands. Over the years, the plot had passed to other families and had been abandoned, like much of the farmland on Isabela. Local residents left it behind because clearing invasive species made staying less appealing than a life on the coast with its more lucrative opportunities in tourism.

Michuy bought the farm with the intention of making it a tourism destination. He planted an organic fruit orchard where he grows bananas, oranges, papayas, passion fruit, avocado, plums, and mangos (all introduced)—the tortoises, he said, particularly love the mangos. Near the orchard was the fenced corral where the farm's star animals live. As we approached, a few chickens skittered past; he serves their fresh eggs for breakfast to visitors who camp overnight. We fed a couple of tortoises a papaya—another favorite. Michuy smiled as he watched the animals and told me he had always dreamed of having a pet tortoise. He used to joke with friends that he would have to capture a wild one, though he knew that would get him in trouble with the park. Instead, a friend who worked at the Darwin station put him in touch with park officials on Isabela. At first they were skeptical that the tortoises would be safe and that tourists would actually come up to an old farm. But eventually, Michuy told me, his face lighting up, the park sent him forty-three tortoises. A small grant from a local development fund established by Lindblad Expeditions, one of the largest and oldest tour companies operating in the islands, helped him build the corrals he needed to contain the tortoises. International senior citizen volunteers come to help him take care of the farm.

Nearly ten years later, the farm was a success. The tortoises grew well there—much faster than at the breeding center. After a few years, the park retrieved thirty-two of the tortoises to repatriate to their native population on Cerro Azul, at the western end of the island. Michuy had not known they would do this; he was annoyed because he had encouraged tourists to "adopt" a tortoise, promising to send them pictures and an update every year to encourage visitors to return or send their friends to see "their" tortoise. But with the remaining animals, enough tourists came that Michuy's family lived comfortably. When I asked what message he thought the animals conveyed to visitors, he told me that people came to the Galápagos looking for a sense of peace and tranquility in nature that was lacking in the places where they lived. Although the planet seemed to be "collapsing," here they could find a place where people cared, where they conserved.[29]

Campo Duro is an oasis for tortoises at the intersection of tourism and conservation, yet not all the animals on Isabela are so well cared for. Quite the opposite. Park guards make weekly patrols on Sierra Negra to dissuade poaching. Periodically they find the remains of butchered tortoises—a total of forty-four between 2006 and 2011. When I asked the island park director why people killed the animals, he told me of a black market for tortoise meat where a midsized tortoise of twenty-five to thirty-five years old sold for about $65. Residents ate tortoise for special occasions or after giving birth. In the highlands, feral game hunters killed tortoises, the park director told me; nearer to the coast, fishers did so. He attributed Isabeleños' tortoise hunting to "tradition" held since people arrived in the archipelago. "But obviously," he said, "many generations have now passed and the park is more than fifty years old, so the prohibitions are not something new, but the people here have not succeeded in understanding this. They think that killing and eating the tortoises is going to do them good. Obviously it may be true, but unfortunately it is no longer allowed and that is what they do not understand."[30] What the park director described as "tradition" was a pattern of hunting tortoises for consumption born of long histories of imperial exploration, penal colonization, and natural history collection.

When I started doing research in the islands, I was asked to help monitor poaching, not by the national park (which had issued my research permit), but by the social scientist at the Darwin station. He suggested I go to Isabela to do an ethnographic study of tortoise poachers, with the end goal of organizing a meeting in which conservationists could confront residents and, through discussion, try to end this practice by persuading them otherwise. For him it was an obvious focus of research—I could address a long-standing conservation problem. For me, it was troubling. Although I was sympathetic to the plight of tortoises, I thought such research would position me as conducting surveillance. How could I gain the trust of island residents, encourage them to open up to me about an illegal practice, and be up front about the aims of the research? I felt I would have to hide my intentions and betray my informants. Nor was I interested in undertaking a study that would reinscribe an already well-tread narrative that positioned residents as threats to iconic species by focusing on a very small minority of the population. Instead, I wanted to understand the history of conservationist policing of human relationships with the tortoises.

At the Darwin station archives, I came across a series of community newsletters issued by the park office in Isabela during the late 1980s. They mostly updated residents on park news, including plans to build a tortoise breeding center on Isabela. Several articles also discouraged tortoise poaching. One issue ran a drawing of a man bringing down his machete on a tortoise, under a

banner that read *"No muerte a los galápagos, cuidalos"*—No death to the tortoises, care for them. The text below reminded residents that they had resided
within a national park since 1959: "Failure to comply with the Laws and Regulations of the National Park Service . . . [is an] evil we do to ourselves and our
Island, because we are killing the touristic and scientific attraction. It is necessary that we all understand this, so we do not give the impression that in Isabela we are thirty years behind." The next issue repeated the message in a
scripted conversation between two fictional residents who discussed the recent killing of six tortoises at La Cazuela on the eastern slopes of Sierra Negra.
One man commented that while on the other islands inhabitants follow the
law and respect the animals, on Isabela "it seems that not even news of the
existence of this law has arrived." The other man countered that not all Isabeleños are like this; it's only the actions of a few *malandrinos* who also rob
farmers of their chickens and young cows. The conversation ends with the
men's exacerbated cry, *"¡Así es mi tierra!"*[31]

The role-play and drawing attempt to educate residents about the existence
of a law, pretending that lack of awareness might account for the killings. Yet
the problem was not that residents did not *know.* Park admonishments not to
kill the tortoises were moral mandates with stakes that extended far beyond
the lives of animals or social shaming. They were reflections of normative
conservationist ideas about the proper order of relationships between humans
and nonhumans. The admonishments accused Isabeleños of destroying precisely that which was valuable about the island for scientists and tourists. They
also suggested that Isabeleños, unlike residents on other islands, were behind
the times in their failure to care for tortoises. The assertion assumed a linear
history of conservation that moves from consumption to care. Failure to conform to ethical assertions about how populations *should* act was read as a lack
of progress to be remedied. Such temporal framings of difference are common in the language of colonialism and development that painted subject
populations as primitive natives in need of guidance from those who knew
better.[32] Although Galápagos residents are not indigenous to the islands, they
fall within a similar categorization when they fail to act according to the norms
of conservationist rule. Within a world already held apart from modernity as
an anachronistic space to be preserved outside of historical time, tortoise
hunters are doubly troubling. Their presence is as "unnatural" as that of other
islanders. But worse, they also actively attacked the prized animals that did
belong.[33] The implicit assertion was that if there must be people in the islands,
then they should respect the conservationist project of protecting the animals.

Yet the tortoise killings continued. Biologists documented twenty-one slain tortoises in 1991, fifteen in 1992, and one in 1993. In 1994—a particularly tense year—park guards found eighty-one butchered tortoises.[34] Conservationists suspected fishers were behind the killings—a symbolic response to escalating tensions over booms in commercial fishing of sea cucumbers in the mid-1990s.[35] Over the next decade, from 1994 to 2005, conservationists found 190 slaughtered tortoises. All but one were found on Isabela, and several were found near Sierra Negra and places frequented by fishers.[36] Many of the tortoises were used for food; park guards even caught fishers with tortoise meat in their cooking pot one year. But park wardens found other intact tortoises dead in the middle of trails, their necks slit. These they took as killings meant to send a message—they were not about a traditional taste for tortoise.[37]

The butchered tortoises are also boundary objects that sit at a more conflictive intersection between conservation and local economies, one shaped by long-running histories of resource use and regulations. The traditions that led people to kill tortoises were not only influenced by the food culture of early settlers. They had been formed through a history of tortoise collection and commodification as old as the settlement of Puerto Villamil itself. They had also been formed by decades of resentment about conservationist rule and repeated contestations about what kinds of life were of value in the archipelago. To explore that, we need to return from the highlands of Sierra Negra to Puerto Villamil to visit the fishers' dock.

Fishers' Dock

Puerto Villamil is small compared with the towns on San Cristóbal and Santa Cruz. Just two thousand people live on Isabela. The community is dominated by a few extended families, some of which can trace ties to the island back seven generations. Although Isabela was one of the first two islands settled, and was once a popular stop for naturalists like Rollo Beck and Charles Townsend, the island community now has fewer connections to the outside world than the two main inhabited islands. While the economies of those two urban sites diversified into government offices (on San Cristóbal) and conservation and tourism (on Santa Cruz), agriculture and fishing remained the dominant livelihoods on Isabela in the late twentieth century. Tourists do visit, but cruise ships do not stop here—the bay is far too shallow. Puerto Villamil is *tranquilo,* with a wide sandy beach dotted with small hotels and more

Figure 7.3 A fisher prepares the day's catch for sale at the fishers' dock in Puerto Ayora, Santa Cruz, a popular stop for tourists who watch the multispecies encounters among fishers, their catch, and the pelicans and sea lions who beg for scraps. Visiting the dock gives a different perspective on fishers than those more commonly portrayed in media coverage of strikes and scientific papers about poaching. (Photo by the author, 2011)

bicycles than cars on the roads. Restaurants that serve Ecuadorian food as well as pizzas and burgers line the main square. It is the kind of place where I have ordered eggs and then watched a kitchen hand run across the street to the market to buy some.

The fishers' dock sits about a half mile east of the town square, next to the dock where tourists arrive in small *lanchas* (ships) from Santa Cruz. In Puerto Ayora, the fishers' dock is a popular tourist attraction on Avenida Darwin near tourist restaurants, hotels, and souvenir shops (fig. 7.3). In the afternoon when the fishers are back with the day's catch of tuna and bacalao grouper, locals stop to buy dinner while tourists snap photos. The scene at the Isabela dock, however, is quieter. Usually it is just the fishers who hang out at the bar across the way at picnic tables nestled in the sand.

I sat there with a team of researchers during my first trip to the islands in 2007, talking to fishers about their views on conservation and Galápagos life. "*Somos los malos de la película*"—we are the bad guys in this movie, they told us. They were frustrated with and angry about this reputation, though it was not wholly undeserved. In 2000, fishers angry over park quotas on the year's lobster fishery set the park director's house on fire, then walked down the street and vandalized the park office. They also took ten tortoises from the breeding center on Isabela and destroyed tortoise egg incubators.[38] Sitting at the nearly empty bar, I found it hard to imagine the uproar in this quiet town, but the men were animated as they spoke, passions rising. Fishers have had a long history of conflict with the national park and Darwin station. In 2004, a strike on Santa Cruz made international headlines when fishers blocked the road to the park and station headquarters for several days. By doing so, they also blocked access to the main tortoise breeding center, and to Lonesome George, prompting a BBC News headline "Galápagos Tortoises Held Hostage."[39] Strikes are a common aspect of labor negotiations and displeasure with governing institutions in Ecuador, but when these protests involved charismatic wildlife, they became sensationalist international news.

Competing ways of valuing nature have long been a source of friction in the islands. The most notorious case remains what the press called the sea cucumber war that erupted in December 1994.[40] It was one in a long history of conflicts in Ecuador about international rights to fish the rich, and lucrative, marine areas around the islands. The national park closed the three-month fishery a month early because the take quota, set at 550,000 animals, already had been surpassed—by more than 6 million animals, according to some reports. These *pepinos*, as they are called in Spanish, were not for local consumption. The fishery boom had been backed by Asian exporters who had come to the islands in the late 1980s after overfishing the Ecuadorian coast. To circumvent laws prohibiting foreign and industrial fishing in Galápagos waters, they provided local fishers with loans for modern equipment and boats and promised to pay high prices for these animals that are a delicacy in Asian markets. Park officials tried to stop further overharvesting by closing the yearly season early. Fishers protested, angry that their opportunity for development had been curtailed. They took control of the Darwin station and park headquarters, keeping workers hostage for three days and threatening tortoises at the breeding center.[41] The strikes were an explicit challenge of conservationist rule and an implicit rejection of the moral ordering of life that privileged the conservation of a Darwinian laboratory over human livelihoods. They were also an expres-

sion of fishers' frustrations about being caught in the crux of two clashing political economic systems: the international fisheries market, to which they had
access, versus conservation and tourism, to which their access was limited.[42]

The tense situation of the mid-1990s nearly led to the islands' inscription on
UNESCO's list of World Heritage Sites "in danger." Instead, the World Heritage
Committee pressured the Ecuadorian government to pass legislation to address
the situation.[43] The result was a comprehensive attempt to reframe the relationship between conservation and development in the islands. In 1998, the
Ecuadorian government passed the Special Law for the Galápagos, written
in consultation with the World Wildlife Fund, the Nature Conservancy, and
Fundación Natura—institutions that had become central players in Ecuadorian
environmental governance when they completed a $10 million "debt-for-nature
swap" with the state in 1989.[44] The result of the swap in the Galápagos was
more funding for conservation as well as a law that established, at least on paper,
conservation and sustainable development as complementary guiding principles for local governance. It was an attempt to make the alliance between conservation and nature tourism established in the 1960s work for local residents who
complained that the islands were managed by external forces and that the benefits of the growing tourism economy did not remain in the islands.[45]

The Special Law responded to tensions associated with industrial fishing
and growing on-island development by establishing special island citizenship
and migratory restrictions, instituting an inspection and quarantine system to
prevent introduction of nonnative species, and reforming the system of tourism permits to promote local ownership. It also increased entrance fees to the
national park and established a mechanism for distributing them among the
park and local institutions so that tourism money would stay in the archipelago instead of going to the central government. The law provided the first
meaningful structure for protecting a fifty-thousand-square-mile marine reserve where industrial fishing was banned and created a participatory management board that brought together representatives from the national park,
Darwin station, fishing industry, and tourism industry to negotiate fishing
calendars and cruise permits.[46] At the center of the law's institutional changes
was a reformulated role for INGALA: an institution previously responsible
for maintaining roads and constructing basketball courts and soccer pitches
would now act as central coordinator of the islands' institutions, responsible
for implementing the law, overseeing migratory controls and island citizenship
permissions, managing tourist visits, and ensuring quarantine procedures on
continent-to-island and interisland transportation.[47]

Yet perhaps the greatest challenge was not improving the technical capacity of core institutions but opening conservationists' understanding of nature. It was at this point, *following* new legislation for a shift toward sustainable development, that biologists made their clearest call for a return to pristine nature. In May 1999, fifty-eight international conservation biologists met in the Galápagos to outline a "Biodiversity Vision" for the future of the archipelago.[48] They feared the islands were poised at a crossroads. Island nature was under threat from development, particularly the booming tourism economy, human population growth, exploitative fishing, and transportation networks that linked islands to each other and to the external world. Although they estimated that the archipelago retained 95 percent of its "original, pre-human diversity of species," they fretted about the future.[49] They had "one last chance" to go "back to Eden" by restoring the islands to their state in 1534, the year before Tomás de Berlanga's ships arrived and the islands entered the records of Western history.[50]

The drama of one final opportunity to restore the islands to an Edenic state reflects core assumptions of conservation biology—chief among them an idealization of untouched nature and threat of looming crisis. Although the idea of pristine wilderness had been roundly critiqued by the late 1990s as an illusory goal for conservation, it continued to be a powerful framework for understanding nature in the Galápagos.[51] The islands were, for conservation biologists who wrote the vision, "a rare remnant of a prehistoric pattern of global biological diversity where great proportions of the world's distinctive and often bizarre species occurred on islands." The Galápagos were "one of our last chances" to save this evolutionary heritage.[52]

The workshop reinscribed the desire to preserve pristine nature that had influenced earlier policies that limited space for local residents. The diverse group of authors said their goal was to harmonize conservation and development, and they acknowledged the need for complementary social analysis. But this fell outside their purview and was never produced. Despite the institutional change called for in the Special Law, a divisive atmosphere remained. In the late 1990s prominent—and extreme—island conservationists looked for ways to privatize the islands and remove residents altogether. A retired Darwin station scientist told me years later that if he had his way, there would not be any people on the islands—but, he said, Ecuador is a democracy, so what could be done?

By killing tortoises and other iconic animals, Galápagos residents took the binary logic that had long structured island governance to its opposite extreme.

They challenged the line drawn to exclude them from visions of the islands as a land of and for tortoises. Where conservationists had suggested removing people from the islands altogether to create a place of ideal pristine nature, hunters targeted iconic animals to assert their right to live in the islands. These residents were not "behind the times"; they were responding to a way of ordering life that made little, if any, room for them.

When I lived in the islands, I did not encounter protests like those of the mid-1990s and early 2000s. But as the fishers at the dock on Isabela showed me and my companions, memories of these conflicts were vibrant. As we discussed Correa's "at-risk" declaration and their struggles with the national park, they brought up an incident that had occurred in the summer of 1994. During a wildfire on the island (thought to be caused by a runaway campfire), several tortoises were airlifted to safety, and one injured tortoise was flown to Florida for treatment.[53] For the fishers, this was a telling example of the park's greater concern for the well-being of wildlife than of human inhabitants. They complained bitterly about the resources spent to protect giant tortoises, citing the injustice of conservationists' rush to care for the animals during times of crisis when residents had difficulty accessing quality health care. A young girl had died shortly before the wildfire because of lack of adequate medical care. It was a story, I later found, that had come up for other ethnographers too—a cultural touch point of indignation.[54]

In a place with as much international acclaim as the Galápagos, iconic animals generate considerable attention. Conservation advocates have long used media to advance their work, shaping public perception of the islands and their challenges. By targeting tortoises, fishers have attempted to turn this attention to their own struggles, although the result is often not successful in the way that they hope. A fisher who had been part of the strikes as a union leader said the oft-repeated charge that they had taken Lonesome George hostage had been misconstrued. They never intended any harm to the tortoises, he told me, but had no other way of being taken seriously.[55] Anthropologist Jill Constantino, who lived on Isabela for several years, recounted a protest that occurred during her fieldwork in 2001. Isabela fishers took several tortoises from the breeding center in the middle of the night, moved them down the road to a makeshift corral in front of the park office in the center of town, and sat themselves down next to the tortoises to draw attention to their plight:

On the first day of the protest, one fisherman—who told me he was trying to be funny—posed with a knife at the throat of a tortoise. This pose and its implicit threat were widely reported in newspapers, magazines, and radio programs. The protest struck an international nerve. When fishermen used Galápagos tortoises in their protest, they hoped to broadcast the absurdity of valuing human/nonhuman relationships over human/human relationships. They implicitly and publicly asked, "Are turtles really more important than fishermen?" The international answer was a resounding "Yes!"[56]

The fishers' sense of humor did not translate. For an international audience conditioned to think of the Galápagos archipelago as a place to be conserved, the fishers positioned themselves as predators by targeting tortoises. They walked into the role that naturalists had ascribed to island colonists for a century. Fishers threaten what Constantino calls the hegemonic moral order of the Galápagos. By asserting themselves and their livelihood concerns, fishers are seen not only as harming marine resources, but also as challenging the dominant understanding of the Galápagos as a living museum of evolution. The hegemonic understanding of nature as global heritage eclipses the voices of residents who dare to question it.[57]

Constantino relates another telling story from when she arrived to teach English in 1995. Looking for information on the community on Isabela, she visited the Darwin station library but found few references. When she asked a scientist why there was so little written on human history in the archipelago, she was told, "Don't you know? We're trying to keep people out!"[58] Local human histories were not worth studying. Fishers do not fit the dominant vision for the Galápagos established by Western scientists when the park was founded. Even though many of their families came to the islands before the mid-twentieth century, their claims to belonging are eclipsed by the hegemony of understandings of the archipelago as an evolutionary landscape. In contrast to residents who saw the tortoises as an important economic resource for tourism, fishers used the tortoises' symbolism to contest the strength of an alliance among scientists, conservationists, and the tourism industry. Iconic species are proxies for dreams of development, but so too are they proxies for frustrations when those dreams are dashed.

Despite all of the animosity, the fishers I spoke with insisted that they were not against conservation. Rather they argued that "there is no conservation without people." How might taking their point seriously change the experiment of ordering nature and society that has shaped the history of conservation?

Island Life

Since the founding of the national park and Darwin station, conservationists have sought to order life in these island laboratories. Three deceptively simple questions were at stake in the experiments that tried to arrange relationships among humans and nature: Who belongs in the archipelago? What should be conserved? And who or what should benefit? These questions have had presumed answers since the 1930s when foreign biologists argued that the Galápagos should be protected as a natural laboratory of Darwinian evolution: scientists, conservationists, and a few tourists belonged; native species should be conserved as proxies of evolution; and the benefit accrued to global publics through the creation of scientific knowledge and the protection of a world patrimony. Island residents were largely irrelevant. When they were considered in later years, they were mainly treated as threats to conservation. But their perspectives have been seldom consulted. As one local official explained, conservationists "propose that if Galápagos has an identity, it is in the history from the writings of Charles Darwin. They think people should have an identity. But the population does not have this idea."[59] Many Galápagos residents are quite proud to live in a World Heritage Site; I often heard a sense of pride and responsibility as people told me that they felt privileged to live in the archipelago. Yet as one fisher told me, island residents did not ask for the Galápagos to be a world patrimony, the people were not consulted, and it is now the people who are blamed for the problems.

The SES model and participatory workshops on park management plans were designed to change this by bringing island residents into the project of conservation. Yet the resulting policy offered them one of two roles. Residents were either *insulares*, who brought continental lifestyles with a colonist mentality; or they were *isleños*, ecologically noble islanders, who could take on appropriate forms of caring for and studying iconic species, and, like Don Michuy or Fredy Cabrera, thus benefit both themselves and the wider project of conservation.[60] These are deeply moral and normative visions that seek to govern how island residents should think and behave.[61] Like the natural laboratory model, this new model also produces a terrain of inclusions and exclusions through its narrowly proscribed vision of who belongs in the Galápagos and who does not. It creates another line ordering relations between society and nature.

The SES model's normative choice between *insular* and *isleño* identities does not take into account the histories of struggle behind the redrawing of this line. The national park's strategies for managing the "proper" relationship among humans and nature have changed considerably over the past fifty years, but change has not been a straightforward progression from fortress conservation models to the kind of collaborative care work on display at Campo Duro. Instead, changes have occurred unevenly in response to local contestations, shifting national policy, and evolving conservation philosophies. Although recent strategies have argued for more inclusive visions for conservation, memories of these conflicts live on, shaping the attitudes people bring to possible collaborations. The conservationists who argued for removing people from the islands have since retired, but their proposals were made in the living memory of many island residents. So too are there conservationists still working in the islands who were held hostage by fishers or otherwise targeted in their protests. While the SES model is based on the premise that the fortress conservation model caused considerable social and ecological problems in the archipelago, the framework does not offer a way of addressing these histories. It misses the point of what was at stake.

Nor do the roles from the SES model bear much in relation to the lived experiences and cultural identities of island residents. The *colono* identity, for example, is celebrated in the islands. The local newspaper on Santa Cruz is named *El Colono* in homage to the difficult lives of early settlers and in recognition that everyone who lives in the islands, like the plants and animals, is a colonist. Many of these colonists would call themselves *galapagueños*, like the Isabela tourism official I met who said she was "*autentica galapagueña*" and could trace her ancestry back to Antonio Gil, the founder of Puerto Villamil. Others, mostly old-timers now, might call themselves *carapachudos*—those who, like the giant tortoises, developed thick shells as they adapted to the harsh conditions of island life. But being *galapagueño* in this sense, even identifying with the tortoises, has little to do with conservationist ethics. Instead, being *galapagueño* is a marker of class status.[62] Since the Special Law of 1998 the term has distinguished those with long ties to the archipelago from recent migrants. The law limited migration and put in place residency-based rights to property ownership. Residency is a valuable commodity that distinguishes the well-established from migrants who are not as socially or economically privileged and are often disparaged by both xenophobic island residents and conservation discourses that fault their presence and their links to their sending communities.[63] Galápagos culture is shaped by numerous social, political, racial, and class divisions—among them fractures and alliances between foreign conservationists and local residents, park guards and those they police, scientists and politicians, *galapagueño* residents and new migrants, Ecuadorians and white *afuereños* (foreigners), those who work in lucrative tourism jobs versus those in support industries.[64] As the multispecies transects show, the people who inhabit these overlapping subject positions have varied economic interests as well as particular understandings of place and nature based on their lived experiences that do not fit into binary models of idealized identity.[65] Struggles over conservation are not between those who destroy and those who conserve nature as the SES model would have it, but among populations who understand the value of nature very differently.[66]

Yet to focus on people's identities and behavior is to miss the fishers' point when they said conservation had to include people. The fishers were not making an argument about how people should act, but rather about how conservation should. By targeting giant tortoises, the fishers were targeting the idea that in the Galápagos it's tortoises all the way down—that the animals need be at the center of life. They challenged the settled answers to the questions of not only who belongs in the islands, but what is conserved and for whom.

The fishers pointed to a theme that came up repeatedly in conversations I had with many people in the archipelago. Fishers do not represent the majority of islanders; indeed, because they do not fit the dominant ideology, they are often stigmatized. Yet their concerns speak to issues of broad political importance. Battles over take quotas for sea cucumbers and lobsters are important questions of fisheries management, but they are also proxies for contestation of conservationist management that draw attention to the broader question of who thrives in the islands. Of core concern to many residents is the success of livelihood strategies and basic quality of life—from the quality of primary and secondary education to clean water and access to health care. In Puerto Ayora, for example, people told me there was no potable water system, inadequate sewerage, no good health care, no good education system—though the town was secure and tranquil, which helped to compensate for such deficiencies. Residents complained of a host of issues to be addressed, from the invasive species that are a typical concern of conservation to issues that are not, including alcoholism, drug abuse, and domestic violence. Water quality was a particular concern. In one awkward moment during an interview, a male German environmentalist warned me and the women I was with not to wash our "private parts" in the shower because the water quality was so bad. (It was good advice: I later read in the local paper that 80 percent of one doctor's cases involved treating female residents who had fungal skin infections.) Except for farmers in the highlands who collect rainwater, everyone in the islands drinks purified water bought on the private market. When I lived in apartments with-out access to rainwater, I learned, like local residents, to dunk my dishes in a solution of bleach after washing them under the tap. Residents' basic concerns about public services highlight stark disparities between the multimillion-dollar tourism and conservation industries and their own quality of life. What might conservation be if these issues were core objects of concern?

Over the past five years, concerns about conserving quality of life have given new dimension to the iconism of giant tortoises. In the summer of 2015, the animals again became part of protests. The national government had pro-posed constitutional reforms, including a new Special Law for the archipelago. The law would reduce the scope of participatory governance structures insti-tuted with the 1998 law and, among other things, phase out wage subsidies that helped offset the high cost of living in the islands. The changes were broadly unpopular in the Galápagos, made more so in the context of a sum-mer of food scarcity and high prices for imported goods because of problems with one of the cargo ships that served the archipelago. More troubling for one community organizer was that the law would effectively centralize state power over the archipelago through presidential political appointments. It had been orchestrated, she told me, by people in Quito who did not really know the Ga-lápagos but saw the archipelago as a place to extract wealth. The plans for one such scheme, the archipelago's first five-star resort, circulated that summer. The $26 million resort was to be a dream space of green luxury backed by a Guayaquil businessman, complete with spa, restaurants, meeting rooms, and a water-treatment facility, as well as fifty-one suites that would cost $800 to $1,000 a night. It would be built at Punta Carola, undeveloped beachfront property on the edge of town on San Cristóbal, privatizing a popular swim-ming spot.[67] As the activist said, "Right now, I see the Galápagos not as a patrimony, not as nature, not as anything. Right now, I see it as a center of tourism where government people want to extract money, nothing more."[68]

The protests responded to these plans, pushing back against extractive cap-italism and moves to centralize control over the archipelago. During a nation-wide strike, crowds marching on Quito raised a giant papier-mâché tortoise in defense of the islands. On social media, an image circulated of the close-up face of a giant tortoise with stripes of paint under its eye in the yellow, blue, and red of the Ecuadorian flag. The tagline read *Fuerza Galápagos*—Galápagos Force. In these contexts the tortoises became a different kind of icon. They were a rallying cry against changes that residents saw as jeopardizing both their livelihoods and the well-being of island nature.[69] They were a call to protect the ability of both wildlife and islanders to thrive in this place that is their home. They were a reminder that the well-being of nature is fundamentally tied to issues of social and environmental justice for island residents.

Not everyone in the Galápagos cares about giant tortoises, either as conservation icons or as symbols of protest. This alternative tour could have featured many other stories of what it is like to live in a natural World Heritage Site. Yet tourism can be an imposition, and many people do not want to open their homes to visitors. I focused on a few of the ways that giant tortoises have become part of experiments in ordering nature and society because the animals are public icons. As charismatic species, giant tortoises are highly valued and sometimes hotly contested. Both materially and symbolically, they inhabit the boundaries that have attempted to distinguish the cultural from the natural, sometimes plowing across these divides themselves, and other times used as objects of transgression by others. Their lives, and their meaning, reflect entangled socio-natural histories of life in the archipelago. Giant tortoises may be living fossils, but they are not prehistoric.

8 · All the Way Down

In which the tortoises are not alone

After a life defined more by fame than by solitude, Lonesome George remains an icon of the Galápagos Islands. Shortly after he died in 2012, his frozen remains were sent to an expert taxidermist at the American Museum of Natural History in New York for preservation. Considerable debate ensued about where his prepped body should repose: Was he a global icon who should go on world tour, or a national symbol who should live in a Quito museum? Islanders, however, fought for him to come home to the Galápagos, and in 2017, after an exhibit at the museum in New York, he was reinstalled at the breeding center where he spent the last four decades of his life. He is now displayed under glass near his old corral, a figure who carries a lineage millions of years old into the future through memory and mourning. He is a metonym for the Galápagos, important in his own right, but more so because of the way he embodied struggles over what kind of place the archipelago is and should be.

As it turns out, though, Lonesome George may not actually be the last of the Pinta species. In 2008, the Galápagos National Park Service mounted an expedition to Volcán Wolf, at the northern end of Isabela, with a team of geneticists from Yale University. They were collecting data from this little-studied and not easily accessible tortoise population to compare with other species, allowing the geneticists to reconstruct past patterns of adaptive radiation as the tortoises dispersed from island to island. Over the past two decades, genetic analyses have revolutionized the study of the history of life, making it possible to peer into the past and see histories of evolution in ways Darwin might only have dreamed about. The Yale team's work has provided a new set of evidence that gives historical depth to the morphologically based taxonomic work of Van Denburgh, Rothschild, and other natural historians—and proves Darwin's hunch that the tortoises from different islands were distinct species. Geneticists have traced a version of the famous branching tree Darwin sketched as he developed his ideas on natural selection that explains the geography of how the giant tortoises dispersed from an initial founder that colonized San Cristóbal more than three million years ago to populate the rest of the islands.

But what conservationists and scientists found on Volcán Wolf was a surprise: several of the tortoises they encountered did not share the dome shape of the majority of the population but were smaller saddle-backs. The team took blood samples from these and other tortoises they found—sixteen hundred in all, which they estimate to be about 20 percent of the total number of tortoises living on the volcano. Back in New Haven, their analysis revealed that the saddle-backed tortoises were not entirely native to the volcano. They had been born there but were hybrid crosses between the native species (*C. becki*) and species from other islands. Seventeen of these hybrid tortoises were Pinta crosses; eighty-four were Floreana hybrids.[1] These were living tortoises from two extinct species, a finding trumpeted in the scientific and popular press. By all accounts, the Floreana tortoises had become extinct in the decade after Darwin's visit in 1835, the last animals presumably casualties of colonists at Asilo de Paz; the Pinta tortoises, with the exception of George, had died off in the mid-twentieth century after centuries of predation, the last individuals falling into crevasses in the lava or perhaps killed by fishers.[2]

The hybrid tortoises did not fit the timeline of evolutionary dispersal by which a founding tortoise, or tortoises, arrived on each of the islands hundreds of thousands of years ago. The genetic maps of species history estimated that the Floreana species was some 850,000 years old, the Pinta species 300,000 years old, and the Volcán Wolf population 280,000 years old. But tracing the crosses' lineages showed that their ancestors had arrived on Volcán Wolf fewer than three hundred years ago.[3] This was not what biologists consider a *natural* dispersal event. Instead, the tortoises more likely had been taken to the volcano by the very sailors from centuries past whose actions are so often lamented today for their exploitation of these animals. Captain David Porter, who in 1822 had compared the tortoises to the "finest veal," told of English ships he captured near Banks Bay on northern Isabela, a popular whaling grounds. Before being captured, the ships' crews had, he regretted, thrown the tortoises they had collected from Santiago overboard to clear the decks for action. "But a few days afterwards, at daylight in the morning,"

Porter found tortoises bobbing in the sea. "We were so fortunate as to find ourselves surrounded by about fifty of them, which were picked up and brought on board, as they had been lying in the same place where they had been thrown over."[4] Porter thought the tortoises unable to swim, but under similar circumstances some animals might have managed to get ashore before being claimed by ships or the depths of the Pacific.

For conservationists who have long strived to maintain the isolation of island populations and the purity of species, the hybrid tortoises were unnatural mixes of species that had evolved on different islands—the product of just the kind of human disturbance they tried to prevent. But the sailors who moved the animals from island to island inadvertently saved the species. The continued existence of these species is a product of the very socio-natural history that nearly caused their extinction. Tracing the genetic footprints of these hybrid tortoises revealed that many of them were directly descended from a "pure" parent from Floreana or Pinta—animals who could well still be alive.

In 2015, a team of park guards and scientists returned to Volcán Wolf and spent ten days searching for the Floreana and Pinta tortoises. Nearly 150 of the 1,300 tortoises they found had the revealing saddle-back shape. They chose 32 of these to take back to the breeding center where they would form the core of a captive breeding program to revive the lost species. Each of these special tortoises, most of which turned out to be Floreana crosses, was air-lifted off the volcano by helicopter to the national park's work ship and then taken to Santa Cruz, where they joined the Floreana tortoises whose eggs I had tossed away. They would be put to work much like Diego and the Española females to revive the lost population, albeit this time with more attention to genetic diversity. Much to everyone's disappointment, however, none of the tortoises the team encountered in 2015 were from Pinta. George's kin had, as he had for so long, evaded capture.

Like George, these wild hybrid tortoises stymied conservationist attempts to restore the natural laboratory. The animals are a reminder that though the islands have been called a laboratory for everything from evolution to ecotourism to climate change, they would be better understood by a different metaphor. Rather than see the islands as an experimental space—with the lab metaphor's implicit connotations of isolation, control, and a scientific manager—the Galápagos Islands are more aptly described by drawing on one of Darwin's own metaphors. He closed the *Origin of Species* with what has become a famous paragraph. "It is interesting to contemplate an entangled bank," he wrote, "clothed with many plants of many kinds, with birds singing on the bushes, with various insects flitting about, and with worms crawling through the damp earth, and to reflect that these elaborately constructed forms, so different from each other, and dependent on each other in so complex a manner, have all been produced by laws acting around us." It is elegant prose that captures the nature of evolution. "There is grandeur in this view of life," he continued, that "from so simple a beginning endless forms most beautiful and most wonderful have been, and are being, evolved."[5] This grandeur is precisely what draws so many people to the Galápagos. But what is missing from the quotation might be the key lesson that the history of the Galápagos can offer us about the nature of life today: the role humans play in shaping this entangled archipelago.

Evolution in the Galápagos archipelago is an entanglement of tortoises, cacti, finches, and ticks, as it is of tectonic plates, volcanoes, and ocean currents. But so too are these forms entangled with histories of exploration and tortoise hunting, with nation-state development and Crusoe-type escapes, with natural history collection and conservation work, with nature tourism and struggles for development and a good life. More so than the geographical imagination of *islands,* the archipelago metaphor—with its emphasis on flows rather than isolation, mobility over disconnection, movement not stasis—allows us to situate the Galápagos in the evolving history of the modern world, understanding how these forces too, and the political struggles that animate them, shape endless forms.

This is the lesson of our journey through the stack of tortoises—the ground is never as solid as we might like to believe. Accepting this means changing our expectations of what the Galápagos should be. The socio-ecological systems model that Galápagos conservationists proposed a decade ago makes headway toward reimagining the archipelago, but its framework is limited by an approach to history as a series of feedback loops that are more at home in ecological theory than in social analysis. It is worth returning to these loops to show how we might think differently about the implications of the past if we do not begin with the assumption of a fall from Eden.

Like most analyses of the social history of the Galápagos, the SES model begins with Tomás de Berlanga and the Western discovery of the archipelago (see fig. 7.1). It is a historical moment that pinpoints the beginning of a humanized world with European presence—discounting the existence, experiences, and views of indigenous peoples who likely made earlier journeys to the archipelago. The implication of beginning with Berlanga, as the figure's downward arrow shows, is that it was all downhill from that point. Choosing this date as the beginning of history sets up a fundamental divide between evolutionary history and social history that posits a prehistoric world of nature as the stage for a history in which all future human actors—regardless of whether they are extractivists, conservationists, tourists, or residents—are a foreign presence. The material effects of this history, the authors of the SES model note, were degrading: the increasing scarcity of natural resources and the introduction of foreign species. While these problems are real, this approach sets up the conservationist dream of returning to a prehistoric Eden, as if, with enough effort and funding, it might be possible to claw our way back up this ladder of history. But history is not a timeline of events, and starting points are not a question of what came first; rather, they are weighted choices that shape the perspective of the analysis that follows.

Letting go of goals to restore the past does not mean ignoring history and turning our vision squarely toward a "novel" future. Such a move would be equally shortsighted. Instead, conservation must reckon with its past, reckon with the particular patterns of entanglement that have made the Galápagos Islands what they are today. This includes histories of European exploration and exploitation that are replayed and revised in nature tourism, as well as the practices of natural history that stripped species of their social and ecological contexts and in doing so shaped understandings of the archipelago as an evolutionary laboratory where nature could be protected from society.

"The past," as William Faulkner famously wrote, "is never dead. It is not even past."[6] In the Galápagos the past is not only alive in our nostalgia, but also is embodied in the giant tortoises and other creatures, ourselves included, through long lineages of evolution. It is evident in the islands' entangled landscapes—where Scalesia and Cedrela shade the same forests, where giant tortoises visit farmland to eat cattle grass and guava, and where sea lions and sea turtles swim alongside snorkelers. Some of these landscapes are more wild, less developed, than others. But they have all changed over the course of the past five hundred years. This is true not only in a material sense, but also in an imaginative one. The long arrow of history and the stacking loops of the SES diagram make it look as if histories of exploration were a long time ago, but these are the very histories that put the Galápagos Islands on the map and that have conditioned Western ways of engaging them ever since. This extractivism was fueled, in part, by the chronicles of European conquistadors, explorers, and naturalists like William Dampier and James Colnett who wrote with awe about the "new world," often painting Latin America, with its strange species and dense jungles, as a world of abundant, unclaimed nature, ripe for the picking.[7] This writing sowed the seeds of both natural history exploration and the tourism economy. The geographical imagination of the remote islands of tortoises as a space of adventure and destination for Europeans and North Americans emerged during this period. So too did the lasting idea that the islands are a storehouse of nature, whether we understand that nature as a resource for a ship's larder or as biodiversity to be protected.

It is essential to recognize the historical production of nature through particular articulations of history and biology. Evolution is not an apolitical, pre-social foundation for the set of epistemological and moral stances that power conservation. Goals to save the islands as a Darwinian museum are an artifact of imperial histories. We cannot selectively remember this past. We cannot memorialize heroes without taking along with them the conditions in which they produced knowledge and how those conditions—like the hunt for natural resources and colonial-era imperialism—remain very much entangled with desire to save evolution in a laboratory.

The SES model marks a second period of Galápagos history as beginning when Ecuador claimed the archipelago in 1832 and colonies transformed the highlands of four islands through agriculture. This is the upswing of the loop—an increase in human presence that caused a downward slide of environmental ruin, prompting foreign scientists to step in. But it was not only colonists who practiced extraction during this period, but natural historians as well. It was only through such practices that they came to understand the tortoises as endemic and endangered; this period is as much one of colonization as it was of the development of the evolutionary understanding of life that has powered conservation in the archipelago. We cannot trace a simple line from Darwin's visit through subsequent histories of collection and classification to the conservation campaigns of the 1930s and 1950s. This suggests that science is itself a progressive, evolving force as well as a monolithic rationality that offers one true take on the nature of life. Instead it is essential that we recognize that changing scientific practices and research questions make nature knowable in particular ways and have tangible political and environmental consequences.

The idea that the Galápagos archipelago is a natural laboratory was just as much a political assertion as a description of the islands' scientific utility. In this laboratory, nature was to be protected according to scientific guidance provided by experts—a technocratic vision for management on which postwar institutions were premised and that powerfully shaped the region during the twentieth century. Evolutionary understandings of life defined a conservationist biopower that identified which species belonged in the natural laboratory and which did not. This perspective reflects the history of isolation that makes island wildlife populations ecologically fragile—easily susceptible to diseases introduced from the outside world and at risk of being outcompeted by foreign species, such as guava or goats that have taken to island life, becoming aggressive invaders that crowd out native species. This understanding informed the desire to preserve nature in accordance with its evolutionary history, or rather, as an artifact of pure nature isolated from human disturbances and the species they introduced.

Imaginations of the Galápagos as a natural laboratory reflect defining characteristics of mid-twentieth-century environmentalism, including reverence for a scientific luminary and a desire to protect the archipelago as the site of an evolutionary origin story, the militarized US science of the postwar period, and the power of neocolonial transnational institutions of environmental governance that emerged at that time. Scientists' desire to protect the archipelago as a natural laboratory reveals a paradox of modern thought: trying to save a place protected beyond the reach of modernity under the watchful eye of scientists— the quintessential agents of modernity—was a project itself steeped in modern logics. Twentieth-century conservation was to put an end to histories of exploitative extraction in the Galápagos Islands and protect them from further degradation. Yet it too ushered in new kinds of extractivism. The creation of the Galápagos National Park was dependent on selling the opportunity to see "Darwin's islands" to a new market of nature tourists in the 1960s and 1970s. This economic approach to conservation too was steeped in modern logics, and continues to be central to conservation strategies today.

Yet we cannot analyze the history of the Galápagos as a progression from centuries of degradation to a period of conservation inspired by Darwin's insights. The history of tortoises is not a progression from use as food and taxonomic specimens to endangered species and tourism icons. Although feasting on tortoises is now prohibited, tourists' aesthetic consumption of the animals bears more in common with these past histories than conservationists would like to believe. The consumption of the tortoises as food, for example, is not a thing of the past, despite conservationists' efforts. And while conservation has saved the species from extinction, it also has fueled a market for selling them that has become more successful than any previous attempt to profit from the islands' natural resources. Rather, the animals' meanings and use values have become layered, much like the rings on their carapaces. So too have these different histories changed the species themselves. The tortoises are not stable entities understood differently through different lenses—be they extractive, scientific, conservationist, or touristic. They are not abstractions. Instead, the animals, the species, are remade as they become part of different world-making projects in the archipelago.[8]

The Galápagos Islands now face the limitations of conservation that has sought to preserve them as museums to Darwin and focused primarily on saving iconic species. Lonesome George was the poster tortoise for this vision. He is now a monument to conservation's past. His death marked the end not of the goal of saving species, but of doing so according to a logic of returning to a more pure past world.

The Galápagos archipelago remains an important microcosm for understanding life. It offers us not a space of pristine isolation, but rather a place in which to think through a new kind of conservation in a world that is not Edenic. The rhetoric of island laboratories has allowed conservation to slip into a dream world in which it is possible to control nature, to keep social life separated from the natural world, to eradicate foreign species, and to control the actions of all visitors. But conservationists need to be aware that their moral imperatives are not universal and the projects they use to achieve them run up against the lifeworlds of others. The major oversight of mid-twentieth-century conservationists was that they failed to allow for the continued existence of the Galápagos human society, the members of which did not understand themselves to be living in a laboratory.

There still are, and will remain, many worlds in the Galápagos. Conservation must ask the difficult questions of what natures and cultures should be conserved, by whom, and for whom. Too often, the proponents of conservation have assumed a unitary vision of the answers to these questions. Whether we conserve tortoises for the sake of nature itself or as umbrellas for saving wider ecosystems so that they might continue to evolve, we are making weighted decisions about what kinds of life matter in the Galápagos. No matter how benevolent we imagine ourselves, how worthy our cause, there is a politics of race and knowledge behind such decisions and assertions of how best to manage nature that must be addressed.

Who should be able to enjoy the Galápagos? Global tourists? Future biologists and ecologists? South Americans now make up nearly a third of visitors to the Galápagos archipelago. Plans to restrict tourism to luxury "green" hotels or to dramatically raise park entrance fees would reduce access to the archipelago for all but the most elite. What of the people who live in the archipelago and who cannot, as one complained to me, even walk down the beach at night at Tortuga Bay to look at the stars? Who benefits from tourism? Market-based mechanisms are not just or effective methods of answering this question. Tourism will continue to be the lifeblood of the Galápagos economy, but policy makers and visitors alike must address the class and racial inequalities of the industry, as well as its environmental consequences. That Galápagos residents are able to earn a decent living through tourism and support industries is essential to the future vitality of the archipelago. People cannot be considered, nor consider themselves, invasive species and at the same time develop a sense of responsibility and care for the place they live.

It is not giant tortoises, but entanglement, all the way down in the Galápagos. The future of conservation rests on recognizing that the fate of endangered species is tied to that of the people who live with them. Yet recognizing this entanglement is only a first step toward reckoning with the histories that have reshaped the Galápagos over the past five hundred years. Tracking tortoise histories shows us that we cannot proceed from a starting point of believing nature is, or should be, separate from human society. Both literally and figuratively, the tortoises heave their weighty bodies across this divide, demonstrating the inseparability of natures and cultures. Life, evolution, flows in ways that are both conditioned by the past and unpredictable, without goal or destination. Evolution is not so much a search for origins, as philosopher Elizabeth Grosz reminds us, but the "elaboration of difference."[9] Conserving life, then, cannot be driven by nostalgia for an inaccessible past world, but must be inspired by hope for a future that is as diverse, as resplendent, and as just as possible.

NOTES

Chapter 1. What We Stand On

1. Modified from the version told by John R. Ross in *Constraints on Variables in Syntax* (1967), via Wikipedia (https://en.wikipedia.org/wiki/Turtles_all_the_way_down).

2. Stephen Hawking, *Brief History*, 1.

3. Animal histories have reached a critical mass over the past decade. Although most of us do not often recognize just how central animals are to our lives, their presence has fundamentally shaped human history. As historian Erica Fudge has written, "the history of animals is a necessary part of our reconceptualization of ourselves as human" ("Left-Handed Blow," 5). This is as true in places like the Galápagos that are known for their wildlife as it is in urban spaces or their rural hinterlands. By investigating relationships among humans and animals, this work collectively has several aims, including challenging anthropocentrism, denaturalizing the supposed divide that separates humans and animals, and examining how humans have exploited other species in the past, often from the perspective of environmentalism or animal welfare. Despite attempts to center animals in historical analysis, they often remain marginalized even in texts in which they are framed as both actors and objects. This is often because of the difficulty of tracing them in historical documents as well as the persistent dominance of anthropocentric understandings of what counts as historical agency. See Wolch and Emel, *Animal Geographies;* Mitchell, "Can the Mosquito Speak?"; Ritvo, "Animal Planet"; Brantz, ed., *Beastly Natures;* Shaw, ed., "Way with Animals"; and Few and Tortorici, eds., *Centering Animals.*

4. The pre-Columbian history of the Galápagos is uncertain. Pedro Sarmiento de Gamboa (*History of the Incas*) wrote that the Galápagos, which he claimed to have discovered in 1567, had previously been visited by Tupac Inca, who called the islands Avachumbi and Ninachumbi (*chumbi* meaning fire) and returned after nine months or

a year with "black people, gold, a chair of brass, and the skin and jawbone of a horse" (136). That the Tupac Inca could have acquired these items in the Galápagos seems doubtful, thus the location of Avachumbi and Ninachumbi has been debated. In the 1950s, explorer Thor Heyerdahl found pottery shards in the archipelago he dated to the pre-Columbian period, although his findings also have been debated. See Heyerdahl and Skjølsvold, "Archaeological Evidence."

5. Darwin, *Journal of Researches*, 378. The phrase was originally John Herschel's. On the tortoises and Pleistocene extinctions, see Auffenberg, "Checklist."

6. Darwin, *On the Origin of Species*, 107.

7. Dorst, "Where Time Stood Still," 30, 32.

8. Darwin, *Journal of Researches*, 377.

9. The phrase "Darwin's finches" was coined by ornithologist Percy Lowe ("Finches") in 1936 and became well known as the title for David Lack's 1947 book *Darwin's Finches*. See Sulloway, "Darwin and His Finches"; "flywheel of evolution" is from Quammen, *Song of the Dodo*, 128. Also see Weiner, *Beak of the Finch;* and Grant and Grant, "Unpredictable Evolution."

10. Simkin, "Geology of Galapagos"; Geist, "On the Emergence."

11. Jackson, *Galapagos*.

12. Cayot, interview with the author, 28 September 2011.

13. Bensted-Smith, ed., *Biodiversity Vision*, 1. Conservationists are a heterogeneous group with various backgrounds and goals. I primarily use the term to refer to the predominately white, northern biologists and ecologists who work in the archipelago and refer to others who work on conservation—such as Ecuadorian scientists, politicians, and national park guards—in more specific terms. In a broad sense, however, "conservationist" is an inclusive category that covers scientists, policymakers, and park guards as well as many island residents and visitors—a diversity that reflects productive tensions within the field.

14. Biology students are taught not to anthropomorphize, that is, ascribe human rationality or emotions to the animals they study. This reminds them of the difference between humans and nonhumans, how humans and animals inhabit and perceive the world differently, and how we humans cannot fully know what precisely is going on behind animals' eyes. But not anthropomorphizing animals can be quite difficult. In trying to understand animals and their behavior, people become emotionally attached to animals, sometimes caring for them, sometimes vilifying them. The conservation enterprise organized around saving endangered charismatic wildlife turns on anthropomorphism, even as it seeks to enroll people to care for wildlife as beings separate from human worlds. In this book, I do not seek to avoid anthropomorphism, but rather to understand the work it does as a tool for crossing species divides, enticing people to care about nature in a particular way, and revealing the broader assumptions about relationships between humans and animals. I refer to the tortoises and other animals as "who" rather than "that" not to personify them, or to make them over in a human

image, but to acknowledge that they are conscious, animate beings and thus actors in the world. On anthropomorphism, see Daston and Mitman, *Thinking with Animals*.

15. Sixty percent of the 168 endemic plant species in the islands are thought to be threatened. A recent count recorded 463 alien insect species—186 more than an inventory eight years before. Tye, "La flora endémica"; Causton et al., "Alien Insects." Figures vary: Gardener, Atkinson, and Rentería, "Eradications and People," estimated that 55–60 percent of the 194 endemic plant species were threatened by goats.

16. Merlen, *Restoring the Tortoise Dynasty*.

17. George had long been overweight and had previously given keepers a scare when he gorged on cacti that fell over in his corral to the point of lethargy. His death, though, was apparently of natural causes, although necropsy revealed that his liver and kidneys were unhealthy. See Nicholls, "Legacy of Lonesome George."

18. Stengers, *Power and Invention*, 60–61.

19. "Biological and geological Eden" is from Dawkins, "Foreword," 6. In a journal, Darwin wrote that he had been "greatly struck from about month of previous March on character of S. American fossils—& species on Galapagos Archipelago.—These facts origin (especially latter) of all my views" (Journal, July 1837); see also Browne, *Charles Darwin Vol. I*; Sulloway, "Darwin and the Galapagos."

20. *Galápagos: The Islands That Changed the World*.

21. Dorst, "Problems of Conservation," 2.

22. Robbins and Moore, "Ecological Anxiety Disorder," 4.

23. Leading conservation biologist Michale Soulé has called the field a "crisis discipline" ("What Is Conservation Biology?"); on ontological divisions and crisis, see Latour, *We Have Never Been Modern*; on crisis and conservation, also see Lowe, *Wild Profusion*.

24. Dorst, "Where Time Stood Still"; on dynamic nature, see, for example, Bennett, *Vibrant Matter*; Massey, *For Space*. On "anachronistic space," see McClintock, *Imperial Leather*, 36; and Fabian, *Time and the Other*.

25. *Galápagos: The Islands That Changed the World*.

26. On environmentalist imaginations of Eden, see Merchant, *Reinventing Eden*.

27. Gregory, *Geographical Imaginations*; as the geographer Jesse Gilley has written, geographical imaginations can "justify actions that make reality conform to what is seen in the imagination, linking the 'real' to the planned, the hoped for, expected, and fantasized." Gilley, "Geographical Imagination," 1222.

28. Baldacchino, "Islands as Novelty Sites"; Deleuze, *Desert Islands*; Grove, *Green Imperialism*; Wallace, *Island Life*; MacArthur and Wilson, *Theory of Island Biogeography*; DeLoughrey, "Myth of Isolates."

29. Islands are a kind of heterotopia, Michel Foucault's term for spaces that are "other": "as perfect, as meticulous, as well arranged as ours is messy, ill-constructed, and jumbled." Foucault's key insight about heterotopias was that by ostensibly sitting outside of modern life, the ordered inverse of a chaotic world, these spaces are actually revealing

sites for reflecting on the organization of modernity. Like Foucault's examples—prisons, cemeteries, museums, and ships—the ideal of a pristine island laboratory offers a space of controlled order and pure nature. Foucault, "Of Other Spaces," 27.

30. See Cronon, "Trouble with Wilderness."

31. Debates about the concept of the Anthropocene and the politics of the term have proliferated in the earth sciences and particularly in the environmental humanities since Paul Crutzen and Eugene Stoermer published their short introductory essay in 2000; indeed, they are the focus of a course I teach. In using the term here, I do not intend to subscribe to the Anthropocene Working Group's identification of the 1950s as the beginning of a geological epoch shaped by a universal human species. Instead, I think that if anything, the Galápagos case points to an age of entanglement among unevenly powerful human populations and multiple forms of natural life wrought through changing histories of European exploration and conquest since the sixteenth century as well as through evolving forms of global capitalism. This approach is more in line with Donna Haraway's concept of the Chthulucene, though I do not follow her resulting concern about overpopulation or science fantasy of ideal forms of human-nonhuman coevolution. The Galápagos Islands have long been ruled by tacit environmentalist desire to limit population growth, with adverse effects for both people and nature, and people there live a daily coexistence with nonhuman species—from giant tortoises to Dengue-carrying mosquitoes. See Haraway, *Staying with the Trouble.* For a critical take on the Anthropocene more broadly, see Bonneuil and Fressoz, *Shock of the Anthropocene.*

32. For philosopher Gilles Deleuze, archipelagos are models of "a world in process" (quoted in Pugh, "Island Movements," 10). Lanny Thompson ("Heuristic Geographies") writes that "archipelagoes are not natural phenomena but rather spatial and historical configurations assembled and reconfigured, shaped largely by imperial and postcolonial processes" (70); see also Massey, *For Space.*

33. Marris, *Rambunctious Garden*; Monbiot, *Feral*; Lorimer, *Wildlife in the Anthropocene.*

34. Stengers, *Power and Invention*, 65.

35. All quotations in this paragraph are from Geertz, "Thick Description," 28, 29.

36. Grenier, *Conservación*; Star and Griesemer ("Institutional Ecology") defined "boundary objects" as things with some degree of stability that travel through various social worlds and in doing so provoke novel articulations between diverse actors; also Fujimura, "Crafting Science." On ecologists' debates about the use of flagship species, see Entwistle, "Flagships for the Future?," 239; Caro and O'Doherty, "On the Use"; Berger, "Population Constraints"; Simberloff, "Flagships"; Walpole and Leader-Williams, "Tourism and Flagship Species."

37. Political ecology and multispecies studies are distinct fields with much to say to one another. Political ecology is strong in explaining how political economic relations shaped by histories of colonialism and imperialism produce particular natures, for

example, through processes of commodification, and how these productions come to be at the heart of conflicts over access to resources. Multispecies studies attend more to the affective, ethical attachments that people have with the natural world and how the processes of relating remake both human and nonhuman nature. Much of the focus of multispecies studies aims to displace the human exceptionalism and anthropocentrism of much of the social sciences and humanities. I follow this work in arguing that animals are not only support for human life or part of the natural backdrop in which we humans live. But because this book is set in a place that has for centuries been valued more for its nonhuman species than for its humans, I seek to use multispecies studies to an opposite aim: to assert that conservation biology must take seriously the lives of the many humans who are entangled with the wildlife it privileges. For an introduction to political ecology, see Robbins, *Political Ecology*. On multispecies work, see Kirksey and Helmreich, "Emergence"; and van Dooren, Kirksey, and Münster, "Multispecies Studies."

38. Latour, *Politics of Nature;* Castree, *Nature;* Whatmore, *Hybrid Geographies.* Instead of a dualistic approach characteristic of modern understandings of a world bifurcated into discrete, interacting realms of Nature and Culture, geographers have theorized a relational understanding of the coproduction of nature and society. For Stephen Hinchliffe, nature is best understood as "enacted": "nature and society make one another (so thus are not independent), but are not necessarily reducible to one another (so thus are not strictly dependent)." Neither nature nor culture can be taken as pregiven, determining forces but emerge in relation to each other. Hinchliffe, *Geographies of Nature,* 9; Andean and Amazonian anthropologists have been leaders in rethinking Nature/Culture dualism through attention to indigenous cosmologies. See in particular de la Cadena, *Earth Beings;* Viveiros de Castro, "Perspectival Anthropology"; Blaser, "Threat of the Yrmo."

39. Bustamante, *Historia de la Conservación Ambiental.*

40. Neumann, *Imposing Wilderness,* 18. Conservation nongovernmental organizations have long been central to the process of value production through tourism. As Dan Brockington and Katherine Scholfield write, they "forg[e] the conditions, discursively and materially, for capital to appropriate aspects or parts of wildlife and nature which had escaped being turned into commodities. In part this is achieved through legitimizing visions of . . . landscapes and wildlife and specific types of nature production." Brockington and Scholfield, "Conservationist Mode," 552. Also see Smith, *Uneven Development;* Brockington, Duffy, and Igoe, *Nature Unbound;* Lowe, *Wild Profusion,* 161; Braun, *Intemperate Rainforest;* and Rutherford, *Governing the Wild.*

41. Taylor, Hardner, and Steward, "Ecotourism and Economic Growth."

42. Quiroga, "Crafting Nature," 129; also Grenier, *Conservación.*

43. Bassett, *Galapagos at the Crossroads;* Watkins and Cruz, "Galápagos at Risk."

44. My historical methodology is inspired by what Hugh Raffles and Jake Kosek have called critical natural history (Raffles, *In Amazonia;* Kosek, *Understories*). I

think of this book as a "history of the present" that uses Foucauldian genealogy and work in historical political ecology. See in particular Davis, *Resurrecting the Granary*; Davis, "Historical Approaches"; and Kull, *Isle of Fire*. Also see Hennessy, "III: Ecological Restoration."

45. Quiroga, "Crafting Nature," 133.

46. "Galapagos Tortoises Held Hostage."

47. Emotional affect is a central strategy of raising awareness and funds for conservation and nature tourism, but affective relationships are always situated in political (and economic) contexts—there is nothing innate about the nature of tortoises or other species that determines human responses to them. Lorimer, "Nonhuman Charisma."

48. Thompson, "When Elephants"; Mitman, "Pachyderm Personalities"; Wolch and Emel, *Animal Geographies*.

49. Tsing, *Friction*.

50. van Dooren, *Flight Ways*, 274.

51. Law, "Enacting Naturecultures"; Latour, *We Have Never Been Modern*; Mol, *Body Multiple*.

52. Nicholls, *Lonesome George*, xviii.

53. Russell, *Evolutionary History*.

54. Here I am paraphrasing Raffles, " 'Local Theory,' " 323.

Chapter 2.　In Darwin's Footsteps

1. The national park service regulates the eighty ships that continually circle the archipelago, all of which follow the same basic itinerary, with preappointed windows for their stops at park-approved visitor sites. The park service does this so that the 150 visitor sites it manages are not overwhelmed—it benefits both island ecosystems and tourists who want to believe they are wholly immersed in a timeless world of pristine nature. Even scientists who might be conducting research nearby are told to stay away from visitor sites and out of sight of tourists. This is a kind of "virtualism": nature tourism is designed to help preserve ecosystems, but also to create a landscape that fits idealized imaginations of nature. Carrier and Macleod, "Bursting the Bubble"; West and Carrier, "Ecotourism and Authenticity"; Honey, "Galápagos Islands"; Britton, "Tourism, Capital, and Place."

2. Grant and Estes, *Darwin in Galápagos*, 16.

3. Darwin, *Journal of Researches*, 384.

4. Keynes, ed., *Charles Darwin's* Beagle *Diary*, 364.

5. Colnett, *Voyage*, 157–158.

6. Berlanga, "Carta a Su Majestad," 540. My translation.

7. Dampier, *Voyage*, 101–102.

8. Davis, *Nimrod of the Sea*, 220.

9. It was Colnett's second fur trading trip. During his first, in 1789, he became instrumental in the disintegration of the "Spanish Lake" when he sailed into Nootka Sound near what is now Vancouver with supplies and Chinese laborers to build a permanent British trading base. But in doing so, he tread into recently claimed Spanish territory and was taken prisoner and held for months at the Spanish naval base at San Blas, Mexico—an episode that brought Britain and Spain to the brink of war. Rights to trade in the Pacific Northwest were eventually settled by conventions that prevented either power from establishing a formal colony on the coast. See Buschmann, Slack, and Tueller, *Navigating the Spanish Lake*, 21; and Gough, "James Colnett."

10. Colnett, *Voyage*, 157–158.

11. Carter, *Road to Botany Bay*, 9.

12. Carter, *Road to Botany Bay*, xxi.

13. Geographer Doreen Massey (*For Space*) called these "traveling imaginations"; also Stepan, *Picturing Tropical Nature*, 14.

14. Pratt, *Imperial Eyes*, 3.

15. Keynes, ed., *Charles Darwin's* Beagle *Diary*, 362.

16. Colnett, *Voyage*, 156. Colnett's account often references those of previous buccaneers and Spanish sailors, which he relied on to find his way through the archipelago, particularly to find places they noted as having freshwater. But although he trusted buccaneers' narratives, Colnett was "not in the habit of giving an implicit faith to Spanish accounts" after his experience at San Blas (ibid., 60).

17. Hall, *Memoir*, 143. Hall stopped in the Galápagos to take measurements along the equator of the earth's magnetism with an invariable pendulum.

18. Hall, *Memoir*, 32.

19. The *Beagle* carried twenty-two chronometers to be sure the crew had a reliable time standard against which to measure their position. Before that, gauging latitude could be done with relative accuracy by using one's naked eyes, an astrolabe, or a quadrant to judge a ship's position in relation to the sun or pole star. It was far more difficult, though, to measure longitude with any accuracy on the basis of the position of the sun or stars. Sailors estimated distance by dead reckoning or using log-lines and sand clocks, tying a line to a log, throwing it overboard, and then measuring with the line how far the ship moved in a set amount of time. The instruments that made accurate chart-making possible—the telescope, a quadrant with a telescope attachment, a filar micrometer, and a pendulum for linear measurement—were inventions of the seventeenth century, but most ships would not have been equipped with them until much later. Fernandez-Armesto, "Maps and Exploration," 747.

20. Melville, *Encantadas*, 7–8.

21. Robert FitzRoy, quoted in Larson, *Evolution's Workshop*, 65.

22. Browne, "Biogeography and Empire"; Drayton, *Nature's Government*; Grove, *Green Imperialism*.

23. Clarke, "Perfectly Natural."

24. Sulloway, "Darwin's Conversion"; Sulloway, "Darwin and the Galápagos"; Sulloway, "Tantalizing Tortoises"; Browne, *Charles Darwin, Vols. I* and *II*.

25. Larson, *Evolution's Workshop*, 9.

26. Melville, *Encantadas*, 6.

27. Melville, *Encantadas*, 9.

28. Jonik, "Melville's 'Permanent Riotocracy,' " 229.

29. Darwin, *Journal of Researches*, 373–374.

30. Melville, *Encantadas*, 87; Keynes, ed., *Charles Darwin's* Beagle *Diary*, 352–353.

31. Darwin, *Journal of Researches*, 374–375.

32. Keynes, ed., *Charles Darwin's* Beagle *Diary*, 25; Secord, "Discovery of a Vocation."

33. Melville, *Encantadas*, 5.

34. Byron, *Voyage of H.M.S. Blonde*, 91–92. On Darwin reading Byron, see Browne, *Charles Darwin, Vol. I*, 297. Byron, a captain in the Napoleonic wars, was in the Pacific to return to Hawai'i the bodies of two monarchs who died of measles while on a state visit to England.

35. It was Byron who had called them "imps of darkness" (*Voyage of H.M.S. Blonde*, 92); Keynes, ed., *Charles Darwin's* Beagle *Diary*, 353.

36. Darwin, *Journal of Researches*, 387.

37. Ecologists now think this behavior has more to do with iguanas' need to thermo-regulate than to escape predators.

38. Ritvo, *Platypus and the Mermaid*.

39. Rogers, *Cruising Voyage*, 191. Rogers remains most famous today as the captain who rescued marooned Scottish sailor Alexander Selkirk from Juan Fernández Island, west of the Chilean coast. Selkirk's experience of living on the island for four years is thought to have inspired Daniel Defoe's *Robinson Crusoe*, a novel that seeped into the imagination of young Charles Darwin.

40. Delano, *Narrative of the Voyages*, 376. Delano is best remembered today as the inspiration for the American captain in Melville's "Benito Cereno," a short story based on Delano's experience boarding a Spanish ship in distress that turns out to have been taken over by revolting slaves. See Grandin, *Empire of Necessity*.

41. Berlanga, "Carta a Su Majestad," 539. My translation.

42. Rogers, *Cruising Voyage*, 191. Rogers, though, estimated that most of the largest land turtles weighed only about one hundred pounds.

43. In the story, Melville (*Encantadas*, 15) recalls listening to the tortoises stomp around the deck from below as he tried to sleep: "Their stupidity or their resolution was so great, that they never went aside for any impediment." In the morning he rose to find that one tortoise had spent the night "butted like a battering-ram against the im-moveable foot of the fore-mast, and still striving, tooth and nail, to force the impossible passage." C. L. R. James reads this behavior, typical of the tortoises, as an allegory for "man struggling through the ages . . . hopelessly, blindly," in a failed civilization that is beyond repair. See James, *Mariners, Renegades, and Castaways*, 115–117.

44. Melville, *Encantadas*, 14.

45. Melville, *Encantadas*, 16.

46. Langdon, ed., *Where the Whalers Went*. According to one study of New England ships' logs, nearly two hundred US whalers alone visited the Galápagos during the nineteenth century.

47. Dolin, *Leviathan*, iii.

48. Dolin, *Leviathan*, 189.

49. Shoemaker, "Whalemeat," 278.

50. Shoemaker, "Whalemeat," 276.

51. Porter, *Journal*, 146.

52. Porter, *Journal*, 162.

53. Morrell, *Narrative*, 125.

54. David Porter quoted in Daughan, *Shining Sea*, 63–64.

55. Porter, *Journal*, 150. He was eventually captured by British warships.

56. Grant and Estes, *Darwin in Galápagos*, 113; Lawson was a Norwegian who had served in the Chilean navy before administering the colony Villamil established on Floreana. See Nordlohne, "The Seven Year Search."

57. Grant and Estes, *Darwin in Galápagos*, 117.

58. Charles H. Townsend—later a major figure in Galápagos tortoise conservation—estimated in 1908 that some twenty thousand Galápagos fur seals (an endemic species, *Arctocephalus galapagoensis*) had been taken between 1870 and 1882, when stocks were generally believed to have been already vastly depleted. See Townsend, "Fur Seals." Also see Soluri, "On Edge."

59. FitzRoy, *Narrative*, 487–488; Grant and Estes, *Darwin in Galápagos*, 98–99.

60. Darwin, *Galapagos Notebook*. Whalers rarely used the name "tortoise," instead calling the animals "terrapin" or, more colloquially, "turpin"—or various spellings of the word: "turpine," "tarpain," "turupin," or "terapen." See Townsend, *Galapagos Tortoises*, 69.

61. Camillas, "Gallipagos Islands." All further quotations from Camillas are from this account.

62. See Larson, *Evolution's Workshop*, 35–59.

63. Townsend, *Galapagos Tortoises*, 59.

64. Townsend, *Galapagos Tortoises*, 84–85.

65. Townsend, *Galapagos Tortoises*, 90.

66. Townsend, *Galapagos Tortoises*, 91.

67. Porter, *Journal*, 162.

68. Darwin, *Charles Darwin's Zoology*, 292.

69. Townsend, *Galapagos Tortoises*, 93.

70. Ferrin, "Place of Wild Tortoises."

71. Camillas, "Gallipagos Islands," 75.

72. Porter, *Journal*, 227.

73. Camillas, "Gallipagos Islands," 75.

74. Porter, *Journal*, 161.

75. Delano, *Narrative of the Voyages*, 378.

76. Delano, *Narrative of the Voyages*, 377.

77. Townsend, *Galapagos Tortoises*, 81, 70.

78. MacFarland, Villa, and Toro, "Galápagos Giant Tortoises."

79. FitzRoy, *Narrative*, 492.

80. FitzRoy, *Narrative*, 492.

81. Braun, *Intemperate Rainforest*.

82. Photographic technology is central to what John Urry calls the "tourist gaze," the visual consumption that is central to the tourist experience. Urry, *Tourist Gaze*.

83. In the case of giant tortoises, they are also about protecting the animals from upper respiratory tract infections.

84. Ferrin, "Place of Wild Tortoises"; Westwood, *Moon Ecuador*, 353.

Chapter 3. What's in a Name?

1. Darwin, *On the Origin of Species*, 107.

2. Darwin, *Journal of Researches*, 394.

3. Darwin, *Journal of Researches*, 394.

4. Other early synonyms included *Testudo californiana, Testudo nigra, Testudo nigrita,* and *Testudo elephantopus,* the latter of which was the only classification based on a type specimen from the Galápagos. Pritchard, "Galápagos Tortoises."

5. Sulloway, "Darwin and His Finches."

6. Darwin, *Journal of Researches*, 384–385.

7. For much of the twentieth century, taxonomists debated whether the island populations are each unique species or whether they are all subspecies belonging to one overarching Galápagos species. Today, geneticists consider the populations to be distinct species, which is the taxonomy I use. (Although genetic analyses have also further complicated tortoise taxonomy, identifying, for example not only one, but two distinct species on Santa Cruz Island.) The tortoises' genus classification has also been debated: in this chapter I use the bionomial names used by the naturalists themselves, including the genus name *Testudo;* in other chapters I use the current genus name *Chelonoidis.* Pritchard, "Galápagos Tortoises"; Poulakakis, Russello, et al., "Unraveling."

8. McOuat, "Species," 477.

9. In the terminology of science studies scholar Bruno Latour, the tortoises were "circulating references" brought to metropole "centers of calculation." See Latour, *Science in Action*.

10. Although they serve to represent a species, type specimens are not necessarily typical of a species' particular features. Ideally, they would clearly show defining characteristics, but this is not always the case. Many early classifications did not have multiple

specimens from a place to compare. Van Denburgh and Rothschild dismissed Quoy and Gaimond's names for Galápagos tortoises because their type was only ten inches long—thus a juvenile tortoise that did not show enough differentiation to be useful once museums had access to more specimens from other places.

11. James, "Collecting Evolution" and *Collecting Evolution*.

12. Van Denburgh, "Expedition," 323.

13. Journal of Joseph R. Slevin of expedition to the Galapagos Islands, Box 10, Joseph Richard Slevin Papers, MSS-429, California Academy of Sciences Library, San Francisco, CA (hereafter CAS Slevin Papers); Van Denburgh, "Expedition," 322.

14. Journal of Joseph R. Slevin; Alan Leviton, interview with the author, 11 February 2011.

15. Larsen, "Equipment," 358.

16. Daston and Galison, *Objectivity*, 19–20.

17. Latour, *Science in Action*, 215. Latour's focus on centers of empire, however, obscures the place-based knowledge of local residents and field scientists that was essential to the production of modern scientific knowledge.

18. Latour, *Science in Action*, 215; Drayton, *Nature's Government*.

19. Although many naturalists, Darwin and Slevin among them, crossed the divide between field collection and specimen-based analysis.

20. Poliquin, *Breathless Zoo*, 118.

21. No Galápagos tortoise, for example, has a nuchal scute—a small section of shell at the cervical spine—typical of most turtles. Van Denburgh, "Preliminary," 4; Van Denburgh, "Expedition," 321.

22. Porter, *Journal*, 151.

23. Porter, *Journal*, 215.

24. Van Denburgh, "Expedition," 260.

25. Linnaeus quoted in Poliquin, *Breathless Zoo*, 116.

26. Bruno Latour argues in *Laboratory Life* that it is by erasing the process of the construction of scientific facts that they seem to so closely match supposedly external natural objects: "The further temptation for the observer, once faced with one set of statements and one reality to which these statements correspond, is to marvel at the perfect match between the scientist's statement and the external reality. . . . To counter this possibility, we offer our observations of the way this kind of illusion is constructed within the laboratory. It is small wonder that the statements appear to match external entities so exactly: they are the same thing" (177). On natural history, see Pratt, *Imperial Eyes*, 5–7; and Tsing, *Friction*, 88–101.

27. Pratt, *Imperial Eyes*, 123–126; Tsing, *Friction*, 89; Raffles, *In Amazonia*, 139–146.

28. Raffles, *In Amazonia*, 135–136: "The extraction of insects from the forest and their reinvention of specimens in the collection demanded persistent, manufactured traces of locality as key components of value at every point. At the same time, scientific practice participated actively in a narrativizing of geography."

29. Günther quoted in Larson, *Evolution's Workshop*, 104.

30. Larson, *Evolution's Workshop*, 103.

31. Baur, "Galápagos," 418.

32. Darwin, "Letter no. 10819."

33. Günther, "Description," 258.

34. Günther quoted in Chambers, *Sheltered Life*, 137.

35. Cheke and Hume, *Lost Land of the Dodo*.

36. After what Lawson and Gould had told him, Darwin believed the islands were home to distinct species. He was further encouraged by a presentation he saw at the Zoological Society of London in 1837. French zoologist Gabriel Bibron discussed his taxonomic work on tortoises—two years previously, Bibron and his collaborator André Marie Constant Dumeril had also decided that the Indian Ocean and Galápagos tortoises were different species. Afterward, Darwin picked his brain about tortoises and was later excited to report that with additional specimens he had examined since his original paper, "The French Bibron says that two species of tortoise come from the Galápagos!!!" Bibron, though, had had few specimens to consult and was offering Darwin only his opinion, not a full scientific analysis.

37. See, for example, Delano's account (*Narrative*) of introducing Galápagos tortoises to Masafuera in 1801. Also see Garman, *Galapagos Tortoises*, on the role of sailors in transporting tortoises among islands.

38. FitzRoy, *Narrative*, 505; Günther, "Description," 253.

39. Darwin, *Journal of Researches*, 384–385.

40. Larson, *Evolution's Workshop*, 103; Darwin, *On the Origin of Species*, 107.

41. At the beginning of the nineteenth century, William Paley's *Natural Theology* (1802) made the case for the widely held "argument from design"—that the properties of nature were evidence of a divine designer.

42. Agassiz had been trained by the French naturalist Georges Cuvier before going to the United States, where he eventually founded the Museum of Comparative Zoology at Harvard. Like his teacher, Agassiz was a lifelong believer in special creation and denied evolution to his death. He thought that after every mass extinction event, God re-created species anew. Waggoner, "Agassiz."

43. Larson, *Evolution's Workshop*, 96.

44. The phrase is itself not old but was coined by John McPhee in 1981 in *Basin and Range*.

45. Rudwick, *Earth's Deep History*; Gould, *Time's Arrow*; Lyell, *Principles*.

46. Quoted in Sevilla, "Darwinians," 54. Recent research has demonstrated that the islands are fewer than five million years old. See Geist, "On the Emergence."

47. Günther, "Gigantic," 297.

48. Günther, "Gigantic," 297.

49. Günther, "Gigantic," 297.

50. Wallace, *Island Life*, 268–269.

51. Baur, "Galápagos," 422, emphasis in original; also Baur, "On the Origin."

52. Baur developed a theory of two types of island chains: those that previously had been linked to continents and showed "harmonic" distribution of organisms, where each island had its particular species; and oceanic islands where the flora and fauna were a mixture of forms that had been accidentally introduced from other places and were thus disharmonic. Baur saw Günther's evidence of discrete tortoise species on each island as evidence that the islands were harmonic and was convinced that a land bridge had once connected the Galápagos to the South American mainland. His subsidence theory offered the "most simple solution" to the distribution of animals in the Galápagos: "All the islands were formerly connected with each other, forming a single large island; subsidence kept on and the single island was divided up into several islands. Every island developed, in the course of long periods, its peculiar races, because the conditions on these different islands were not absolutely identical. Thus it has been made probable that the Galápagos are of continental origin" ("Galápagos," 422). Today, naturalists recognize the islands as a "disharmonic" taxonomic landscape where the improbability of colonization has resulted in an unusual collection of species compared with continental areas. See Tye et al., "Outstanding," 16.

53. Baur, "Galápagos," 418.

54. Quoted in Larson, *Evolution's Workshop*, 114.

55. Darwin, "On the action."

56. Porter, *Journal*, 136.

57. Van Denburgh, "Expedition," 367.

58. Garman, *Galápagos Tortoises*, 262.

59. Beebe, *Galápagos*, 228.

60. Stewart, "Further Observations," 200.

61. Gerlach, Muir, and Richmond, "First Substantiated Case."

62. Although . . . Baur's insistence that the islands were once connected to each other has been supported by recent research on changing sea levels. See Nicholls, "How Sea Level."

63. Baur quoted in Larson, *Evolution's Workshop*, 113.

64. Günther, "Description," 258, quoted in Larson, *Evolution's Workshop*, 104.

65. Garman, "Reptiles," 76.

66. All were from the Bay Area, including the Stanford-Hopkins expedition of 1898, sent under the direction of Stanford University president David Starr Jordan, a former student of Agassiz who championed evolution. See Larson, *Evolution's Workshop*, 119–143.

67. See Rothschild, *Dear Lord Rothschild*; James, *Collecting Evolution*; and Larson, *Evolution's Workshop*.

68. Pitelka, "Rollo Beck."

69. Rothschild, *Dear Lord Rothschild*, v.

70. Rothschild, *Dear Lord Rothschild*, 2; on Karl Jordan, see Johnson, *Ordering Life*.

71. Rothschild, "On Giant Land Tortoises."

72. Barrow, "Specimen Dealer."

73. Haraway, "Teddy Bear Patriarchy"; Rutherford, *Governing the Wild*; Drayton, *Nature's Government*; Endersby, *Imperial Nature*.

74. Johnson, *Ordering Life*.

75. Walter Rothschild to Albert Günther, 13 February 1897, A. Günther Collection, Box 16, Letters L. W. Rothschild, 1884–1898, Natural History Museum at Tring, permission of the Trustees of the Natural History Museum, London (hereafter Günther Collection, BMNH Tring Museum).

76. Quoted in Larson, *Evolution's Workshop*, 113.

77. Walter Rothschild to Albert Günther, 27 December 1897, Günther Collection, BMNH Tring Museum.

78. Although Tring was very much a personal project for Rothschild, he was generous with his collections (see Jordan, "Obituary")—to a point. When the Natural History Museum's keeper, George Boulenger, made a formal description and taxonomic designation for one of the Galápagos tortoises Rothschild had acquired, Rothschild snapped that Boulenger had no right to name one of his animals and threatened to cut the British Museum out of his will. Boulenger apologized, pleading ignorance.

79. Rothschild and Hartert, "Review," 120. This was not the case, as Beck took nine more on his next trip in 1901; R. H. Beck to E. Hartert, 8 July 1901, Tring Museum Correspondence, Box 165, Natural History Museum at Tring, United Kingdom (hereafter BMNH Tring Museum Correspondence).

80. Beck, "In the Home," 172.

81. Rothschild, *Dear Lord Rothschild*, 201.

82. Rothschild, "Description," 119.

83. R. H. Beck to E. Hartert, 8 July 1901, Tring Museum Correspondence.

84. R. H. Beck, quoted at length in Murphy, *Oceanic Birds*, 4. Rothschild was secretive about his expeditions; a perceived unjustness about this relationship was part of the reason. He explained to Günther that he did "not tell people my collecting concerns, except friends like you who do not misunderstand me." He was concerned about backlash over how the trip was financed because his "5 collectors [were] not paid six pence except the profit on the specimens after I have had a see." W. Rothschild to A. Günther, 23 December 1897, Günther Collection, BMNH Tring Museum.

85. R. H. Beck to E. Hartert, 10 September 1901, BMNH Tring Museum Correspondence.

86. Beck, "In the Home," 161.

87. Beck, "In the Home," 168–169.

88. Beck, "In the Home," 162.

89. Beck, "In the Home," 168.

90. Beck, "In the Home," 168.

91. In 1827, Richard Harlan ("Description") wrote of the abundance of tortoises, citing Porter's journal: Notes "great profusion" of the tortoises—"Four boats were despatched every morning, to bring in a stock of tortoises, and returned at night, bringing with them from twenty to thirty each, averaging about 60 lbs; and in four days we had as many as we could conveniently stow. They were piled up on the quarter deck for a few days, in order that they might have time to discharge the contents of the intestines, which are considerable; after which, they were stowed away below, like any other provision" (291).

92. Drayton, *Nature's Government*; Haraway, "Teddy Bear Patriarchy"; Tyrell, *Crisis*.

93. Today, conservationists debate the extent to which the Cal Academy's scientific collecting contributed to the endangerment of tortoises and other Galápagos wildlife and whether scientists should have known better. See James, *Collecting Evolution*, 59–60.

94. These specimens also became the foundation of the Cal Academy when, while Beck's team was in the archipelago, the Great Fire of 1906 destroyed the museum and its collections. It is a case that clearly demonstrates that since Linnaeus, the existence of natural science institutions has rested squarely on the backs of specimens and the men (and few women) who collected and preserved them. In the case of the Cal Academy, saving specimens meant saving science as well. See James, *Collecting Evolution*.

95. Van Denburgh, "Preliminary Description," 3.

96. Speculation haunted the Fernandina tortoise, which Van Denburgh named *phantastica*, for decades; had it truly been the last of its species?, or might Beck, alone on the island, have invented the tale of skinning the lone animal by moonlight to enhance the value of his collections? For decades, searches of the island turned up no tortoises. But in February 2019, park guards and biologists on an expedition to Fernandina found a single adult female tortoise in a remote part of the island, as well as scat presumably left by other tortoises. This species, thought to have been extinct for more than a century, was indeed still alive. See "Giant Tortoise"; James, *Collecting Evolution*, 3–4.

97. DeSola, "Liebespiel," 80.

98. Andrei, "Nature's Mirror," 117–118.

99. Townsend, "New Information."

100. Townsend, "Impending Extinction."

101. See Raby, *American Tropics*.

102. R. H Beck to Senor Gil, 6 March 1903, Box 3, Correspondence, 1899–1903, Rollo H. and Ida M. Beck Collection, MSS.036, California Academy of Sciences Library, San Francisco, CA.

103. DeSola, "Notes," 54.

104. "Hardy animals": Townsend, "Impending Extinction," 55; "the only hope of keeping the stock alive": Townsend, "Galápagos Islands Revisited," 165, quoting an article published in *Nature*, 4 August 1928.

105. Townsend, "Giant Tortoise."

106. Townsend, "Giant Tortoise."

107. See Hennessy, "Producing 'Prehistoric' Life."

108. On trusteeship, see Li, *Will to Improve.*

Chapter 4. The Many Worlds at World's End

1. Latorre, *Curse,* 1.

2. Latorre, *Curse,* 1, 3. Full capitalization in the original.

3. Latorre, *Curse,* 2.

4. Von Hagen, *Ecuador and the Galápagos,* 215.

5. Von Hagen erected the Darwin bust at Wreck Bay on San Cristóbal Island, but Darwin first landed nearby at Cerro Tijeretas (Frigatebird Hill) on 16 September 1835. See Grant and Estes, *Darwin in Galápagos,* 81.

6. Von Hagen, *Ecuador the Unknown,* 96. A prolific travel writer and self-promoter, he made sure that prominent naturalists heard about his efforts in Ecuador. He spent six months traveling the archipelago with his wife—a circumscribed version of the Darwin Memorial Expedition he had wanted to take, but could not finance, that would have followed the *Beagle*'s course along the entire South American coast. Following his stay, protecting the Galápagos became a personal mission; he spent the next two years trying to strengthen legislative protections and rally support for a research station. See Barrow, *Nature's Ghosts,* 179.

7. Otterman, *Clinker Islands,* 43.

8. Larson, *Evolution's Workshop,* 108.

9. The short timespan did not detract from the success of the expedition: Henry Fairfield Osborn wrote in the book's Foreword, "in less than one hundred actual hours on land, the *Noma* party accomplished results—artistic, photographic, observational— which are entirely without rival." Beebe, *World's End,* iv.

10. Beebe, *World's End,* 145.

11. Beebe, *World's End,* 60–61, 57.

12. Gillis, *Islands of the Mind.*

13. Beebe, *World's End,* 57.

14. Beebe, *World's End,* 301, 295–308.

15. They were received in Guayaquil by the British consul, then sent on a cattle boat to Panama, where the taxi driver worked dredging the canal before catching yellow fever and escaping to the United States; Beebe, *World's End,* 295–308. With somewhat different details, von Hagen also recounts the story of the *Alexandra* in his books (*Ecuador and the Galápagos Islands,* 236–247; and *Ecuador the Unknown,* 145–154). He first heard the tale in 1935 from an Icelandic settler on Santa Cruz, Finsen, who had a copy of the ship's log, published in Norway as *Mandskapet fra Bark Alexandra,* ved Alf Harbitz, Steen'ske Bogtrykkeri og Forlag (Kristiania, 1915).

16. Rose, "Last Raid," 194.

17. Rose, "Last Raid," 179.

18. Beebe, *World's End*, 123; the tortoises would go to Charles Townsend's zoo colonies.

19. See Adams, *Against Extinction*, on the transition from zoo-based to in situ conservation as the dominant paradigm over the twentieth century. But also Braverman, *Wild Life*, on contemporary tensions between in situ and ex situ animal conservation.

20. On the history of habitat and conservation, see Alagona, *After the Grizzly*.

21. Parts of this section appear in Hennessy, "Politics of a Natural Laboratory"; also see Lewis, "Negotiating for Nature," 42–52.

22. Kohler, *Landscapes and Labscapes*.

23. Quoted in Barrow, *Nature's Ghosts*, 176.

24. Quoted in Larson, *Evolution's Workshop*, 166.

25. Kohler, *Landscapes and Labscapes*.

26. The phrase is early-twentieth-century German preservationist Hugo Conwentz's term for pristine places and objects that stood as symbols of nature of important patriotic, aesthetic, scientific, or general value. Conwentz's arguments for protecting *Naturdenkmal* were particularly influential in Britain. See Conwentz, *Care of Natural Monuments*. Discussed in Ross, *Ecology and Power*, 243.

27. Callicott and Nelson, *Great New Wilderness Debate*; Kupper, *Creating Wilderness*; Kupper, "Nature's Laboratories?"

28. Kohler, *Landscapes and Labscapes*, 149. In a Latin American context, Gregory Cushman looks to Alexander von Humboldt as the ancestor of early- to mid-twentieth-century naturalists' environmental protection efforts, including the creation of Venezuela's first national park in 1937 under the leadership of Henri Pittier (see Cushman, "Humboldtian Science"). Humboldt did not visit the Galápagos during his travels in South America, and naturalists working in the islands, such as Theodor Wolf and Misael Acosta Solís, looked instead to Darwin.

29. Weiner, *Beak of the Finch*; Lack, however, was not as charmed by the experience of doing research on the islands as Beebe had been—perhaps because he had no yacht: the Galápagos were "scarcely a residential paradise": "The biological peculiarities are offset by an enervating climate, monotonous scenery, dense thorn scrub, cactus spines, loose sharp lava, food deficiencies, water shortage, black rats, fleas, jiggers, ants, mosquitoes, scorpions, Ecuadorian Indians of doubtful honesty, and dejected, disillusioned European settlers." Not all field biologists were adventurous, outdoorsy types. Quoted in Larson, *Evolution's Workshop*, 167.

30. Larrea, *El Archipiélago de Colón*, 216–217; Acosta Solís, "La Protección," 89–105.

31. Cuvi, "La Institucionización."

32. Acosta Solís, *Galápagos*; Acosta Solís, "Problems of Conservation."

33. The Galápagos case fits neither the regional timeline nor political history of national park creation that Emily Wakild describes in "Panorama of Parks." She argues that Latin American park creation in the early twentieth century was predominantly

nationalist and found "little reason to believe parks were external creations imposed by a 'global conservation aristocracy' " (246). In Mexico, for example, the forty national parks created before 1940 were a response to postrevolution calls for both social change and nature protection—"tangible representations of how revolutionaries nationalized their national territory" (Wakild, *Revolutionary Parks,* 10). Yet this is not the case in the Galápagos, where international scientific elites—working with Ecuadorian diplomats and scientists, to be sure—had a powerful role in shaping the creation of the national park and Darwin station. See also Lewis, "Negotiating for Nature," and Cushman, "Road to Survival," on the history of US conservationists in Latin America during the 1940s. On Ecuadorian histories of conservation, see Bustamante, *Historia de la Conservación Ambiental.*

34. Cuvi, "La Institucionización," 114; as Cuvi points out, it is not the case, however, that conservation in Ecuador began only with the second phase of foreign scientists interested in the Galápagos in the late 1950s. That narrative obscures the efforts of Acosta Solís and other early domestic conservation advocates in the 1930s and 1940s. In 1940, Acosta Solís was central to the creation of the first Ecuadorian institution dedicated to conservation, El Instituto Ecuatoriano de Ciencias Naturales, an organization founded to connect "ciencia y la patria," in whose journal Acosta Solís converted to the role of "defensora de las riquezas naturales" (ibid., 113).

35. Raby, *American Tropics,* 216; see also McCook, *States of Nature;* De Bont, *Stations in the Field;* and Vetter, "Labs in the Field?"

36. De Bont, *Stations in the Field,* 38.

37. "Les parcs nationaux"; De Bont, "World Laboratory."

38. The protection of "strict wilderness reserves" was included as one of the categories of conservation in the 1940 Convention on Nature Protection and Wild Life Preservation in the Western Hemisphere, written primarily by Harold Coolidge and Alexander Wetmore (assistant secretary at the Smithsonian), with input from conservationists across Latin America. See Lewis, "Negotiating for Nature." Yet Emily Wakild argues that the convention had little effect in Latin America, where more than half the national governments failed to initially sign or ratify it ("Panorama of Parks," 257). She argues that given the temporal distance between the overthrow of Spanish and Portuguese empires and national park creation in Latin America (predominately in the twentieth century), regional conservation cannot be considered expressly colonial, as it was in much of southern Africa and South Asia. Yet this perspective misses both the neocolonialism of transnational institutions that were instrumental in creating the Galápagos park and how ideologies of nature shaped during the colonial period informed "post"-colonial understandings of conservation and resource use as well as class and racial hierarchies that have structured park politics throughout the region. See Cushman, "Struggle over Airways," on South America; and Ross, *Ecology and Power,* more generally.

39. Dorsey, *Dawn of Conservation Diplomacy;* Barrow, *Nature's Ghosts;* Ross, *Ecology and Power,* 264–271; Adams, *Against Extinction.*

40. Barrow, *Nature's Ghosts*, 178.

41. Moore, "Protection and Conservation."

42. J. Huxley to R. Moore, 7 April 1936, British Museum of Natural History (BMNH) DF206/160 British Association Galápagos Islands Committee: correspondence with J. S. Huxley 1935–1937; BMNH DF206/158 British Association Galápagos Islands Committee: minutes and circular letters 1935–1937, Permission of the Trustees of the Natural History Museum, London (hereafter BMNH).

43. Acosta Solís, *Galápagos*, 75.

44. Von Hagen to H. Swarth, 27 June 1935, DF206/158 British Association Galápagos Islands Committee: minutes and circular letters 1935–1937; Barrow, *Nature's Ghosts*, 179. The "International Wild Life Committee" (i.e., the American Committee), after so much involvement in the first decree, declined to participate (von Hagen, *Ecuador the Unknown*, 218). The British consul in Quito, Stafford London, reported to the home office that the Guayaquil committee was something von Hagen had organized for diplomatic purposes rather than as an active body. See Stephen Gaselee (Foreign Office) to Tate Regan, FRS, 20 August 1936, BMNH DF1004/361 Expeditions: Galápagos Islands 1935–1960, BMNH.

45. Von Hagen to O. J. R. Howarth, Secretary of the British Association for the Advancement of Science, n.d., BMNH DF206/159 British Association Galápagos Islands Committee: letters of O. J. R. Howarth, Secretary, 1935–1937, BMNH.

46. Sevilla Perez, "Galápagos Islands"; Luna, *Historía Política Internacional*.

47. Velasco Ibarra had served just a year before he dissolved the legislature and began jailing opponents. Páez also established a repressive dictatorship before being ousted by General Alberto Enríquez Gallo. Pineo, *Ecuador and the United States*, 102.

48. Lewis, "Negotiating for Nature," 46–47. The creation of a nature reserve as a way to draw a frontier space under national control was a strategy that other South American governments also used. See Freitas, "Park for the Borderlands"; and Wakild, "Protecting Patagonia."

49. In the United States, Kingsley Nobel, curator of herpetology at the American Museum of Natural History, told Huxley that von Hagen "is, as you suspect, a promoter rather than a scientist. There are so many things we do not understand about his arrangements that the Museum has been very cautious in backing up his various schemes." K. Nobel to J. Huxley, 23 January 1937, BMNH DF206/160 British Association Galápagos Islands Committee: correspondence with J. S. Huxley 1935–1937, BMNH.

50. As Mark Barrow recounts, he was accused of issuing bad checks, forging a Mexican government official's signature to get financial backing for an expedition (*Nature's Ghosts*, 181).

51. Von Hagen to J. Huxley, 20 February 1937, BMNH DF206/158 British Association Galápagos Islands Committee: minutes and circular letters 1935–1937, BMNH. Upon returning to California, von Hagen was apprehended by the FBI and interned in

a concentration camp in 1941 because of his German heritage. His 1949 book *Ecuador and the Galápagos Islands* includes the transcript of a letter from the director of naval intelligence thanking him for supplying the US Navy with information on the Galápagos during his 1935 stay. Perhaps von Hagen included it (not mentioned in *Ecuador the Unknown*, published in London) as a way of clarifying his allegiances. He also noted a suggestion that the US navy collaborate with the British Association's Galápagos Committee (which Huxley organized) to establish a research station in the archipelago: "A radio, a vessel, and constant zoological surveillance of the islands might have given the navy precisely what it wanted: observation, a check on vessels moving in the islands, and a building of up of basic intelligence, so that if the time came to act, the navy would possess vital information." The navy, however, did not take von Hagen's offer. See von Hagen, *Ecuador and the Galápagos Islands*, 259–260. Coolidge too recognized security concerns and framed the 1940 convention to respond to German and Japanese expansion in Latin America. See Cushman, "Struggle over Airways."

52. Black, *Archipiélago del Ecuador*, 7; Larrea, *El Archipiélago de Colón*, 207–237.

53. He also suggested the introduction of barbasco, a plant then grown on the Santa Elena peninsula, for export to US pharmaceutical companies that used it to make steroids. Acosta Solís, *Observado Fitológicamente*, 59–68.

54. Sevilla Perez, "Galápagos Islands"; Luna, *Historía Política Internacional*.

55. "Al Gobierno del Ecuador en aquel feraz territorio que por primera vez nace a la luz de la sociedad," quoted in Luna, *Historía Política Internacional*, 6; the official report also recognized the presence of one Juan Johnson, an old inhabitant of the island, though it gave no description of his story.

56. Luna, *Historía Política Internacional*, 63–70.

57. "Dice que acogió la empresa con todo el ardor que me inspiraba el progeso del país cuando me convencía que al aumentar su territorio y población le aumentaba nuevas riquezas," cited in Luna, *Historía Política Internacional*, 66.

58. Latorre, *El Hombre*, 78–79, 81; Luna, *Historía Política Internacional*. The Ecuadorian government did not outlaw debt peonage, known in Ecuador as *concertaje*, until 1918, when liberal governments sought to increase the mobility of labor to provide workers for thriving lowland export industries. See Lyons, *Remembering the Hacienda*, 19. The colonial-era *hausipungo* labor system that tied indigenous laborers to haciendas was not abolished until the Agrarian Reform Law of 1964.

59. Latorre, *El Hombre*, 82–84. As Ricardo Salvatorre and Carlos Aguirre wrote, penitentiaries offered the "promise of a double reconstruction: that of the moral subject by means of isolation (reflection and remorse) and moral suasion; and that of the *Homo economicus* through the operation of redemptive work and the system of incentives and penalties associated with it" ("Birth of the Penitentiary," 27). The Latin American experience, though varied, was largely shaped by relations of personal dependence and a slower market orientation among the working classes than in North America and Europe. Palmer, "Confinement," 231; Astudillo, "Environmental Historical Archaeology."

60. Luna, *Historía Política Internacional*, 6. On Galápagos, see Astudillo, "Environmental Historical Archaeology"; on Latin American prisons, see Salvatore and Aguirre, *Birth of the Penitentiary*, and more broadly, Foucault, *Discipline and Punish*. On Ecuadorian agrarian organization: Duncan and Rutledge, eds., *Land and Labour;* and Lyons, *Remembering the Hacienda.*

61. Foote, "Race, State and Nation," 276.

62. Keynes, ed., *Charles Darwin's* Beagle *Diary,* 356.

63. FitzRoy, *Narrative,* 491.

64. Keynes, ed., *Charles Darwin's* Beagle *Diary,* 355.

65. Keynes, ed., *Charles Darwin's* Beagle *Diary,* 355.

66. DeLoughrey, *Routes and Roots,* 13.

67. Friday, for Joyce, was the "trusty savage," the "symbol of the subject races"; from a lecture on Daniel Defoe at Trieste in 1912, quoted in Maganiello, *Joyce's Politics,* 109. Gilles Deleuze critiques Defoe's portrayal of Friday as "docile towards work, happy to be a slave, and too easily disgusted by cannibalism. Any healthy reader would dream of seeing him eat Robinson" (*Desert Islands and Other Texts,* 12).

68. Darwin, *Journal of Researches,* 376.

69. Keynes, ed., *Charles Darwin's* Beagle *Diary,* 355.

70. Latorre, *El Hombre,* 102; Striffler, *In the Shadows.*

71. Heflin, *Herman Melville's Whaling Years,* 90–105.

72. Melville, *Encantadas,* 53.

73. The most provocative detail of this story for Ecuadorian readers may be that Melville says the island is a Peruvian territory. Melville, *Encantadas,* 50, 51 (story 47–54).

74. Melville, *Encantadas,* 53.

75. Lundh, *Galápagos,* n.p.

76. Latorre argues that the narrative of colonies undone by criminal laborers is overblown, noting that most of the early Galápagos colonies also included sizable contingents of free men and families. Other writers attribute unrest to the uneven gender composition of colonies that had few women. But it is futile to try to isolate a direct cause; issues of class, gender, and racial inequity, as well as the sheer difficulty of labor and social life on the island, all contributed to unrest. Latorre, *Curse,* 40–41; Mann, *Yachting on the Pacific,* 37.

77. Foote, "Race, State and Nation."

78. See Jonik, "Melville's 'Permanent Riotocracy' "; also James, *Mariners, Renegades and Castaways,* 19.

79. I use the term "postcolonial" to reference the end of formal Spanish rule, but not to suggest a "stage in which colonial domination had been economically, politically, and culturally erased and/or transcended" (Moraña, Dussel, and Jáuregui, "Colonialism and Its Replicants," 11). To do so would be to misunderstand the cultural and political-economic history of Latin America and the continued influences of what Aníbal

Quijano ("Coloniality of Power") called "coloniality." Instead, I follow Catherine Walsh's assertion that "the postcolonial . . . is a perspective, a lens for reading the dynamic, experiential, and shifting nature of coloniality, its new fields and dispositions of power, its relations, practices struggles and resistances" ("[Post]Coloniality in Ecuador," 509, quoting Walsh, "Staging Encounters"). I concur with Walsh's argument that "postcoloniality in Ecuador is a lived experience" ("[Post]Coloniality in Ecuador," 516) and further argue that the debates on postcoloniality in Latin America would be enriched by more grounded attention to how the materiality of nature figures in these lived experiences, and how colonial histories reworked ontologies of nature-culture relationality.

80. Deleuze, *Desert Islands and Other Texts*, 12–13.

81. Captain David Porter told Watkins's story—at least as he had heard it—at length in his *Journal of a Cruise*. It has since been repeated many times, including in Rose, "Last Raid."

82. Porter, *Journal*, 141.

83. Melville, *Encantadas*, 79.

84. Cushman, *Guano*.

85. In that position, he continued trying to make the Galápagos profitable, for Ecuador and for himself, by planning ventures with US businessmen. One such contract was to use the islands as a fueling station for US ships in the Pacific, but it was not realized. He also served, at various points, as secretary of state (under President Urbina) and as governor of Manabí province before his death in 1866.

86. He also reported that Bindloe, Wennman, Hood, James, Indefatigable, Culpepper, Ronde Rock, and Tower had smaller deposits.

87. "En la primera fila de las naciones más opulentas," quoted in Luna, *Historía Política Internacional*, 94.

88. Briones and other escapees were taken to Guayaquil and executed. Much of the story of what happened was told by the cooper on the ship *George Howland*, a Mr. Peacock, who survived the ordeal, having been locked in the ship's hull. Lundh, *Galápagos*.

89. Luna, *Historía Política Internacional*, 81.

90. "Act to Authorize."

91. Luna, *Historía Política Internacional*, 95–97.

92. A definitive report on the islands' guano did not come until geologist Theodor Wolf voyaged to the islands twenty years later. Wolf found that unlike the Peruvian islands to the south and closer to the coast, the Galápagos did not have significant guano reserves, something he deemed unsurprising given the yearly winter rains that washed the large islands. Only the small, drier islets had considerable guano, but those deposits were too limited in size to be profitably mined (Wolf, *Memoria*, 17; Sevilla Perez, "Galápagos Islands," 32).

93. Mann, *Yachting on the Pacific*, 14.

94. Wolf, *Memoria*, 15.

95. McCook, *States of Nature*, 5.

96. Braun, "Producing Vertical Territory"; McCook, *States of Nature*. The Jesuits' interest in state territory extended only so far. Wolf never traveled to the Amazon— indeed, on the map he published in 1892, the region was struck through with text reading "Uncharted regions inhabited by savage Indians!" As historian Ana Sevilla Perez has noted, the region was home to indigenous populations that threatened the idea of the nation-state and was repeatedly claimed by Peru, making it a frontier of terra incognita. Sevilla Perez, "La incompleta nacionalización."

97. The Catholic Church saw Wolf's enthusiasm for evolution as anti-Catholic. Tension between science and religion plagued his tenure at the Polytechnic. The Ecuadorian church denied him permission to visit the islands, prompting Wolf to renounce his professorship and Jesuit vows, leaving behind privileged support from García Moreno's government and his life in Quito. Starting over in Guayaquil, he managed to scrape together funding for a Galápagos expedition in 1875 and traveled to the islands on Valdizán's ship. While Wolf was in the archipelago, García Moreno was assassinated by liberal political rivals. Upon his return to the continent, Wolf was named state geologist by García Moreno's successor. In this position, Wolf journeyed to the archipelago again in 1878 with the primary mission of revisiting the question of whether the islands had mineable guano reserves. Sevilla Perez, "Galápagos Islands."

98. Wolf, *Memoria*, 27–29 (quotation on 28); Sevilla Perez, "Galápagos Islands."

99. Latorre, *Curse*, 49–59.

100. Quoted in Luna, *Historía Política Internacional*, 127.

101. Church, "Report upon Ecuador," 54.

102. As Ecuadorian representative in France, Flores Jijón had himself offered the islands as collateral for protection of the state in 1865, but Napoleon rejected the deal. Luna, *Historía Política Internacional*, 121, 127–129.

103. Astudillo, "Environmental Historical Archaeology," 48.

104. Ecuadorian naturalist and writer Nicolás Martínez visited Isabela (which he called Albemarle, following the English tradition) in 1906 and described the process of skinning cattle in Puerto Villamil. Colonists killed many more animals than they could eat and so threw carcasses into the sea. Martínez, *Impresiones*, 140–141.

105. Latorre, *El Hombre*, 199–211.

106. Latorre, *Manuel J. Cobos*.

107. Latorre, *El Hombre*, 143; Astudillo ("Environmental Historical Archaeology") discusses whether the settlement El Progreso under Cobos is better characterized as a hacienda or a plantation, building on the distinction in size of markets and capital investment introduced by Eric Wolf and Sidney Mintz in 1957. Drawing primarily from Philip Curtin's definition of a plantation system, Astudillo concludes that El Progreso was a hybrid system that reflected a transitional period in the Ecuadorian economy in which the stratified Spanish colonial social hierarchy shaped republican models for a capitalist, export-oriented economy that supplied a global market with natural tropical commodities. See also Wolf and Mintz, "Haciendas"; and Curtin, *Rise and Fall*.

108. They also imported bamboo as an important construction material. Astudillo, "Environmental Historical Archaeology."

109. Fernando Astudillo's historical ecological research corroborates Mann's description, finding deforestation of the Scalesia zone on San Cristóbal and conversion to grassland shortly after colonization. "Environmental Historical Archaeology," 200.

110. Van Denburgh, "Expedition," 243.

111. Latorre, *El Hombre*, 202; also Martínez, *Impresiones*.

112. Roberts, *El Ecuador*.

113. Astudillo, "Environmental Historical Archaeology," 193–194; Latorre, *Manuel J. Cobos*; Mann, *Yachting on the Pacific*, 29.

114. Pineo, *Social and Economic Reform*, 11.

115. Astudillo, "Environmental Historical Archaeology," 203; Latorre, *Manuel J. Cobos*.

116. Latorre, *El Hombre*, 183–187.

117. Latorre, *Manuel J. Cobos*.

118. Somewhat perversely, in an effort to reform Ecuadorian prisons, the state penal code of 1906, adopted in the wake of the El Progreso scandal, legalized the deportation of criminals and use of penal labor in agricultural colonies in the Galápagos—practices that had been central to state sovereignty in the archipelago since 1832. Salvatore and Aguirre, "Birth of the Penitentiary," 11.

119. Foote, "Race, State and Nation."

120. Wittmer, *Floreana*, 16.

121. Pinchot, *To the South Seas*, 134–135.

122. Quoted in Latorre, *Curse*, 120–121.

123. The following years saw an "invasion" of US tuna boats in Galápagos waters, caused in part by the Mexican government reducing its licenses for tuna vessels because of declining stocks. In response, the boats ventured farther south to the Galápagos, where the Ecuadorian government had little oversight. In 1932–1933, Van Camp took $37 million worth of tuna from the archipelago, 70 percent of its annual catch. Ecuador saw little benefit from this haul—one of the large US tuna ships, the *Chicken of the Sea*, paid only $823 for a year's fishing license but took 250 tons of tuna in just one trip. The Ecuadorian government was supportive of fisheries development but felt it was not recouping its share of the trade. This kicked off the first "tuna war" over access to the seas, which lasted from 1929 to 1941. The second tuna war, 1952–1980, began with the Declaration of Santiago, in which Ecuador, Peru, and Chile claimed sovereignty over the sea two hundred miles from the coast—a claim contested by the United States, which recognized the states' territorial rights as extending only twelve miles offshore. Latorre, *El Hombre*, 292–295; Luna, *História Política Internacional*.

124. Otterman, *Clinker Islands*, 91–101.

125. Mann, *Yachting on the Pacific*, 11.

126. These stories are now the subject of the 2014 documentary *The Galápagos Affair: Satan Came to Eden*, by Dan Geller and Dayna Goldfine. The film is based on the

1983 book of the same title by John Treherne published in New York by Random House.

127. Ritter, "Adam and Eve," 409, 411; Strauch, *Satan Came to Eden,* 11, 12.

128. Ritter, "Adam and Eve," 409, 411, 414.

129. Wittmer, *Floreana,* 27.

130. Wittmer, *Floreana,* 64.

131. Burton, "Capt. Hancock's."

132. Just what happened on Floreana remains a mystery; Margret Wittmer, the last witness, died in the islands in 2000. But it has been the topic of much speculation. For firsthand accounts, see Wittmer, *Floreana,* and Strauch, *Satan Came to Eden.* These memoirs are quoted extensively in the 2014 documentary *The Galápagos Affair.* See also Latorre, *Curse,* 127–185.

133. Ritter, "Adam and Eve," 412.

134. Conway and Conway, *Enchanted Islands.*

135. Wittmer, *Floreana,* 19.

136. J. T. Howell, botanist on the *Zaca,* Templeton-Crocker expedition, March–September 1932, letter to L. DeStefano, 7 March 1978, on his recollections of the Ritters in Galápagos Boxes, Group 6, California Academy of Sciences Library, San Francisco, CA.

137. Conway and Conway, *Enchanted Islands,* 98.

138. Conway and Conway, *Enchanted Islands,* 96, 91; Ritter, "Satan Walks."

139. They were all examples of what Foucault ("Of Other Spaces") called heterotopias: "counter-sites, a kind of effectively enacted utopia in which the real sites, all the other real sites that can be found within the culture, are simultaneously represented, contested, and inverted" (24).

140. Thompson, "Heuristic Geographies," 67.

141. Larson, *Evolution's Workshop,* 175. The United States had laid the groundwork for the base over several previous years. As early as 1934, the War Department had suggested acquiring the islands to President Roosevelt, though he rejected it as conflicting with his Good Neighbor Policy. Luna, *Historía Política Internacional,* 242–260; Lewis, "Negotiating for Nature," 48. The US imposition of bases in the Galápagos and in Salinas, on the Ecuadorian coast, was also part of a neocolonial battle over control of South American airways during the period that spanned World War I and World War II. See Cushman, "Struggle over Airways," 200.

142. G. C. Hodges, "The Pacific's Key to Panama," *Sunset: The Pacific Monthly* 39 (1917): 36, in Box 2, Folder 8, CAS Slevin Papers. On US concerns about Ecuador's control over the archipelago, particularly after Japanese ships sought fishing licenses from Ecuador, see E. Trent, "Bay Group Seeks Galápagos Rights," *San Francisco Chronicle,* 21 April 1929, in Box 3, CAS Slevin Papers; Parks and Rippy, "Galápagos Islands." For an Ecuadorian perspective, see Latorre, *El Hombre,* 314; and Luna, *Historía Política Internacional,* 237–241.

143. Harrison, "Study of the U.S. Air Forces'."

144. Manuel María Borrero, "Informe de la Comision de Relaciones Exteriores," 18 December 1944, Documentos de la Occupación del Archipelago de Galápagos, 1940–1944, Tomo I, G.3.4.1, Archivo del Ministerio de Relaciones Exteriores, Quito, Ecuador, 33 pp.

Chapter 5. Making a Natural Laboratory

1. Portions of this chapter were previously published in Hennessy, "Mythologizing Darwin's Islands." Also see Larson, *Evolution's Workshop*, 194–196.

2. Acosta Solís was one of three Ecuadorians, though many more Ecuadorian scientists and students joined the trip in the Galápagos. Participants list, Box 7, Galápagos Islands Collection, MSS-181, California Academy of Sciences Library, San Francisco, CA (hereafter CAS Galápagos Collection).

3. "El hecho."

4. Bowman, ed., *Galápagos;* also Perlman, "Good Life."

5. The convention was resigned in 1989 and 2016, each time for an additional twenty-five years.

6. On sailing in the *Beagle*'s wake, Perlman, "Galapagos Quest Is On"; "natural laboratory of evolution," quoting Robert Usinger, in "UC Island Safari Complete": "Our field work in the Galapagos is finished and our members are already comparing notes and digesting findings and starting to write scientific papers. . . . Our original idea that many scientists could profitably combine their efforts to study this natural laboratory of evolution has been realized beyond our fondest expectations."

7. Huxley, "Charles Darwin," 3, 9; Schmidt, "Huxley Packs House." On the political stakes of Huxley's rhetoric and the creation of the myth that Darwin discovered evolution in the islands, see Hennessy, "Mythologizing Darwin's Islands"; and Sulloway, "Darwin and the Galapagos."

8. Dorst, "Where Time Stood Still," 28, 30.

9. Dorst, "Where Time Stood Still," 28.

10. During the war, the Smithsonian and the American Committee attempted to align their desire for a research station with US military endeavors. President Roosevelt supported the effort with a memorandum to his secretary of state saying, "I would die happy if the State Department could accomplish something" to protect the Galápagos, but their efforts were unsuccessful—the military was uninterested in studies of evolutionarily odd species. Roosevelt quoted in Larson, *Evolution's Workshop*, 176.

11. Eibl-Eibesfeldt, *Galápagos* (1961), 21.

12. Woram, "Who Killed the Iguanas?"

13. Eibl-Eibesfeldt, *Galápagos* (1961), 177–178.

14. Eibl-Eibesfeldt, *Galápagos* (1961), 177.

15. Eibl-Eibesfeldt, *Galápagos* (1961), 176.

16. Pineo, *Ecuador*, 128–129.

17. Otterman, *Clinker Islands*, 172–173.

18. R. Lévêque to J. Dorst, 30 June 1960, Unsorted Correspondence, Charles Darwin Research Station Archive, Puerto Ayora, Galápagos, Ecuador (hereafter CDRS Archive).

19. Ross, *Ecology and Power*.

20. Quoted in Larson, *Evolution's Workshop*, 180.

21. Quoted in Larson, *Evolution's Workshop*, 180. Huxley's vision was much criticized, and UNESCO's mission was later distanced from his evolutionary humanism. Indeed, he later regretted his foundational pamphlet. Yet the document remains helpful for understanding the organization's early years as well as Huxley's interest in protecting the Galápagos as a kind of world monument of evolution. Sluga, "UNESCO"; Anker, *Imperial Ecology*.

22. Larson, *Evolution's Workshop*.

23. Anker, *Imperial Ecology*.

24. President Harry Truman's Four Points speech in 1949 outlined what Arturo Escobar (*Encountering Development*) has called this "new imperialism," ushering in an age of development and US leadership to advance the progress of the "underdeveloped" world. See also Goldman, *Imperial Nature*.

25. Sluga, "UNESCO," 396–397; Ross, *Ecology and Power*; De Bont, "World Laboratory."

26. Drayton, *Nature's Government*; also Cushman, *Guano*.

27. *Estudios Sobre Galápagos, Abril 1956–Enero 1957*, Archivo Reservado, Departamento de Fronteras, G.3.4.4, Ministerio de Relaciones Exteriores, Quito; Luna, *Historia Política Internacional*, 320–323. See also Heyerdahl and Skjølsvold, "Archaeological Evidence," though attribution of the source of the pottery shards has been contested.

28. The men were chosen because of their recent experience in the islands. At the time, both were still in their twenties. Bowman had just finished his dissertation at Berkeley and had started as an assistant professor of biology at San Francisco State University, where he taught until he retired in 1988. Eibl was starting his career in animal ethology at the Max Planck Institute for Behavioral Physiology in Westphalia, Germany. Bowman made fifteen research trips to the Galápagos during his career and was an active participant in the Charles Darwin Foundation for many years. Eibl went on to found the field of human ethology, was professor of zoology at the Ludwig Maximilian University in Munich, and later headed the department of human ethology at the Max Planck Institute.

29. See Ross, *Ecology and Power*, 370; Mitman, *Reel Nature*; on Galápagos documentaries, see Larson, *Evolution's Workshop*.

30. Behrman, "On the Trail of Darwin," 6.

31. Bowman, "Treasure Islands," 23, 24.

32. Barnett, "Fantastic Galápagos."

33. Bowman, "Treasure Islands," 23.

34. Bowman, *Report*, 12.

35. Bowman, "Treasure Islands," 23.

36. Bowman, "Treasure Islands," 24.

37. Eibl-Eibesfeldt, *Galapagos* (1961), 51; also Bowman, *Report*, 1.

38. Eibl-Eibesfeldt, *Galapagos* (1961), 52.

39. Bowman, "Treasure Islands," 22.

40. Eibl-Eibesfeldt also took home four juvenile tortoises, whose growth he tracked over two years. Eibl-Eibesfeldt, *Galapagos* (1961), 60–61.

41. Eibl-Eibesfeldt, *Galapagos* (1961), 63.

42. Eibl-Eibesfeldt, *Galapagos* (1961), 63.

43. Behrman, "On the Trail of Darwin," 6; Bowman, *Report*.

44. The original plan had been to co-author a joint report, but following the mission, relations between the two soured. The conflict between them lasted for years—even to the extent that they were never concurrently on the Charles Darwin Foundation board. See Hennessy, "Mythologizing Darwin's Islands."

45. Bowman, *Report*, 36.

46. Hal Coolidge later recounted to Bowman the politicking that went into this resolution: "I also recall the preparation of the resolution at the International Zoological Congress in London where Van Straelen and Ripley played a strong role in getting the resolution about the Station adopted. It was not customary for Zoological Congresses to adopt resolutions and it took a great deal of lobbying effort to get this resolution put through. There was no question about enthusiastic support for it, but we had to overcome some stubborn bureaucratic and procedural objections." H. Coolidge to R. Bowman, 26 August 1978, Robert Irwin Bowman Papers MSS-062, Box 6, California Academy of Sciences Library, San Francisco, CA (hereafter CAS Bowman Papers).

47. Also involved were Jean Dorst, director of the French Natural History Museum; Hal Coolidge; Sir Peter Scott, who later led the World Wildlife Fund; S. Dillon Ripley; and Jean-Georges Baer, then president of the IUCN. Bowman was also on the committee, serving as secretary for the Americas.

48. "Embajador Ecuatoriano," 6.

49. Jorge Espinosa (Government of Ecuador Ministry of Foreign Affairs) to Luther Evans (Director-General of UNESCO), 19 February 1958; A. Balinski (Ecuador Permanent Representative to UNESCO) to R. Galindo (Chief, UNESCO Bureau of Relations with Member States), 30 May 1958; R. Galindo to J. Dorst, 12 November 1958, 551.46 A5/01 (866) AMS/TA Marine Sciences—Research Station, Part II: from 1.6.57, Galápagos Isles, Ecuador, Part. Prog. & TA, UNESCO Archives, Paris.

50. R. Bowman to N. Rothman, 21 November 1960, Box 8, CAS Bowman Papers.

51. S. D. Ripley to R. Bowman, 4 November 1958, Box 8, CAS Bowman Papers.

52. R. Bowman to A. Eglis, 26 November 1960, Box 8, CAS Bowman Papers.

53. R. Bowman to N. Rothman, 21 November 1960, Box 8, CAS Bowman Papers.

54. B. Bowman to H. Coolidge, 31 October 1959, Box 6, CAS Bowman Papers.

55. I. Eibl-Eibesfeldt to T. Grivet 12-1-1958, 557.46 (866) AMS/Galápagos, UNESCO Archives. Bowman may have been particularly sensitive to US-European politics because of his strained relationship with Eibl-Eibesfeldt.

56. The first Ecuadorian director was appointed in 2016. The national park, on the other hand, has from its inception been a state-run institution fully staffed by Ecuadorians.

57. H. Coolidge to R. Bowman, 6 February 1960, Box 6, CAS Bowman Papers. This was a potentially serious threat, considering that Bowman's research focused on Galápagos finches.

58. R. Bowman to H. Coolidge, 11 February 1960, Box 6; R. Bowman to N. Rothman 1 May 1961, Box 8; both CAS Bowman Papers.

59. The decree read, in part:
"The Charles Darwin Research Station is hereby empowered to determine the reserved zones or natural monuments, without restriction of area, on the following islands: Santa Cruz, Isabela, Española, Santa Fé, and others . . .

". . . to determine which indigenous species of flora and fauna at present threatened with extinction must have priority for conservation.

". . . [is] authorized to take all steps considered suitable, with the corresponding support of military and civil authorities, for the control and extermination of animals, either native or introduced, that have actually become a menace and are affecting the maintenance of suitable environmental conditions required for the conservation and perpetuation of the fauna and flora of the islands.

". . . Any type of uncontrolled colonization for farming purposes, burning and exploitation of the trees for the supply of lumber and charcoal is henceforth prohibited in those areas so determined by the Charles Darwin Research Station."
1959 Decree: Junta Militar de Gobierno, Executive Decree No. 523, Box 3, CAS Bowman Papers. For more on the negotiations, see Hennessy, "Mythologizing Darwin's Islands."

60. Mitchell, *Rule of Experts;* Cushman, *Guano;* Escobar, *Encountering Development;* Goldman, *Imperial Nature;* Li, *Will to Improve;* McCook, *States of Nature.* Progress was not a creation of the mid-twentieth century but an idea that, as historian Richard Drayton argues, stemmed from the late eighteenth century, when an "imperialism of improvement" promised that nature would be best governed by "those who understood nature's laws" (*Nature's Government,* xv). In the early twentieth century, this promise became coupled with naturalists' mounting fears of extinction to power a global drive for conservation that has over the past century resulted in the creation of more than two hundred thousand protected areas that include 15 percent of the world's terrestrial area and 3 percent of its oceans. On extinction fears, see Barrow, *Nature's Ghosts;* and Kolbert, *Sixth Extinction.*

61. Gilley, "Geographical Imagination," 1223.

62. Latour and Woolgar, *Laboratory Life;* Knorr Cetina, "The Couch."

63. Gordillo, *Stories,* 135–140, 158–177; Latorre, *El Hombre,* 344–356.

64. Faris, Catton, and Larsen, "Galápagos Expedition"; Hillinger, "Galápagos Ho!"; Otterman, *Clinker Islands,* 211–230; Bocci, "Invasive Life," 71–79.

65. Harrsch, "Filiate Science Antrorse."

66. Quoted in Faris, Catton, and Larsen, "Galápagos Expedition," 50.

67. Faris, Catton, and Larsen, "Galápagos Expedition," 51.

68. Quoted in Bocci, "Invasive Life," 77.

69. Faris, Catton, and Larsen, "Galápagos Expedition," 50.

70. Hillinger, "Galápagos Ho!"

71. Hillinger, "Galápagos Ho!"

72. R. Bowman to A. Eglis, 26 November 1960, Box 3, CAS Bowman Papers.

73. Hillinger, "Galápagos Ho!"

74. Harrsch admitted to racist views against African Americans and Mexicans. Faris, Catton, and Larsen, "Galápagos Expedition," 49.

75. Hillinger, "U.S. Colony."

76. Hillinger, "U.S. Colony."

77. Sundberg, "Conservation Encounters"; the concept was originally developed by Pratt, *Imperial Eyes.*

78. "I could say hello, *muchas gracias,* but not much else." Raymond Lévêque, interview with Matthew James, 8 March 2000.

79. R. Lévêque to J. Dorst, 27 July 1960, Unsorted Correspondence, CDRS Archive.

80. Raymond Lévêque interview; R. Lévêque to J. Dorst, 5 September 1960, Unsorted Correspondence, CDRS Archive.

81. R. Lévêque to Jean Dorst, 27 July 1960, Unsorted Correspondence, CDRS Archive. It is unclear whether island colonists would have received nature documentaries as Lévêque intended, for he also told Dorst: "It's bad, all these guys turning up in the islands to make films or give media reports. Especially if they're like Zuber. This poor guy found a way to make the whole archipelago hate him within 8 months. Not surprising, once you hear how he behaved during, and after . . . the mountain of dirtbags and idiots who assist with his photos won't bring him any luck either. In the meantime, requests for information about the Galápagos are pouring in, and new settlers, as well." On the centrality of wildlife film-making to mid-twentieth-century conservation efforts, see Mitman, *Reel Nature.*

82. Raymond Lévêque, interview.

83. Lévêque did not have the illustrious career of other naturalists involved with the station project in its early years. He published several ornithological papers after his stay but only returned to the Galápagos years later as a guide for Swiss tourists.

84. R. Lévêque to J. Dorst, 17 June 1960, Unsorted Correspondence, CDRS Archive.

85. R. Lévêque to J. Dorst, 17 June 1960, Unsorted Correspondence, CDRS Archive; Simkin, "Geology," 78.

86. Eibl-Eibesfeldt, *Galápagos* (1961), 22.

87. See Striffler, *In the Shadows;* and Soluri, *Banana Cultures.*

88. Striffler, *In the Shadows.*

89. R. Bowman to H. Coolidge, 27 May 1960, quoting a letter from Lévêque, Box 6, CAS Bowman Papers.

90. R. Bowman to A. Eglis, 26 November 1960, Box 8, CAS Bowman Papers (emphasis in original).

91. Callicott and Nelson, *Great New Wilderness Debate.*

92. Acosta Solís, "Problems of Conservation," 284; on his conservation philosophy in regard to state development more broadly, see Cuvi, "Misael Acosta-Solís."

93. Acosta Solís, "Problems of Conservation," 283, 282.

94. Koford, "Economic Resources," 289.

95. Bonifaz, "El Ecuador," 3.

96. Perry, *Island Days,* 39–40.

97. Koford also suggested that sheep would be a more suitable replacement for "more destructive and relatively useless feral goats" ("Economic Resources," 289).

98. Also see Brockington and Scholfield, "Conservationist Mode."

99. R. Bowman to N. Rothman, 21 November 1960, Box 8, CAS Bowman Papers (emphasis in original).

100. "En las Islas Galápagos," 4.

101. Grimwood and Snow, "Recommendations," 1. They also recommended allowing fishing to continue with permits, the appointment of a chief park warden and eighteen guards, and that local residents be encouraged to develop, "as rapidly as possible," facilities to cater to tourists (3).

102. Mountfort, "Problems of Tourism."

103. Dawson, ed., *Economic Feasibility,* 3, 26.

104. Brockington, *Fortress Conservation.*

105. De Bont, "World Laboratory"; also Kupper, "Nature's Laboratories?"

106. Kramer, "Conservacion" (my translation).

107. "Ley de reforma," 1; quotation from Blankstein and Zuvekas, "Agrarian Reform," 79.

108. Striffler, *In the Shadows,* 119–127; Blankstein and Zuvekas, "Agrarian Reform." The most significant effect of the first agrarian reform legislation was to formally end the *hausipungo* system of land tenancy, similar to sharecropping, that tied indigenous populations in the Sierra to *patrones*—and that was then widely seen as a "precapitalist" labor relation that blocked efficient development of market-oriented agriculture. The second push toward agrarian reform of the 1970s was more responsive to peasant organizing but also predominately served a desire for modernization.

109. Castillo, Oral History Interview, 37.

110. Jacoby, *Crimes Against Nature;* Carruthers, *Kruger National Park;* Neumann, *Imposing Wilderness.* For an example of a national park used by an authoritarian state to organize territory, and its tensions with agrarian reform, see Frederico Freitas's account of farmer evictions from Iguaçu National Park in Brazil, "Ordering the Borderland." Also discussed there is the relationship between the military and conservation: "The institutionalization of the park system in the years that followed (1970s) suggests the evictions helped catalyse a nature state mandate as a component of the expanded government powers introduced during the Brazilian military regime" (161).

111. On conservation territories, see Zimmerer, *Globalization,* 8. Also see Brosius, Tsing and Zerner, eds., *Communities and Conservation;* West, Igoe, and Brockington, "Parks and Peoples"; West, *Conservation;* Adams and Hutton, "People, Parks and Poverty."

112. On conflicts that arose from tensions between colonization and conservation in South America, see Amend and Amend, eds., *National Parks Without People?;* and Klubock, *La Frontera.* As Wakild has shown in Mexico (*Revolutionary Parks*), early-twentieth-century park creation was done in part to meet social justice objectives, offering a different story, albeit not one that was without conflict. See also Jacoby, *Crimes Against Nature;* West, *Conservation;* and Scott, *Weapons of the Weak.*

113. J. Dorst to R. Lévêque, 21 July 1960, Unsorted Correspondence, CDRS Archive.

114. R. Lévêque to J. Dorst, 17 June 1960, Unsorted Correspondence, CDRS Archive.

115. R. Lévêque to J. Dorst, 5 September 1960, Unsorted Correspondence, CDRS Archive.

116. Perry, *Island Days,* 36–38.

117. Carruthers, *Kruger National Park;* also Neumann, *Imposing Wilderness;* West, *Conservation.*

118. Sundberg, "Conservation Encounters," 239.

119. José Villa, interview with the author, 27 September 2011. Further quotations from Villa are from this interview.

120. Kramer, "Conservación de los Recursos Naturales."

121. Since the archipelago became part of Ecuador, it had alternately been an independent province and subsumed within the province of Guayas.

122. Kramer, "Conservación de los Recursos Naturales."

123. Valdivia, Wolford, and Lu, "Border Crossings."

124. Wolford, Lu, and Valdivia, "Environmental Crisis."

125. Grenier, *Conservación Contra Natura,* 391.

Chapter 6. Restoring Evolution

1. Bacon, "Tortoise Goes Home."

2. At least, those fifteen were all that naturalists could find at the time; scientists have since told me of seeing an old, unmarked tortoise on Española, likely an "original" inhabitant.

3. Parts of this chapter were previously published in Hennessy, "Producing 'Prehistoric' Life."

4. Place-based strategies became leading methods of saving species around the world in the late twentieth century. Although zoo-based breeding remains an important strategy for carrying on species from threatened habitats, a focus on place-based conservation now dominates attempts to save Galápagos wildlife. On the relationships between ex situ and in situ management of wildlife, see Braverman, *Wild Life* and *Zooland.*

5. Howard Snell, interview with the author, 30 January 2012. Further quotations from Snell are from this interview.

6. Linda Cayot, interviews with the author, 24 and 28 September 2012. Further quotations from Cayot are from these interviews.

7. James Gibbs, personal communication with the author, 24 October 2018.

8. Bensted-Smith, ed., *Biodiversity Vision,* 1.

9. Grosz, *Time Travels,* 37.

10. Geographers and others have developed Foucault's concept of biopower as an approach to understanding conservationist management of nonhuman life. For Foucault, biopower was a productive power centrally concerned with the management of populations, to "make live or let die." See Biermann and Mansfield, "Biodiversity, Purity, and Death"; Biermann and Anderson, "Conservation"; Lorimer and Driessen, "Bovine Biopolitics"; and Lorimer, *Wildlife in the Anthropocene.* These reworkings of Foucault's theories are helpful for analyzing how conservationists apply ecological and evolutionary understandings of life as strategies for managing the islands and their species. Yet biopower does not go far in examining the lived experiences through which ecological knowledge is gained and applied, the agency of nonhumans, or the intimacies and affective relationships that develop between conservationists and the nonhuman species with which they work.

11. In Foucault's words, Darwin "found that population was the medium between the milieu and the organism, with all the specific effects of population: mutations, eliminations, and so forth. So in the analysis of living beings it is the problematization of the population that makes possible the transition from natural history to biology. We should look for the turning point between natural history and biology on the side of population" (*Security, Territory, Population,* 78).

12. Grosz, *Nick of Time,* 32.

13. Crosby, *Ecological Imperialism* and *Columbian Exchange;* Robbins and Moore, "Ecological Anxiety Disorder."

14. For Foucault the power to "make die or let live" was an earlier form of sovereign power that stood in contrast to biopower. In the realm of conservation management of nonhuman life, the temporal distinction does not hold; conservation of species populations depends equally on both biopower and sovereign power, differentially applied to divide life-forms into those that conservationists seek to make flourish, whose life is

managed through productive biopower, to those that threaten this valued life and should be made to die.

15. van Dooren, "Breeding Cranes."

16. Carlos Ochoa is a pseudonym.

17. Carlos Ochoa, interview with the author, 1 March 2012; Van Denburgh, "Expedition," 243.

18. Dorst, "Problems," 1.

19. José Villa, interview with the author, 27 September 2011. Further quotations from Villa are from this and other interviews.

20. Perry, *Island Days*.

21. Perry, *Island Days*, 59.

22. MacFarland, "Giant Tortoises," 644.

23. MacFarland, "Giant Tortoises," 640.

24. MacFarland, "Giant Tortoises," 640.

25. MacFarland, Villa, and Toro, "Galápagos Giant Tortoises"; Kramer, "Galápagos: Island Under Siege," 188; Eibl-Eibesfeldt, *Survey*, 20–21.

26. Perry, *Island Days*, 57; MacFarland, "Giant Tortoises," 646; Fausto Llerena, interview with the author, 21 August 2011.

27. Perry, *Island Days*, 46.

28. Dawson, "Cacti," 212.

29. Darwin, *Journal of Researches*, 478.

30. It was the kind of takeover that geographer Alfred Crosby would soon call "ecological imperialism" as he described early explorers' introduction of European animals and microbes—horses, pigs, and smallpox among them—that helped them conquer the Americas.

31. Don Fausto was honored in 2015 with a species epithet when geneticists classified two separate populations of tortoises on Santa Cruz as distinct species; scientists named the new population *C. donfaustoi*. The animals live near Cerro Fatal on the eastern part of the island, northeast of the park and station headquarters. Poulakakis, Edwards, et al., "Description."

32. Márquez, Cayot, and Rea, *La Crianza*.

33. van Dooren, "Breeding Cranes," 291–294.

34. As I went about my work that day, I still felt torn—the *hembra*'s weeks of nurturing the eggs, her careful digging of a nest, were for naught. She would not miss them— tortoises are not social animals; they do not live in herds or rear their young. But I felt wasteful—why pitch eggs into the *monte*? Why couldn't someone take them home to eat?, I asked. Daniel (a pseudonym) gave me a strange look, as if to say, have you learned nothing here? Obviously we were not supposed to be eating tortoises!

35. Cayot and Morillo, "Rearing."

36. Milinkovitch, Monteyne, et al., "Genetic," 334; Milinkovitch, Kanitz, et al., "Recovery"; Deamer, "Did 1 Sexually Active."

37. Tom Fritts, interview with the author, 29 January 2012.

38. Grigioni, "La estimulación sexual." Linda Cayot, interview with the author, 29 September 2011.

39. Others suggested electro-ejaculation, a technique commonly used for captive breeding of large mammals. But Grigioni warned against this; the technique is rarely used on tortoises and could have potentially caused George serious damage or killed him—a risk no one in the Galápagos was willing to take. See Pritchard, "Further Thoughts."

40. Kramer, "Galápagos Conservation," 4.

41. van Dooren, "Breeding Cranes," 293.

42. van Dooren, "Invasive Species," 290.

43. Haraway, When Species Meet, 80.

44. Washington Tapia, interview with the author, 22 February 2012; author's translation from Spanish. Further quotations from Tapia are from this and other interviews.

45. Lavoie et al., Thematic Atlas, 10.

46. Washington Tapia, interview with the author, 22 February 2012.

47. Anonymous interview with the author, 31 October 2011.

48. Project donor institutions included the Galápagos National Park and Darwin station, government of Ecuador, German government (KfW), Inter-American Development Bank, World Wildlife Fund, United Nations Foundation, AECI (Spanish Agency for International Development Cooperation), and USAID, along with private sector companies. Coello and Saunders, "Final Project Evaluation."

49. Lavoie et al., Thematic Atlas, 6; project leader quoted in Bocci, "Tangles," 424.

50. Anonymous interview with the author, 31 October 2011; Romer, "In the Galápagos Islands."

51. Campbell et al., "Increasing," 737.

52. Rose, "Judas Work," 63.

53. Campbell, "Manipulation," 4.

54. Campbell, "Manipulation," 9.

55. Lavoie et al., Thematic Atlas, 10.

56. Bocci, "Tangles," 438.

57. Coello and Saunders, "Final Project Evaluation," 2011; Bocci, "Tangles."

58. See Nicholls, "Invasive Species"; and Cabrera and Garcia, "Reports."

59. van Dooren, "Invasive Species," 293–294.

60. Haraway, Staying with the Trouble and When Species Meet.

61. Tapia, Málaga, and Gibbs, "Giant Tortoises Hatch."

62. James Gibbs, interview with author, 2 October 2011; Gibbs et al., "Demographic Outcomes." The hope for 100 percent survival is a response driven by the sense of emotional attachment that conservation depends on—it's not an expectation that reflects the reality of natural reproduction. Indeed, low survivorship, or "recruitment" as ecologists call it, is likely the reason female tortoises lay multiple eggs in multiple nests each year.

63. Gibbs et al., "Demographic Outcomes."

64. Gibbs, Márquez, and Sterling, "Role"; Blake, Wikelski, et al., "Seed Dispersal."

65. Gibbs and Woltz, "Pilot Survey."

66. "Nativeness is not a sign of evolutionary fitness or of a species having positive effects," Davis et al., "Don't Judge," 153.

67. Blake, Wikelski, et al., "Seed Dispersal," 6.

68. Nicholls, "Invasive Species"; Gardener, Atkinson, and Rentería, "Eradications."

69. Quoted in Vince, "Embracing Invasives," 1383.

70. Davis et al., "Don't Judge," 153–154.

71. Hobbs, "Novel Ecosystems."

72. Botkin, "Naturalness."

73. Botkin, "Naturalness," 264.

74. Curnutt, "Guide," 1756.

75. Robbins and Moore, "Ecological Anxiety Disorder," 3–4.

76. Linda Cayot, interview with author, 28 September 2011.

77. Hamann, "On Vegetation," 138.

78. Anonymous interview with the author, 31 October 2011.

79. Knafo et al., "Sterilisation"; Rivera et al., "Sterilisation."

80. Quoted in "Lonesome George."

81. Anonymous interview with the author, 16 December 2011.

82. Peter Pritchard, interview with the author, 12 June 2012.

83. Donlan et al., "Pleistocene"; Donlan, "Restoring"; Shapiro, *How to Make a Mammoth*.

84. James Gibbs, interview with the author.

85. Hunter et al., "Equivalency."

86. Tapia, "Return to Española Island."

87. "Giant Tortoise."

88. Koop et al., "Introduced"; Fessel and Tebbich, *"Philornis downsi"*; Fundación Charles Darwin (FCD) y WWF–Ecuador, *Atlas de Galápagos*, 166–167.

Chapter 7. Laboratory Life

1. Tapia, Ospina, et al., "Toward a Shared Vision."

2. González et al., "Rethinking the Galapagos Islands," 1. The model was developed through a project coordinated by the national park along with scholars from the Universidad Autónoma de Madrid and two Ecuadorian universities: the Universidad Andina Simón Bolívar and Universidad San Francisco de Quito. The project was titled "Galápagos como sistema socioecológico: Nuevas estrategias y prioridades para la conservación de la biodiversidad y el desarrollo humano en el archipiélago, en el marco del cambio global y la teoría de los sistemas complejos adaptativos." Representatives of other local institutions, including the Galápagos National Institute (INGALA), the Charles Darwin

Foundation, and the government also participated. Financing was provided from Grupo Santander, el Centro de Estudios para America Latina, la Universidad Autónoma de Madrid, and la Agencia Española de Cooperación Internacional. See Tapia, Ospina, et al., "Toward a Shared Vision."

3. González et al., "Rethinking the Galapagos Islands"; Tapia, Novoa, et al., "Entendiendo Galápagos"; Berkes, Colding, and Folke, eds., "Introduction." Three of the fundamental assumptions of socio-ecological systems are that linear models are insufficient for understanding a world of uncertain change at multiple scales, that qualitative analysis is a necessary complement to quantitative data, and that multiple perspectives are necessary to understand complex change and system dynamics. From a social perspective, history, as well as political and economic contexts, was crucial for understanding change and planning for sustainability. See Berkes and Folke, "Linking," 6–7.

4. Cote and Nightingale, "Resilience Thinking"; Nelson, "Resilience"; Clarke and Hansen, eds., *Emergence and Embodiment;* Walsh-Dilley, Wolford, and McCarthy, "Rights for Resilience."

5. Grenier, *Conservación,* 391.

6. González et al., "Rethinking the Galapagos Islands," 7.

7. Dean, *With Broadax and Firebrand;* Miller, *Environmental History of Latin America.*

8. González et al., "Rethinking the Galapagos Islands," 7.

9. See, for example, Mitchell, "Work of Economics."

10. The 2002 Biodiversity Vision had expressed a similar vision for island life in which a sustainable Galápagos archipelago was one where "a small, well educated, healthy human population co-exists with nature, uses resources sparingly and works constantly to control alien species. The people would have their own, distinctive way of life, appropriate to oceanic islands that evolved in isolation from man and are consequently so vulnerable to human presence. They would accept restrictions and responsibilities and enjoy to the full the privilege of living in one of the most special natural environments on Earth." Bensted-Smith, ed., *Biodiversity Vision,* iii.

11. Like the biopower that has governed conservationist management of endemic and introduced species, conservationists also sought to manage what Foucault called the "conduct of conduct," ordering relationships between people and the environment such that people, as David Scott wrote, citing Jeremy Bentham, "following only their own self-interest, will do as they ought" ("Colonial Governmentality," 202). For Foucault, government was a constellation of multiple modes of power aimed at managing populations, including disciplinary powers of surveillance as well as governmental management. Governmental management of human populations in the Galápagos involves both attempts to create ideal citizens who act in appropriate ways toward the environment and surveillance and policing. Yet Foucault has often been critiqued for presenting government as a totalizing power; to emphasize the limits of government,

I draw from analyses of resistance that draw from the work of Antonio Gramsci. See, for example, Li, *Will to Improve;* and West, *Conservation.* On governmentality, see Foucault, *Security, Territory, Population;* Agrawal, "Environmentality"; and Rutherford, "Green Governmentality."

12. My point is that the kind of rich, embedded understanding of the challenges of Galápagos conservation is not gained through models that seek to simplify life, but through the kind of fieldwork that takes one into the thick of multispecies entanglements. What we need is more thick description—of the natural history kind, of which Darwin was a master, and of the ethnographic sort that can help us understand how people make decisions and how their everyday lives are embedded in and productive of nature. As the anthropologist Celia Lowe has argued (*Wild Profusion*), nature is made at the intersection of people with unique social histories and plants and animals with unique evolutionary histories. How they come together is the focus of multispecies ethnography. This is not typical of how we have been taught to understand scientific knowledge—it is neither objective nor universal. Instead, this knowledge is situated, specific. My story is subjective, made through my particular embodied experiences, the stories that others have told me, and the features of the landscape that I have been called to notice. This is not knowledge representative of all life in the archipelago, but rather is illustrative of the nature and politics of dense entanglements. Yet the stories I tell here and throughout the book are not anecdotal. They are ethnographic; together they add up to more than a series of poignant tales. What we give up in seeking comprehensive knowledge, we gain in depth of understanding that is essential for figuring our way through socio-environmental troubles.

13. I also draw from conversations with Joe Bryan, from the "cultural transect" Jake Kosek takes in his 2006 book *Understories* (228–254), and from Anna Tsing's focus on the "arts of noticing" (*Mushroom at the End of the World,* 17).

14. The airport is LEED Gold certified, was built using steel tubes recycled from oil drilling in the Ecuadorian Amazon, and is powered by wind turbines and solar energy. See "Aeropuerto."

15. Kramer, Informe comprensivo, 31.

16. Cuvi and Guijarro, "Una trayectoria."

17. Cepeda, "Conservación."

18. Tapia, Ospina, et al., "Toward a Shared Vision," 15.

19. See Robbins, *Political Ecology,* 16–17.

20. "Plan Maestro," 21.

21. Kramer, Informe comprensivo, 27–28.

22. For example, in 1984 UNESCO nominated the islands to its Man-in-the-Biosphere program, but its first progress report in 1999 found that the archipelago had been managed primarily as a classic national park that did not fulfill the Biosphere Reserve criteria. Heslinga, "Regulating Ecotourism."

23. Williams, *Country and the City;* Cronon, *Nature's Metropolis;* Cronon, "Kennecott Journey."

24. See the project's website, www.gianttortoise.org. It is funded in part by the Max Planck Institute for Ornithology in Germany, where Blake's team works with ornithologist Martin Wileski, who runs the Department of Migration and Immuno-ecology and researches animal movement in species around the world.

25. "Salad bowl" is from Blake, "Slow Motion," 78; Yackulic, Blake, and Bastille-Rousseau, "Benefits"; Blake, Yackulic, et al., "Migration."

26. Benitez-Capistros et al., "Emergent."

27. For a time, Blake ran a website where one could monitor in real time the movements of a few tortoises with special satellite transmitters. See Haraway, "Cyborg Manifesto."

28. Fairhead and Leach, *Science, Society and Power*.

29. Wilfrido Michuy, interview with the author, 18 October 2011.

30. Park director, interview with the author, 17 October 2011.

31. "Boletín Informativo del Comite Interinstitucional de Desarrollo y Conservación del Canton Isabela," Febrero 1989, Nro. 4, n.p., CDF History Box, CDRS Archives. My translation.

32. Fabian, *Time and the Other*; Escobar, *Encountering Development*; Gupta, *Postcolonial Developments*.

33. McClintock, *Imperial Leather*, 36.

34. Grenier (*Conservación*) notes that tortoise killings in the late 1980s and early 1990s coincided with major conservation efforts, including the founding of the Galápagos Marine Reserve in 1986. Cayot and Lewis ("Recent Increase") point out that the numbers of found tortoises also depends on the number of field excursions taken in a given year. In the early 1990s, La Cazuela, where thirty of the eighty-one tortoises were found, was the most-visited site.

35. See, for example, Grove, "Crisis in the Galápagos"; also Bremner and Perez, "Case Study."

36. Márquez et al., "Human-Caused."

37. Márquez et al., "Human-Caused." When tortoises are killed for food, their plastrons (undershells) are usually slit or removed. Park wardens say the state of the dead body is an obvious indicator of how and why a tortoise died. Tortoises are not the only animals found slaughtered in what conservationists interpret as acts of protest. In January 2008, park rangers found fifty-three dead sea lions on Pinta ("Sea Lions Massacred"). Sea lion killings are not uncommon—their penises are sold to Asian markets—but these had had their skulls crushed, apparently either acts of vengeance or fishers' attempt to defend themselves and their catch from aggressive animals.

38. Bassett, *Galapagos at the Crossroads*.

39. It is not only fishers who strike in the Galápagos. Several months later national park guards closed the road to their own headquarters in protest of the actions of a park director whom they thought was allying with the fishers against the interests of conservation.

40. See Grenier, *Conservación*, 317–336.

41. Bremner and Perez, "Case Study"; Camhi, "Industrial Fisheries."

42. In response to the strikes, several programs over the past two decades have attempted to help fishers transition into small-scale tourism, with limited success. See Engie and Quiroga, "Emergence," and Engie, "Adaptation."

43. Heslinga, "Regulating Ecotourism."

44. That deal was one of many neoliberal reforms that attempted to address the economic crash and crisis of social welfare caused by structural adjustment programs mandated by the International Monetary Fund that privatized Ecuadorian state industries in the early 1980s. The swap encouraged conservation by funding local NGOs with money raised by foreign NGOs that bought some of Ecuador's foreign debt on secondary markets where the price was low because of the high likelihood that the state would default. In exchange for the canceled debt, the state agreed to promote conservation. See also Harvey, *Brief History of Neoliberalism*.

45. Grenier (*Conservación*) defines the central axis of the conflict of the 1990s as the radicalization of "selective tourism" and the explosion of the sea cucumber fishery. He thinks the government in the mid-1990s effectively smothered the productive sectors, leading to intense conflict. Pablo Ospina (*Galápagos* and "Migraciones") found that the sea cucumber conflict led to a sense of Galápagos community as residents united against outside control, but then migration was also growing rapidly and proved too heterogeneous to cement a sense of Galápagos identity.

46. For analysis, see Heylings and Bravo, "Evaluating Governance."

47. Historian Pablo Ospina (*Galápagos*) identified three central problems of governability: (1) the great density of public institutions; (2) the control of these institutions by local inhabitants and their leaders—extensive social ties make institutions impotent in their ability to enforce laws; and (3) the plurality of mechanisms of political representation by which the local leaders extend their authority and make sure that they have the control of public institutions, from family relationships to union leaderships. The line between political and civil society is exceptionally blurry in the Galápagos because of the back-and-forth of leaders between institutions and unions.

48. Sponsored by the Charles Darwin Foundation and the World Wildlife Fund, attendees included Galápagos specialists, including staff from the national park and Darwin station, as well as visiting scientists from Ecuador, the United States, and Europe. It was a largely insular community of scientists and policymakers, many of whom had worked on Galápagos conservation for years. Although not addressed extensively at this workshop, the concept of biodiversity has provided a strong lens for the political organization of naturalists and conservationists since it was coined in the 1980s. See Takacs, *Idea of Biodiversity*; Raby, *American Tropics*, traces the history of the concept as it developed through the work of largely US biologists at tropical research stations during the twentieth century. Anthropologist Celia Lowe (*Wild Profusion*) argues that the biodiversity discourse was built on earlier ideas of environmental risk, but that "biodiversity was

not so much a solution to the problems of environmental risk, however, as its problematization. It instigated a new form of critical inquiry into the relationship between
entities conceived of as 'nature' and 'the human.' Thrust into the light was, on the one
hand, nature, understood as the linkages between genetic variation, species populations, communities and ecosystems, and land and marinescapes and, on the other
hand, humanity, with its ability to instigate what biologist Michael Soule has termed
the 'sixth great extinction' " (4). These two poles of biodiversity and humanity have been
central to how Galápagos policymakers frame local conflicts.

49. Bensted-Smith, ed., *Biodiversity Vision*, 1: the "95 percent" estimation has since
become among the most-cited statistics about the islands, used to celebrate the achievements of conservation and to argue for continued protection of this nearly-pristine archipelago. See Watkins and Cruz, "Galápagos At Risk." But the fact is not as clear-cut
as it seems. In discussing the need for extended taxonomic work on Galápagos species,
Bungartz et al. ("Neglected") point out that while between 18,000 and 25,000 Galápagos species have been identified, as much as half of the islands' biodiversity remains
unknown. Since 2008, the Darwin station has kept an online inventory of all species
in the archipelago. In 2012, the total count was 11,305—6,484 of which were accepted
names, the rest discarded as duplicates or misidentifications. The 95 percent then refers to the presence of reported species. But, the authors caution, "Even if the above
cited 95% estimate were correct for known taxa, the CDF database unequivocally illustrates that the majority of species-numerous groups, all providing essential ecosystem services, have almost entirely been neglected!" (137).

50. Bensted-Smith, ed., *Biodiversity Vision*, 1. On the 1534 baseline as a goal for conservation efforts, authors wrote: "The Galápagos is probably the only remaining large
oceanic archipelago in the world where:

1. We can still accurately quantify the distributions and conditions of terrestrial biodiversity prior to human settlement, and

2. Virtually the full complement of biodiversity is still recoverable, and

3. It could be possible to restore the distributions of almost all terrestrial
biodiversity to the conditions prior to human settlement. Therefore, an approxi-
mation to the original pre-settlement state of Galápagos terrestrial biodiversity
remains a valid, definable benchmark: an ultimate goal. To achieve it would take
much more than 50 years and great advances in science and conservation, nevertheless we should not lose sight of it, as we struggle now to cope with the
hordes of invading alien species. That was the feeling of the of participants in the
Biodiversity Workshop, who described this ultimate goal as "the restoration of
the populations and distributions of all extant native biodiversity and of natural
ecological/evolutionary processes to the conditions prior to human settlement."
If this extremely ambitious goal were one day to be achieved, it would represent
the pinnacle of accomplishment in conservation biology—the restoration of the

biological nature of the Galápagos Islands almost to the conditions of 1534." Bensted-Smith, ed., *Biodiversity Vision,* 48.

51. Cronon, "Wilderness."

52. Bensted-Smith, ed., *Biodiversity Vision,* 16.

53. Márquez, Gordillo, and Tupiza, "The Fire of 1994." A station herpetologist I asked about the incident said the tortoise had been accidentally injured as park guards attempted to protect it from the fire. Conservationists had done their best to keep this incident out of news coverage of the wildfire, though.

54. Grenier, *Conservación;* Ospina, "El Hada del Agua."

55. Interview with the author, 5 March 2012.

56. Constantino, "Tortoise Soup," 96.

57. As Celata and Sanna ("Post-Political Ecology") write, "When one principle, such as the protection of biological diversity, acquires a universal character, or when a place is designated as world heritage, the scope of social struggles is strictly limited for two reasons: in the first place because the conflict prevents a consensual solution to the problem, and in the second place because disagreement can be judged as insulting with respect to the 'bigger' problem to which it refers" (987).

58. Constantino, "Tortoise Soup," 90.

59. Interview with the author, 22 May 2007.

60. The authors called the Galápagos a "humans-*with*-nature" model rather than a "humans-*in*-nature" situation, as they characterized most socio-ecological systems, because the absence of an indigenous human population had "precluded the coevolution of cultural and natural forces." This idea implicitly suggests that the co-evolution of an indigenous society with its environment would have been sustainable—an idea roundly critiqued by anthropologists and ecologists as the myth of the "ecologically noble savage." Calls for cultural change in islanders' lifestyles reflect a common trope of environmentalism that frames controversies as revolving around Edenic narratives in which an original, pristine nature is lost through some culpable human act that results in environmental degradation and moral jeopardy. Hames, "Ecologically"; Nadasdy, "Transcending"; Redford, "Ecologically"; Slater, "Amazonia."

61. Agrawal, "Environmentality."

62. Ospina, *Galápagos.*

63. There is also diversity among urban and rural populations in terms of wealth, community identity (many agriculturalists identify with continental communities of origin such as Loja or Salasaca), and migration status. For example, while most fishers are permanent residents, many farm owners employ undocumented migrants who are willing to work for wages significantly less than what is demanded by Galápagos residents. On the Galápagos, see Ospina, *Galápagos,* and Bocci, *Invasive Life;* more broadly, Bebbington, "Organizations and Intensifications."

64. Quiroga, "Crafting Nature"; Ospina and Facloni, eds., *Galápagos.*

65. See, for example, Cairns, Sallu, and Goodman, "Questioning"; Valdivia, Wolford, and Lu, "Border Crossings."

66. Ospina, "El Hada del Agua."

67. Correa's government had declared a moratorium on hotel construction in the Galápagos in 2013, but the following year, the Governing Council of the Galápagos lifted it to allow 20 hotels it had selected—some including more than 100 rooms and a golf course. Santos, "Galápagos al límite."

68. Interview with the author, 23 July 2015.

69. Alvear and Lewis, "Nota Actualizada."

Chapter 8. All the Way Down

1. I told the story of the Floreana tortoises in Hennessy, "Molecular Turn"; also see Poulakakis, Glaberman, et al., "Historical DNA"; Garrick et al., "Genetic Rediscovery"; and Beheregaray and Caccone, "How We Rediscovered."

2. Van Denburgh, "Expedition," 243; Pritchard, "Galápagos Tortoises," 53–54.

3. Garrick et al., "Genetic Rediscovery."

4. Porter, *Journal of a Cruise*, 151–152.

5. Darwin, *On the Origin of Species*, 489–490.

6. Faulkner, *Requiem for a Nun*, 73.

7. Pratt, *Imperial Eyes*; Raffles, *In Amazonia*. On extractivism and environmental history, see Dean, *With Broadax and Firebrand*; Soluri, *Banana Cultures*; Striffler, *In the Shadows*; Mintz, *Sweetness and Power*; Robins, *Mercury, Mining and Empire*; and Cushman, *Guano*.

8. Scholars call this "pluriversality"—a concept offered as a corrective to the presumed universality of Western worldviews and scientific rationality. The concept foregrounds the idea that "reality 'is *done* and *enacted* rather than observed'" and is thus multiple (Annemaire Mol, quoted in Collard, Dempsey, and Sundberg, "Manifesto," 328). See also Blaser and de la Cadena, "Pluriverse."

9. Grosz, *Time Travels*, 38.

BIBLIOGRAPHY

Acosta Solís, Misael. *Galápagos: Observado Fitológicamente*. Quito: Imprenta de la Universidad Central, 1937.

———. "Problems of Conservation and Economic Development of the Galápagos." In *The Galápagos: Proceedings of the Galápagos International Scientific Project*, edited by Robert Bowman, 282–285. Brussels: Palais des Academies, 1966.

———. "La Protección de las Islas Galápagos y el Mar Territorial." In *Los Recursos Naturales del Ecuador y su Conservación*. Tercera Parte, Tomo I. México, D.F.: Instituto Panamericano de Geografía y Historia, 1965.

"An Act to Authorize Protection to Be Given to Citizens of the United States Who May Discover Deposites of Guano." *United States Statutes at Large*, Thirty-Fourth Congress, Sess. I, Ch. 164, 1855, 119–120. http://legisworks.org/sal/11/stats/STATUTE-11-Pg119.pdf.

Adams, William. *Against Extinction: The Story of Conservation*. London: Earthscan, 2004.

Adams, William, and J. Hutton. "People, Parks and Poverty: Political Ecology and Biodiversity Conservation." *Conservation and Society* 5 (1997): 147–183.

Aeropuerto Ecológico Galápagos. "About Us." http://www.ecogal.aero/en/about-us.

Agrawal, Arun. "Environmentality: Community, Intimate Government, and the Making of Environmental Subjects in Kumaon, India." *Current Anthropology* 46, no. 2 (2005): 161–190.

Alagona, Peter. *After the Grizzly: Endangered Species and the Politics of Place*. Berkeley: University of California Press, 2013.

Alvear, Cecilia, and George Lewis. "Nota Actualizada 13 de Junio 2015: Continúan Protestas en Galápagos y desde Milán el Presidente Correa denuncia 'violencia.'" *Galápagos Digital*, 12 June 2015.

Amend, Stephan, and Thora Amend, eds. *National Parks Without People? The South American Experience*. Gland, Switzerland: IUCN—World Conservation Union, 1995.

Andrei, Mary Anne. "Nature's Mirror: How the Taxidermists of Ward's Natural Science Establishment Reshaped the American Natural History Museum and Founded the Wildlife Conservation Movement." PhD diss., University of Minnesota, 2006.

Anker, Peder. *Imperial Ecology: Environmental Order in the British Empire, 1895–1945.* Cambridge, MA: Harvard University Press, 2001.

Astudillo, Fernando. "Environmental Historical Archaeology of the Galápagos Islands: Paleoethnobotany of Hacienda El Progreso, 1870–1904." PhD diss., Simon Fraser University, 2017.

Auffenberg, W. "Checklist of Fossil Land Tortoises (Testudinidae)." *Bulletin of the Florida State Museum. Biological Sciences* 18 (1974): 121–251.

Bacon, J. "A Tortoise Goes Home." *Zoonooz* 51, no. 2 (1978): 4–7.

Baldacchino, Godfrey. "Islands as Novelty Sites." *Geographical Review* 97, no. 2 (2007): 165–174.

Barnett, Lincoln. "The Fantastic Galapagos: Darwin's Treasure of Wildlife." *Life,* 8 September 1958, 56–76.

Barrow, Mark. *Nature's Ghosts: Confronting Extinction from the Age of Jefferson to the Age of Ecology.* Chicago: University of Chicago Press, 2009.

———. "The Specimen Dealer: Entrepreneurial Natural History in America's Gilded Age." *Journal of the History of Biology* 33 (2000): 493–534.

Bassett, Carol Ann. *Galapagos at the Crossroads: Pirates, Biologists, Tourists, and Creationists Battle for Darwin's Cradle of Evolution.* Washington, DC: National Geographic, 2009.

Baur, George. "The Galapagos Islands." *Proceedings of the American Antiquarian Society,* October 1891, 418–423.

———. "On the Origin of the Galápagos Islands." *American Naturalist* 25, no. 291 (March 1891): 217–229.

Bebbington, Anthony. "Organizations and Intensifications: Campesino Federations, Rural Livelihoods and Agricultural Technology in the Andes and Amazonia." *World Development* 24, no. 7 (1996): 1161–1177.

Beck, R. H. "In the Home of the Giant Tortoise." *Seventh Annual Report of the New York Zoological Society,* 1 April 1903, 160–174.

Beebe, William. *Galápagos: World's End.* New York: G. P. Putnam's Sons, 1924.

Beheregaray, Luciano, and Adalgisa Caccone. "How We Rediscovered 'Extinct' Giant Tortoises in the Galápagos Islands—and How to Save Them." Conversation, 7 January 2016. https://theconversation.com/how-we-rediscovered-extinct-giant-tortoises-in-the-Galápagos-islands-and-how-to-save-them-52073.

Behrman, D. "On the Trail of Darwin in the Galápagos." *UNESCO Futures* 247 (July 1957): 6–7.

Benitez-Capistros, Francisco, Giorgia Camperio, Jean Hugé, Farid Dahdouh-Guebas, and Nico Koedam. "Emergent Conservation Conflicts in the Galápagos Islands: Human-Giant Tortoise Interactions in the Rural Area of Santa Cruz Island." *PLoS ONE* 13, no. 9 (2018): e0202268. https://doi.org/10.1371/journal.pone.0202268.

Bennett, Jane. *Vibrant Matter: A Political Ecology of Things.* Durham, NC: Duke University Press, 2010.

Bensted-Smith, R., ed. *A Biodiversity Vision for the Galapagos Islands: Based on an International Workshop of Conservation Biologists in Galapagos in May 1999.* Puerto Ayora, Galapagos: Charles Darwin Foundation and World Wildlife Fund.

Berger, J. "Population Constraints Associated with the Use of Black Rhinos as Umbrella Species for Desert Herbivores." *Conservation Biology* 11 (1997): 69–78.

Berkes, Fikret, Johan Colding, and Carl Folke, eds. "Introduction." In *Navigating Social-Ecological Systems: Building Resilience for Complexity and Change,* 1–29. Cambridge: Cambridge University Press, 2008.

Berkes, Fikret, and Carl Folke. "Linking Social and Ecological Systems for Resilience and Sustainability." In *Linking Social and Ecological Systems: Management Practices and Social Mechanisms for Building Resilience,* edited by Fikret Berkes and Carl Folke, 1–25. Cambridge: Cambridge University Press, 1998.

Berlanga, Tomás de. "Carta a Su Majestad." *Colección de documentos inéditos, relativos al descubrimiento, conquista y organización de las antiguas posesiones españolas de América y Oceanía. Tomo XLI, Cuaderno II.* Madrid: Imprenta de Manuel G. Hernandez, 1884, 538–544.

Biermann, Christine, and Robert M. Anderson. "Conservation, Biopolitics, and the Governance of Life and Death." *Geography Compass* 11 (2017). https://doi.org/10.1111/gec3.12329.

Biermann, Christine, and Becky Mansfield. "Biodiversity, Purity, and Death: Conservation Biology as Biopolitics." *Environment and Planning D: Society and Space* 32 (2014): 257–273.

Black, Juan. *Archipiélago del Ecuador.* Quito: Fundación Charles Darwin para las Islas Galápagos y World Wildlife Fund, 1973.

Blake, Stephen. "Slow Motion." *BBC Wildlife,* June 2013, 74–79. http://www.gianttortoise.org/linked/bbc_wildjune2013_uncorrected_last_draft_blake.pdf.

Blake, Stephen, Anne Guézou, Sharon L. Deem, Charles B. Yackulic, and Fredy Cabrera. "The Dominance of Introduced Plant Species in the Diets of Migratory Galápagos Tortoises Increases with Elevation on a Human-Occupied Island." *Biotropica* 47, no. 2 (2015): 246–258.

Blake, Stephen, Martin Wikelski, Fredy Cabrera, Anne Guézou, Miriam Silva, E. Sadeghayobi, Charles B. Yackulic, and Patricia Jaramillo. "Seed Dispersal by Galápagos Tortoises." *Journal of Biogeography* 39, no. 11 (2012): 1961–1972.

Blake, Stephen, Charles B. Yackulic, Martin Wikelski, Washington Tapia, James P. Gibbs, Sharon Deem, Fredy Villamar, and Fredy Cabrera. "Migration by Galapagos Giant Tortoises Requires Landscape-Scale Conservation Efforts." In Galapagos Report 2013–2014, 144–150. Puerto Ayora, Galapagos, Ecuador: Galapagos National Park Directorate, Governing Council of Galapagos, Charles Darwin Foundation, and Galapagos Conservancy. http://www.gianttortoise.org/linked/galapagosreport_2013-2014-20-blake-144-150.pdf.

Blankstein, Charles S., and Clarence Zuvekas. "Agrarian Reform in Ecuador: An Evaluation of Past Efforts and the Development of a New Approach." *Economic Development and Cultural Change* 22, no. 1 (October 1973): 73–94.

Blaser, Mario. "The Threat of the Yrmo: The Political Ontology of a Sustainable Hunting Program." *American Anthropologist* 111, no. 1 (2009): 10–20.

Blaser, Mario, and Marisol de la Cadena. "Pluriverse." In *A World of Many Worlds*, edited by Marisol de la Cadena and Mario Blaser, 1–22. Durham, NC: Duke University Press, 2018.

Bocci, Paolo. "Invasive Life: Illegal Immigrants and Invasive Species on the Galápagos Islands, Ecuador." PhD diss., University of North Carolina at Chapel Hill, 2017.

———. "Tangles of Care: Killing Goats to Save Tortoises on the Galápagos Islands." *Cultural Anthropology* 32, no. 3 (2017): 424–449.

Bonifaz, Cristóbal. "El Ecuador y Las Islas Galápagos." *Noticias de Galápagos* 1 (July 1963): 1–3.

Bonneuil, Christophe, and Jean-Baptiste Fressoz. *The Shock of the Anthropocene*. Translated by David Fernbach. London: Verso, 2016.

Botkin, Daniel B. "The Naturalness of Biological Invasions." *Western North American Naturalist* 61, no. 3 (2001): 263–264.

Bowman, Robert, ed. *The Galápagos: Proceedings of the Galápagos International Scientific Project*. Brussels: Palais des Academies, 1966.

———. *Report on a Biological Reconnaissance of the Galápagos Islands During 1957*. Paris: United Nations Educational, Scientific, and Cultural Organization (UNESCO), 1960.

———. "Treasure Islands of Science." *Americas* 10, no. 12 (1958): 18–24.

Brantz, Dorothee, ed. *Beastly Natures: Animals, Humans, and the Study of History*. Charlottesville: University of Virginia Press, 2010.

Braun, Bruce. *Intemperate Rainforest: Nature, Culture, and Power on Canada's West Coast*. Minneapolis: University of Minnesota Press, 2002.

———. "Producing Vertical Territory: Geology and Governmentality in Late Victorian Canada." *Ecumene* 7 (2000): 7–46.

Braverman, Irus. *Wild Life: The Institution of Nature*. Palo Alto, CA: Stanford University Press, 2015.

———. *Zooland: The Institution of Captivity*. Palo Alto, CA: Stanford University Press, 2012.

Bremner, Jason, and Jaime Perez. "A Case Study of Human Migration and the Sea Cucumber Crisis in the Galápagos Islands." *Ambio* 31, no. 4 (2002): 306–310.

Britton, S. "Tourism, Capital, and Place: Towards a Critical Geography of Tourism." *Environment and Planning D: Society and Space* 9 (1991): 451–478.

Brockington, Dan. *Fortress Conservation: The Preservation of the Mkomazi Game Reserve, Tanzania*. Bloomington: Indiana University Press, 2002.

Brockington, Dan, R. Duffy, and J. Igoe. *Nature Unbound: Capitalism and the Future of Protected Areas*. London: Earthscan, 2008.

Brockington, Dan, and K. Scholfield. "The Conservationist Mode of Production and Conservation NGOs in Sub-Saharan Africa." *Antipode* 42 (2010): 551–575.

Brosius, Peter, Anna Lowenhaupt Tsing, and Charles Zerner, eds. *Communities and Conservation: Histories and Politics of Community-Based Management.* Walnut Creek, CA: AltaMira, 2005.

Browne, Janet. "Biogeography and Empire." In *Cultures of Natural History: From Curiosity to Crisis,* edited by Nicholas Jardine, James Secord, and Emma Spary, 305–321. Cambridge: Cambridge University Press, 1996.

———. *Charles Darwin: A Biography. Volume I: Voyaging.* Princeton, NJ: Princeton University Press, 1995.

———. *Charles Darwin: A Biography. Volume II: The Power of Place.* Princeton, NJ: Princeton University Press, 2004.

Bungartz, Frank, Frauke Ziemmeck, Nathalia Tirado, Patricia Jaramillo, Henri Herrera, and Gustavo Jiménez-Uzcátegui. "The Neglected Majority: Biodiversity Inventories as an Integral Part of Conservation Biology." In *The Role of Science for Conservation,* edited by Matthias Wolf and Mark Gardener, 119–142. London: Routledge, 2012.

Burton, M. J. "Capt. Hancock's Thrilling Discoveries of Strange Animals and Human Exiles." *San Francisco Chronicle,* 11 April 1937, Box 3, Joseph R. Slevin Collection, MSS.429, California Academy of Sciences Library, San Francisco.

Buschmann, Rainer, Edward Slack, and James Tueller. *Navigating the Spanish Lake: The Pacific in the Iberian World, 1521–1898.* Honolulu: University of Hawai'i Press, 2014.

Bustamante Ponce, Teodoro. *Historia de la Conservación Ambiental en Ecuador: Volcanes, Tortugas, Geólogos y Políticos.* Quito: FLACSO Ecuador y Editorial Abya Yala, 2016.

Byron, George Anson. *Voyage of H.M.S. Blonde to the Sandwich islands, in the years 1824–1825: Captain the Right Hon. Lord Byron, Commander.* Edited by Maria Calicott. London: John Murray, 1826.

Cabrera, Wilson, and Omar Garcia. "Reports from the Front: Personal Accounts from National Park Field Staff." In *Galápagos: Preserving Darwin's Legacy,* edited by Tui De Roy, 213–219. Richmond Hill, Ontario: Firefly, 2009.

Cairns, Rose, Susannah Sallu, and Simon Goodman. "Questioning Calls to Consensus in Conservation: A Q Study of Conservation Discourses on Galápagos." *Environmental Conservation* 41, no. 1 (2014): 13–26.

Callicott, J. Baird, and Michael P. Nelson. *The Great New Wilderness Debate.* Athens: University of Georgia Press, 1998.

Camhi, Merry. "Industrial Fisheries Threaten Ecological Integrity of the Galápagos Islands." *Conservation Biology* 9, no. 4 (August 1995): 715–724.

Camillas. "Gallipagos Islands—No. 2: A Sailor's Epistle Home, Albemarle Island, Jan. 5, 1847, Notes of a Terrapin Hunt." In *The Seamen's Friend: Vol. V, No. X.* Honolulu: Hawaii (1847): 74–75.

Campbell, Karl. "Manipulation of the Reproductive System of Feral Goats (*Capra hircus*) to Increase the Efficacy of Judas Goats: Field Methods Utilising Tubal Sterilisation, Abortion, Hormone Implants and Epididymectomy." PhD diss., University of Queensland, 2006.

Campbell, Karl, Greg S. Baxter, Peter John Murray, and Bruce E. Coblentz. "Increasing the Efficacy of Judas Goats by Sterilisation and Pregnancy Termination." *Wildlife Research* 32 (2005): 737–743.

Caro, T. M., and Gillian O'Doherty. "On the Use of Surrogate Species in Conservation." *Conservation Biology* 13 (1999): 805–814.

Carrier, James, and Donald Macleod. "Bursting the Bubble: The Socio-Cultural Context of Ecotourism." *Journal of the Royal Anthropological Institute* 11, no. 2 (June 2005): 315–334.

Carruthers, Jane. *The Kruger National Park: A Social and Political History.* Scottsville, South Africa: University of KwaZulu-Natal Press, 1995; reprint 2013.

Carter, Paul. *The Road to Botany Bay: An Exploration of Landscape and History.* Minneapolis: University of Minnesota Press, 2010.

Castillo, Blanca Vargas de. Oral History Interview. In *Desde las Islas Encantadas: Historias de Vida de Colonos en Galápagos,* by Pablo Ospina, 17–65. Quito: UNDP, Corporación Editora Nacional, 2005.

Castree, Noel. *Nature.* New York: Routledge, 2005.

Causton, C., S. B. Peck, B. J. Sinclair, L. Roque-Albelo, C. J. Hodgson, and B. Landry. "Alien Insects: Threats and Implications for Conservation of Galápagos Islands." *Annals of the Entomological Society of America* 99, no. 1 (2006): 121–143.

Cayot, Linda, and Ed Lewis. "Recent Increase of Killing of Giant Tortoises on Isabela Island." *Noticias de Galápagos* 54 (November 1994): 2–7.

Cayot, Linda, and Germán E. Morillo. "Rearing and Repatriation of Galápagos Tortoises: *Geochelone nigra hoodensis,* a Case Study." Paper presented at Conservation, Restoration, and Management of Tortoises and Turtles—An International Conference, July 1993, Purchase, New York.

Celata, Filippo, and Venere Stefania Sanna. "The Post-Political Ecology of Protected Areas: Nature, Social Justice and Political Conflicts in the Galápagos Islands." *Local Environment* 17, no. 9 (2012): 977–990.

Cepeda, Fausto. "La conservación de los recursos naturales renovables." *Trama* 49 (July 1989): 5–7.

Chambers, Paul. *A Sheltered Life: The Unexpected History of the Giant Tortoise.* Oxford: Oxford University Press, 2006.

Cheke, Anthony, and Julian Hume. *Lost Land of the Dodo: An Ecological History of Mauritius, Réunion, and Rodrigues.* London: T & AD Poyser, 2008.

Church, George Earl. "Report upon Ecuador." United States Congress, Senate Committee on Foreign Relations, Subcommittee on Disarmament, Message from the President of the United States, Transmitting, in Compliance with Senate Resolution of December 18, 1882, the Report of Mr. George Earl Church upon Ecuador. Washington, DC, 1883.

Clarke, Bruce, and Mark B. N. Hansen, eds. *Emergence and Embodiment: New Essays on Second-Order Systems Theory*. Durham, NC: Duke University Press, 2009.

Clarke, Jay. "Perfectly Natural: Exploring the Galápagos Islands and Encountering the Friendly Wildlife Is a Once-in-a-Lifetime Experience." *Saint Paul Pioneer Press*, 19 November 1995.

Coello, Segundo, and Alan Saunders. "Final Project Evaluation, Control of Invasive Species in the Galápagos Archipelago." ECU/oo/G31. Prepared for the Global Environment Fund: United Nations Development Programme, Ministry of the Environment, 2011.

Collard, Rosemary-Claire, Jessica Dempsey, and Juanita Sundberg. "A Manifesto for Abundant Futures." *Annals of the Association of American Geographers* 105, no. 2 (2015): 322–330.

Colnett, James. *A Voyage to the South Atlantic and round Cape Horn into the Pacific Ocean, for the Purpose of Extending the Spermaceti Whale Fisheries, and Other Objects of Commerce, by Ascertaining the Ports, Bays, Harbours, and Anchoring Births, in Certain Islands and Coasts in Those Seas at Which the Ships of the British Merchants Might Be Refitted: Undertaken and Performed by Captain James Colnett, of the Royal Navy, in the Ship Rattler*. London, 1798.

Constantino, Jill. "Tortoise Soup for the Soul: Finding a Space for Human History in Evolution's Laboratory." In *The Anthropology of Extinction: Essays on Culture and Species Death*, edited by Genese M. Sodikoff, 89–102. Bloomington: Indiana University Press, 2011.

Conway, Ainslie, and Frances Conway. *The Enchanted Islands*. New York: Putnam, 1947.

Conwentz, Hugo Wilhelm. *The Care of Natural Monuments with Special Reference to Great Britain and Germany*. Cambridge: Cambridge University Press, 1909.

Cote, M., and A. Nightingale. "Resilience Thinking Meets Social Theory: Situating Social Change in Socio-Ecological Systems (SES) Research." *Progress in Human Geography* 36, no. 4 (2012): 475–489.

Cronon, William. "Kennecott Journey: The Paths Out of Town." In *Under an Open Sky: Rethinking America's Western Past*, edited by William Cronon, George A. Miles, and Jay Gitlin, 28–51. New York: W. W. Norton, 1993.

———. *Nature's Metropolis: Chicago and the Great West*. New York: W. W. Norton, 1997.

———. "The Trouble with Wilderness; or, Getting Back to the Wrong Nature." In *Uncommon Ground: Toward Reinventing Nature*, edited by William Cronon, 69–90. New York: W. W. Norton, 1995.

Crosby, Alfred W. *The Columbian Exchange: Biological and Cultural Consequences of 1492*. Westport, CT: Praeger, 1972.

―――. *Ecological Imperialism: The Biological Expansion of Europe, 900–1900.* Cambridge: Cambridge University Press, 2004 (originally published 1986).

Curnutt, John L. "A Guide to the Homogenocene." *Ecology* 81, no. 6 (2000): 1756–1757.

Curtin, Philip D. *The Rise and Fall of the Plantation Complex: Essays in Atlantic History.* New York: Cambridge University Press, 1998.

Cushman, Gregory. *Guano and the Opening of the Pacific World: A Global Ecological History.* Cambridge: Cambridge University Press, 2013.

―――. "Humboldtian Science, Creole Meteorology, and the Discovery of Human-Caused Climate Change in South America." *Osiris* 26 (2011): 19–44.

―――. "The Road to Survival." In *Guano and the Opening of the Pacific World: A Global Ecological History,* by Gregory Cushman, 243–281. Cambridge: Cambridge University Press, 2013.

―――. "The Struggle over Airways in the Americas, 1919–1945: Atmospheric Science, Aviation Technology, and Neocolonialism." In *Intimate Universality: Local and Global Themes in the History of Weather and Climate,* edited by James R. Fleming, Vladimir Jankovic, and Deborah Coen, 175–222. Sagamore Beach, MA: Science History Publications, 2006.

Cuvi, Nicolás. "La Institucionización del conservacionismo en el Ecuador (1949–1953): Misael Acosta Solís y el Departamento Forestal." *ProcesoS, Revista Ecuatoriana de Historía* 22 (2005): 107–129.

―――. "Misael Acosta Solís y el conservacionismo en el Ecuador, 1936–1953." *Scripta Nova* 9, no. 15 (June 2005): 107–129.

Cuvi, Nicolás, and David Guijarro. "¿Una trayectoria hacia la insustentabilidad? La movilidad terrestre en la isla Santa Cruz, Galápagos." *Revista Transporte y Territorio* 15 (2016): 216–240.

Dampier, William. *A Voyage to New Holland, &c. in the Year, 1699: Wherein Are Described. . . .* London, 1703.

Darwin, Charles. *Charles Darwin's Zoology Notes and Specimen Lists from H.M.S. Beagle.* Edited by R. D. Keynes. Cambridge: Cambridge University Press, 2000.

―――. *Galapagos Notebook.* Edited by Gordon Chancellor and John van Wyhe. Available at Darwin Online. The Complete Work of Charles Darwin Online, edited by John van Wyhe. http://darwin-online.org.uk. Last modified 4 November 2013.

―――. Journal, July 1837, 13 recto. Available at Darwin Online. The Complete Work of Charles Darwin Online, edited by John van Wyhe. http://darwin-online.org.uk.

―――. *Journal of Researches into the History and Geology of the Countries Visited During the Voyage of the H.M.S. Beagle Round the World under the Command of Capt. Fitz Roy, R.N.* London: John Murray, 1845.

―――. "Letter no. 10819." To J. D. Hooker, 28 January 1877. Darwin Correspondence Project. http://www.darwinproject.ac.uk/DCP-LETT-10819.

―――. *Narrative of the Surveying Voyages of His Majesty's Ships Adventure and Beagle Between the Years 1826 and 1836, Describing Their Examination of the Southern Shores of South America, and the Beagle's Circumnavigation of the Globe. Volume III: Journal and Remarks, 1832–1835.* London: Henry Colburn, 1839.

———. "On the action of sea-water on the germination of seeds." *Journal of the Proceedings of the Linnean Society of London (Botany)* 1 (6 May 1856): 130–140. Available at Darwin Online. The Complete Work of Charles Darwin Online, edited by John van Wyhe. http://darwin-online.org.uk.

———. *On the Origin of Species by Means of Natural Selection, or the Preservation of Favoured Races in the Struggle for Life.* 1st ed. London: John Murray, 1859. Available at Darwin Online. The Complete Work of Charles Darwin Online, edited by John van Wyhe. http://darwin-online.org.uk.

Daston, Lorraine, and Peter Galison. *Objectivity.* Brooklyn, NY: Zone Books, 2007.

Daston, Lorraine, and Gregg Mitman. *Thinking with Animals: New Perspectives on Anthropomorphism.* New York: Columbia University Press, 2005.

Daughan, George C. *The Shining Sea: David Porter and the U.S.S. Essex During the War of 1812.* New York: Basic Books, 2013.

Davis, Diana K. "Historical Approaches to Political Ecology." In *The Routledge Handbook of Political Ecology,* edited by Tom Perreault, Gavin Bridge, and James McCarthy, 263–276. Abingdon: Routledge, 2015.

———. *Resurrecting the Granary of Rome: Environmental History and French Colonial Expansion in North Africa.* Athens: Ohio University Press, 2007.

Davis, Mark, et al. "Don't Judge Species on Their Origins." *Nature* 474 (June 2011): 153–154.

Davis, William M. *Nimrod of the Sea, Or the American Whaleman.* Boston: Charles E. Lauriat, 1926.

Dawkins, Richard. "Foreword." *Galápagos: The Islands That Changed the World,* by Paul Stewart. New Haven: Yale University Press, 2008.

Dawson, E. Yale. "Cacti in the Galápagos Islands." In *The Galápagos: Proceedings of the Galápagos International Scientific Project,* edited by Robert Bowman, 209–214. Brussels: Palais des Academies, 1966.

Dawson, Michael, ed. *Economic Feasibility of Galapagos Tours: Report to Comision de Valores, Corporacion Financiera Nacional.* Cambridge, MA: Arthur D. Little, 1968.

De Bont, Raf. *Stations in the Field: A History of Place-Based Animal Research, 1870–1930.* Chicago: University of Chicago Press, 2015.

———. "A World Laboratory: Framing the Albert National Park." *Environmental History* 22 (2017): 404–432.

de la Cadena, Marisol. *Earth Beings: Ecologies of Practice Across Andean Worlds.* Durham, NC: Duke University Press, 2015.

Deamer, Kacey. "Did 1 Sexually Active Tortoise Really Save His Species?" Live Science, 27 September 2016. https://www.livescience.com/56277-sexually-active-giant-tortoises-save-species.html.

Dean, Warren. *With Broadax and Firebrand: The Destruction of the Brazilian Atlantic Forest*. Berkeley: University of California Press, 1995.

Delano, Amsa. *Narrative of the Voyages and Travels, in the Northern and Southern Hemispheres: Comprising Three Voyages Around the World; Together with a Voyage of Survey and Discovery, in the Pacific and Oriental Islands*. Boston: E. G. House, 1817.

Deleuze, Gilles. *Desert Islands and Other Texts, 1953–1974*. Los Angeles: Semiotext(e), 2004.

DeLoughrey, Elizabeth. "The Myth of Isolates: Ecosystem Ecologies in the Nuclear Pacific." *Cultural Geographies* 20, no. 2 (2012): 167–184.

———. *Routes and Roots: Navigating Caribbean and Pacific Island Literatures*. Honolulu: University of Hawai'i Press, 2009.

DeSola, C. Ralph. "The Liebespiel of *Testudo vandenburghi*, a New Name for the Mid-Albemarle Island Galapagos Tortoise." *Copeia* 1930, no. 3 (September 1930): 79–80.

———. "Notes on the Sex Determination in a Species of the Galapagos Tortoise." *Copeia* 171 (April–June 1929): 54–55.

Dolin, Eric Jay. *Leviathan: The History of Whaling in America*. New York: W. W. Norton, 2007.

Donlan, C. Josh. "Restoring America's Big Wild Animals." *Scientific American*, June 2007, 72–77.

Donlan, C. Josh, et al. "Pleistocene Rewilding: An Optimistic Agenda for Twenty-First Century Conservation." *American Naturalist* 168, no. 5 (2006): 660–681.

Dorsey, Kirkpatrick. *The Dawn of Conservation Diplomacy: U.S.-Canadian Wildlife Protection Treaties in the Progressive Era*. Seattle: University of Washington Press, 1998.

Dorst, Jean. "Problems of Conservation in the Galapagos Islands." Humid Tropics Research Programme, UNESCO, 23 June 1961. https://unesdoc.unesco.org/ark:/48223/pf0000153302.

———. "Where Time Stood Still: The Galapagos Islands and Their Prehistoric Creatures." *UNESCO Courier* 14, no. 9 (1961): 28–32. https://unesdoc.unesco.org/ark:/48223/pf0000064239.

Drayton, R. *Nature's Government: Science, Imperial Britain and the "Improvement" of the World*. New Haven: Yale University Press, 2000.

Duncan, Kenneth, and Ian Rutledge, eds. *Land and Labour in Latin America: Essays on the Development of Agrarian Capitalism in the Nineteenth and Twentieth Centuries*. Cambridge: Cambridge University Press, 1977.

Eibl-Eibesfeldt, Irenäus. *Galápagos*. New York: Doubleday, 1961.

———. "Galapagos: Wonders of a Noah's Ark off the Coast of Ecuador." *UNESCO Courier* 11, no 1 (1958): 19–23.

———. *Survey on the Galapagos Islands*. International Union for the Conservation of Nature, 1959. https://unesdoc.unesco.org/ark:/48223/pf0000156448?posInSet=1&queryId=bd79761c-cb27-4886-8f26-456d4779aa65.

"El hecho mas grande desde la visita de Darwin será el Congreso de Galápagos." *El Comercio.* 19 January 1964.

"Embajador Ecuatoriano en París expuso en UNESCO importancia de Galápagos." *Diario del Ecuador,* 29 October 1961.

"En las Islas Galápagos." *El Comercio,* 16 October 1961.

Endersby, James. *Imperial Nature: Joseph Hooker and the Practices of Victorian Science.* Chicago: University of Chicago Press, 2008.

Engie, Kim. "Adaptation and Shifting Livelihoods in the Small-Scale Fisheries of the Galápagos Marine Reserve, Ecuador." PhD diss., University of North Carolina–Chapel Hill, 2015.

Engie, Kim, and Diego Quiroga. "The Emergence of Recreational Fishing in the Galápagos Marine Reserve: Adaptation and Complexities." In *The Galapagos Marine Reserve: A Dynamic Social-Ecological System,* edited by Judith Denkinger, Luis Vinueza, and Carlos F. Mena, 203–226. New York: Springer Science + Business Media, 2014.

Entwistle, A. "Flagships for the Future?" *Oryx* 34 (2000): 239–240.

Escobar, Arturo. *Encountering Development: The Making and Unmaking of the Third World.* Princeton, NJ: Princeton University Press, 1995.

"Estudios Sobre Galápagos, Abril 1956–Enero 1957." Archivo Reservado, Departamento de Fronteras, G.3.4.4, Ministerio de Relaciones Exteriores, Quito, Ecuador.

Fabian, Johannes. *Time and the Other: How Anthropology Makes Its Object.* New York: Columbia University Press, 1983.

Fairhead, James, and Melissa Leach. *Science, Society and Power: Environmental Knowledge and Power in West Africa and the Caribbean.* Cambridge: Cambridge University Press, 2003.

Faris, Robert E. L., William R. Catton Jr., and Otto N. Larsen. "The Galápagos Expedition: Failure in the Pursuit of a Contemporary Secular Utopia." *Pacific Sociological Review* 7, no. 1 (Spring 1964): 48–54.

Faulkner, William. *Requiem for a Nun.* New York: Random House, 1951.

Fernandez-Armesto, Felipe. "Maps and Exploration in the Sixteenth and Early Seventeenth Century." In *The History of Cartography, Volume 3: Cartography in the European Renaissance,* edited by David Woodward, 738–759. Chicago: University of Chicago Press, 2007.

Ferrin, Lynn. "The Place of Wild Tortoises." *IMAGE* 21 (October 1990): 24–30, in Robert Bowman Papers, Box 3, California Academy of Sciences Library, San Francisco.

Fessel, Birgit, and Sabine Tebbich. "*Philornis downsi:* A Recently Discovered Parasite on the Galápagos Archipelago—A Threat for Darwin's Finches?" *Ibis* 144, no. 3 (July 2002): 445–451.

Few, Deborah, and Zeb Tortorici, eds. *Centering Animals in Latin American History.* Durham, NC: Duke University Press, 2013.

FitzRoy, Robert. *Narrative of the Surveying Voyages of His Majesty's Ships Adventure and Beagle Between the Years 1826 and 1836, Describing Their Examination of the Southern Shores of South America, and the Beagle's Circumnavigation of the Globe. Volume II: Proceedings of the Second Expedition, 1831–36, under the command of Captain Robert Fitz-Roy, R.N.* London: Henry Colburn, 1839.

Foote, Nicola. "Race, State and Nation in Early Twentieth Century Ecuador." *Nations and Nationalism* 12, no. 2 (2006): 261–278.

Foucault, Michel. *Discipline and Punish: The Birth of the Prison.* New York: Vintage, 1995.

———. "Of Other Spaces." *Diacritics* 16 (Spring 1986): 22–27.

———. *Security, Territory, Population: Lectures at the Collège de France 1977–1978.* Edited by Michel Snelleart. New York: Picador, 2009.

Freitas, Frederico. "Ordering the Borderland: Settlement and Removal in the Iguaçu National Park, Brazil, 1940s–1970s." In *The Nature State: Rethinking the History of Conservation,* edited by Wilko Graf von Hardenberg, Matthew Kelley, Claudia Leal, and Emily Wakild, 158–175. Oxon: Routledge Earthscan, 2017.

———. "A Park for the Borderlands: The Creation of the Iguaçu National Park in Southern Brazil, 1880–1940." *Revista de Historia IberoAmericana* 7, no. 2 (2014): 65–88.

Fudge, Erica. "A Left-Handed Blow: Writing the History of Animals." In *Representing Animals,* edited by Nigel Rothfels, 3–18. Bloomington: Indiana University Press, 2003.

Fujimura, Joan. "Crafting Science: Standardized Packages, Boundary Objects, and 'Translation.'" In *Science as Practice and Culture,* edited by Andrew Pickering, 168–211. Chicago: University of Chicago Press, 1992.

Fundación Charles Darwin (FCD) y WWF–Ecuador. *Atlas de Galápagos, Ecuador: Especies Nativas y Invasoras.* Quito: FCD y WWF–Ecuador, 2018. https://www.darwin foundation.org/images/research/Atlas_de_Galapagos_Ecuador.pdf.

Galápagos: The Islands That Changed the World. London: BBC Video, 2007.

"Galapagos Tortoises Held Hostage." BBC News, last updated 27 February 2004. http://news.bbc.co.uk/2/hi/americas/3491658.stm.

Gardener, Mark R., Rachel Atkinson, and Jorge Luis Rentería. "Eradications and People: Lessons from the Plant Eradication Program in Galapagos." *Restoration Ecology* 18, no. 1 (2010): 20–29.

Garman, Samuel. *The Galapagos Tortoises.* Cambridge, MA: Museum of Comparative Zoology, 1917 (*Memoirs of the Museum of Comparative Zoology at Harvard College* 30, no. 4 [1917]: 261–290).

———. "The Reptiles of the Galapagos Islands. From the Collections of Dr. Geo. Baur." *Bulletin of the Essex Institute* 24 (1892): 73–87.

Garrick, Ryan C., et al. "Genetic Rediscovery of an 'Extinct' Galápagos Giant Tortoise Species." *Current Biology* 22, no. 1 (January 2012): R10–R11.

Geertz, Clifford. "Thick Description: Toward an Interpretive Theory of Culture." In *The Interpretation of Cultures*, 28–29. New York: Basic Books, 1973.

Geist, Dennis. "On the Emergence and Submergence of the Galápagos Islands." *Noticias de Galápagos* 56 (March 1996): 5–9.

Gerlach, Justin, Catharine Muir, and Matthew D. Richmond. "The First Substantiated Case of Trans-Oceanic Tortoise Dispersal." *Journal of Natural History* 40, nos. 41–43 (2006): 2403–2408.

"Giant Tortoise Believed Extinct for 100 Years Found in Galápagos." *The Guardian*, 21 February 2019. https://www.theguardian.com/environment/2019/feb/21/giant-tortoise-believed-extinct-for-100-years-found-in-galapagos.

Gibbs, James P., E. A. Hunter, K. T. Shoemaker, W. H. Tapia, and L. J. Cayot. "Demographic Outcomes and Ecosystem Implications of Giant Tortoise Reintroduction to Española Island, Galapagos." *PLoS ONE* 9, no. 10 (2014): e110742. https://doi.org/10.1371/journal.pone.0110742.

Gibbs, James P., Cruz Márquez, and Eleanor J. Sterling. "The Role of Endangered Species Reintroduction in Ecosystem Restoration: Tortoise-Cactus Interactions on Española Island, Galápagos." *Restoration Ecology* 16 (2008): 88–93.

Gibbs, James P., and Hara Woltz. "A Pilot Survey of the Central Colony of the Waved Albatross *Phoebastria irrorata* on Española Island." *Galápagos Research* 67 (October 2010): 18–20.

Gilley, Jessey. "Geographical Imagination." In *Encyclopedia of Geography*, edited by Barney Warf. 6 vols. http://sk.sagepub.com/reference/geography/n477.xml.

Gillis, John. *Islands of the Mind: How the Human Imagination Created the Atlantic World*. New York: Palgrave, 2009.

Goldman, Michael. *Imperial Nature: The World Bank and Struggles for Social Justice in the Age of Globalization*. New Haven: Yale University Press, 2006.

González, José A., Carlos Montes, José Rodríguez, and Washington Tapia. "Rethinking the Galapagos Islands as a Complex Social-Ecological System: Implications for Conservation and Management." *Ecology and Society* 13, no. 2 (2008): 1–26. https://www.ecologyandsociety.org/vol13/iss2/art13/.

Gordillo, Jacinto. *Stories from 44 Years in the Galápagos Islands*. Quito: Abya-Yala, 2011.

Gough, Barry M. "James Colnett." In *Dictionary of Canadian Biography*, vol. 5 (1801–1820). Toronto: University of Toronto/Université Laval, 1983. http://www.biographi.ca/en/bio/colnett_james_5E.html.

Gould, Stephen Jay. *Time's Arrow, Time's Cycle: Myth and Metaphor in the Discovery of Geological Time*. Cambridge, MA: Harvard University Press, 1987.

Grandin, Greg. *Empire of Necessity: Slavery, Freedom, and Deception in the New World*. New York: Metropolitan Books, 2014.

Grant, Peter R., and Rosemary Grant. "Unpredictable Evolution in a 30-Year Study of Darwin's Finches." *Science* 296, no. 5568 (2002): 707–711.

Grant, Thalia K., and Gregory B. Estes. *Darwin in Galápagos: Footsteps to a New World.* Princeton, NJ: Princeton University Press, 2009.

Gregory, Derek. *Geographical Imaginations.* Cambridge, MA: Wiley-Blackwell, 1994.

Grenier, Christophe. *Conservación Contra Natura: Las Islas Galápagos.* Quito: Abya-Yala, 2007.

Grigioni, Sveva. "La estimulación sexual del Solitario Jorge: Un estudio en comportamiento." Puerto Ayora, Galápagos: Charles Darwin Research Station, 1993.

Grimwood, Ian, and David Snow. "Recommendations on the Administration of the Proposed National Park of the Galapagos Archipelago and the Development of Its Tourist Potential." Unpublished manuscript, 1966. Charles Darwin Research Station Library, Puerto Ayora, Galápagos.

Grosz, Elizabeth. *The Nick of Time: Politics, Evolution and the Untimely.* Durham, NC: Duke University Press, 2004.

———. *Time Travels: Feminism, Nature, Power.* Durham, NC: Duke University Press, 2005.

Grove, Jack. "Crisis in the Galapagos." *Iguana Times* 4, no. 2 (June 1995): 37–39.

Grove, Richard. *Green Imperialism: Colonial Expansion, Tropical Island Edens and the Origins of Environmentalism, 1600–1860.* Cambridge: Cambridge University Press, 1994.

Guha, Ramachandra, and Juan Martinez-Alier. *Varieties of Environmentalism: Essays North and South.* Oxford: Earthscan, 1997.

Günther, Albert. "Description of the Living and Extinct Races of Gigantic Land-Tortoises." *Philosophical Transactions of the Royal Society of London* 165 (1875): 251–284.

———. "The Gigantic Land Tortoises of the Mascarene and Galápagos Islands III." *Nature* 12 (August 1875): 296–297.

Gupta, Akhil. *Postcolonial Developments: Agriculture in the Making of Modern India.* Durham, NC: Duke University Press, 1998.

Hall, Basil. *Memoir on the Navigation of South America, to Accompany a Chart of That Station.* London: Printed by H. Teape, for the Hydrographical Office of the Admiralty, 1825.

Hamann, Ole. "On Vegetation Recovery, Goats and Giant Tortoises on Pinta Island, Galapagos, Ecuador." *Biodiversity and Conservation* 2, no. 2 (1993): 138–151.

Hames, Raymond. "The Ecologically Noble Savage Debate." *Annual Review of Anthropology* 36 (2007): 177–190.

Haraway, Donna. "A Cyborg Manifesto: Science, Technology, and Socialist-Feminism in the Late Twentieth Century." In *Simians, Cyborgs and Women: The Reinvention of Nature,* 149–181. New York: Routledge, 1991.

———. *Staying with the Trouble: Making Kin in the Chthulucene.* Durham, NC: Duke University Press, 2016.

———. "Teddy Bear Patriarchy: Taxidermy in the Garden of Eden, New York City, 1908–1936." In *Primate Visions: Gender, Race, and Nature in the World of Modern Science,* 26–58. New York: Routledge, 1990.

———. *When Species Meet.* Minneapolis: University of Minnesota Press, 2008.

Harlan, Richard. "Description of a Land Tortoise, from the Gallapagos Islands, commonly known as the 'Elephant Tortoise.'" *Journal of the Academy of Natural Sciences of Philadelphia* 5 (1825–1827): 284–292.

Harrison, Paul H. "Study of the U.S. Air Forces' Galapagos Islands Base." John Woram, Las Encantadas: Human and Cartographic History of the Galápagos Islands, 28 October 1947. http://galapagos.to/TEXTS/USAF1947.HTM.

Harrsch, Don. "Filiate Science Antrorse." John Woram, Las Encantadas: Human and Cartographic History of the Galápagos Islands, n.d. http://galapagos.to/TEXTS/HARRSCH.HTM.

Harvey, David. *A Brief History of Neoliberalism.* Oxford: Oxford University Press, 2007.

Hawking, Stephen. *A Brief History of Time.* New York: Bantam, 1988.

Heflin, Wilson. *Herman Melville's Whaling Years.* Nashville: Vanderbilt University Press, 2004.

Hennessy, E., and Amy McCleary. "Nature's Eden? The Production and Effects of 'Pristine' Nature in the Galápagos Islands." *Island Studies Journal* 6, no. 2 (2011): 131–156.

Hennessy, Elizabeth. "The Molecular Turn in Conservation: Genetics, Pristine Nature, and the Rediscovery of an Extinct Species of Galápagos Giant Tortoise." *Annals of the Association of American Geographers* 105, no. 1 (January 2015): 87–104.

———. "Mythologizing Darwin's Islands." In *Darwin, Darwinism and Conservation in the Galápagos Islands: The Legacy of Darwin and Its New Applications,* edited by Diego Quiroga and Ana Sevilla, 65–90. Basel: Springer International, 2017.

———. "The Politics of a Natural Laboratory: Claiming Territory and Managing Life in the Galápagos Islands." *Social Studies of Science* 48, no. 4 (2018): 483–506.

———. "Producing 'Prehistoric' Life: Conservation Breeding and the Remaking of Wildlife Genealogies." *Geoforum* 49 (October 2013): 71–80.

———. "III: Ecological Restoration and Tortoise Histories in the Galápagos Islands." In "Historical Geographies of, and for, the Present," edited by Levi Van Sant, Elizabeth Hennessy, and Mona Domosh. *Progress in Human Geography* (October 2018): 7–10.

Heslinga, Jane. "Regulating Ecotourism in Galápagos: A Case Study of Domestic-International Partnerships." *Journal of International Wildlife Law and Policy* 6, no. 1 (2003): 57–77.

Heyerdahl, Thor, and Arne Skjølsvold. "Archaeological Evidence of Pre-Spanish Visits to the Galápagos Islands." *Memoirs of the Society for American Archaeology,* no. 12. Supplement to *American Antiquity* 22, no. 2, pt. 3 (October 1956): 71 pages.

Heylings, Pippa, and Manuel Bravo. "Evaluating Governance: A Process of Understanding How Co-management Is Functioning, and Why, in the Galapagos Marine Reserve." *Ocean and Coastal Management* 50 (2007): 174–208.

Hillinger, Charles. "Galapagos Ho! Utopian Dream Lures Families to Island." *Los Angeles Times,* 9 October 1960, E1.

———. "U.S. Colony in Galapagos Nearing Tragic End as People Leave Isles." *Los Angeles Times,* 26 December 1960, B1.

Hinchliffe, Stephen. *Geographies of Nature.* London: SAGE, 2007.

Hobbs, Richard J. "Novel Ecosystems: Theoretical and Management Aspects of the New Ecological World Order." *Global Ecology and Biogeography* 15, no. 1 (2006): 1–7.

Holling, C. S. "Understanding the Complexity of Economic, Ecological and Social Systems." *Ecosystems* 4 (2001): 390–405.

Honey, Martha. "The Galápagos Islands: Test Site for Theories of Evolution and Ecotourism." In *Ecotourism and Sustainable Development: Who Owns Paradise?,* 121–159. Washington, DC: Island Press, 2008.

Hunter, E. A., J. P. Gibbs, L. J. Cayot, and W. Tapia. "Equivalency of Galápagos Giant Tortoises Used as Ecological Replacement Species to Restore Ecosystems Functions." *Conservation Biology* 27, no. 4 (2013): 701–709.

Huxley, Julian. "Charles Darwin: Galápagos and After." In *The Galápagos: Proceedings of the Galápagos International Scientific Project,* edited by Robert Bowman, 3–9. Brussels: Palais des Academies, 1966.

———. *Evolution: The Modern Synthesis.* London: George Allen & Unwin, 1942.

Jackson, Michael H. *Galapagos: A Natural History.* Calgary: University of Calgary Press, 1985.

Jacoby, Karl. *Crimes Against Nature: Squatters, Poachers, Thieves and the Hidden History of American Conservation.* Berkeley: University of California Press, 2001.

James, C. L. R. *Mariners, Renegades, and Castaways: The Story of Herman Melville and the World We Live In.* London: Allison & Busby, 1985 (originally published 1953).

James, Matthew. *Collecting Evolution: The Galápagos Expedition That Vindicated Darwin.* New York: Oxford University Press, 2017.

———. "Collecting Evolution: Vindication of Charles Darwin by the 1905–1906 Galápagos Expedition of the California Academy of Sciences." *Proceedings of the California Academy of Sciences* 4, no. 61 (2010): 197–210.

Johnson, Kristin. *Ordering Life: Karl Jordan and the Naturalist Tradition.* Baltimore: Johns Hopkins University Press, 2012.

Jonik, Michael. "Melville's 'Permanent Riotocracy.'" In *A Political Companion to Herman Melville,* edited by Jason Frank, 229–258. Lexington: University Press of Kentucky, 2014.

Jordan, Karl. "Obituary: Lord Rothschild, F.R.S. (1868–1937)." *British Birds* 31 (1938): 146–148.

Keynes, R. D., ed. *Charles Darwin's* Beagle *Diary.* Cambridge: Cambridge University Press, 2001.

Kirksey, S. Eben, and Stefan Helmreich. "The Emergence of Multispecies Ethnography." *Cultural Anthropology* 25 (2010): 545–576.

Klubock, Thomas. *La Frontera: Forests and Ecological Conflict in Chile's Frontier Territory.* Durham, NC: Duke University Press, 2014.

Knafo S. E., S. J. Divers, S. Rivera, L. J. Cayot, W. Tapia Aguilera, and J. Flanagan. "Sterilisation of Hybrid Galapagos Tortoises (*Geochelone Nigra*) for Island Restoration. Part 1: Endoscopic Oophorectomy of Females Under Ketamine-Medetomidine Anaesthesia." *Veterinary Record* 168, no. 2 (2011). doi: 10.1136/vr.c6520.

Knorr Cetina, Karin. "The Couch, the Cathedral, and the Laboratory: On the Relationship Between Experiment and Laboratory in Science." In *Science as Practice and Culture,* edited by Andrew Pickering, 113–138. Chicago: University of Chicago Press, 1992.

Koford, Carl. "Economic Resources of the Galápagos Islands." In *The Galápagos: Proceedings of the Galápagos International Scientific Project,* edited by Robert Bowman, 286–290. Brussels: Palais des Academies, 1966.

Kohler, Robert E. *Landscapes and Labscapes: Exploring the Lab-Field Border in Biology.* Chicago: University of Chicago Press, 2002.

Kolbert, Elizabeth. *The Sixth Extinction: An Unnatural History.* New York: Henry Holt, 2014.

Koop, Jennifer A. H., Peter S. Kim, Sarah A. Knutie, Fred Adler, and Dale H. Clayton. "An Introduced Parasitic Fly May Lead to Local Extinction of Darwin's Finch Populations." *Journal of Applied Ecology* 53 (April 2016): 511–518.

Kosek, Jake. *Understories: The Political Life of Forests in Northern New Mexico.* Durham, NC: Duke University Press, 2006.

Kramer, Peter. "Conservacion de los Recursos Naturales de las Islas Galápagos." PNUD/AT/UNESCO/ECU/68/13, Project 18 Sept 1970 until July 1972. Revision date: 14 June 1971.

———. "Galápagos: Ecology and Conservation: September 1970–December 1973." Paris: UNESCO, 1975.

———. "The Galápagos: Islands Under Siege." *Ambio* 12, no. 3/4 (1983): 186–190.

———. "Galápagos Conservation: Present Position and Future Outlook." *Noticias de Galápagos* 22 (1974): 3–5.

———. Informe comprensivo, cubriendo el periodo de Septiembre de 1970 a Diciembre de 1972, por petición del Director de la Division de Investigación de Recursos Naturales, To: Director de la Oficina de Relaciones de los Estado Miembros, UNESCO, Paris, France, 1972, 36 pages. CDF History Box, Archives of the Charles Darwin Research Station.

Kull, Christian. *Isle of Fire: The Political Ecology of Landscape Burning in Madagascar.* Chicago: University of Chicago Press, 2004.

Kupper, Patrick. *Creating Wilderness: A Transnational History of the Swiss National Park.* New York: Berghahn Books, 2014.

———. "Nature's Laboratories? Exploring the Intersection of Science and National Parks." In *National Parks Beyond the Nation: Global Perspectives on "America's Best*

Idea," edited by A. Howkins, J. Orsi, and M. Fiege, 114–132. Norman: University of Oklahoma Press, 2016.

Lack, David. *Darwin's Finches: An Essay on the General Biological Theory of Evolution.* Gloucester, MA: Peter Smith, 1947.

Langdon, Robert, ed. *Where the Whalers Went: An Index to the Pacific Ports and Islands Visited by American Whalers (and Some Other Ships) in the 19th Century.* Canberra: Pacific Manuscripts Bureau, Research School of Pacific Studies, Australian National University, 1984.

Larrea, Carlos. *El Archipiélago de Colón.* Quito: Casa de la Cultura Ecuatoriana, 1960.

Larsen, Anne. "Equipment for the Field." In *Cultures of Natural History,* edited by N. Jardine, J. A. Secord, and E. C. Spary, 358–377. New York: Cambridge University Press, 1996.

Larson, Edward. *Evolution's Workshop: God and Science on the Galapagos Islands.* New York: Basic Books, 2001.

Latorre, Octavio. *The Curse of the Giant Tortoise: Tragedies, Crimes and Mysteries in the Galápagos Islands.* Quito: National Cultural Fund, 2005 (originally published 1990).

———. *El Hombre en las Islas Encantadas: La historia humana de Galapagos.* Quito: [O. Latorre], 1999.

———. *Manuel J. Cobos, Emperador de Galápagos.* Quito: Fundación Charles Darwin, 1991.

Latour, Bruno. *Politics of Nature: How to Bring the Sciences into Democracy.* Cambridge, MA: Harvard University Press, 2004.

———. *Science in Action: How to Follow Scientists and Engineers Through Society.* Cambridge, MA: Harvard University Press, 1987.

———. *We Have Never Been Modern.* Cambridge, MA: Harvard University Press, 1993.

Latour, Bruno, and Steve Woolgar. *Laboratory Life: The Construction of Scientific Facts.* Princeton, NJ: Princeton University Press, 1979.

Lavoie, C., F. Cruz, G. V. Carrion, K. Campbell, C. J. Donlan, S. Harcourt, and M. Moya. *The Thematic Atlas of Project Isabela: An Illustrative Document Describing, Step-by-Step, the Biggest Successful Goat Eradication Project on the Galápagos Islands, 1998–2006.* Puerto Ayora, Galápagos: Charles Darwin Foundation, 2007.

Law, John. "Enacting Naturecultures: A Note from STS." Centre for Science Studies, Lancaster University, Lancaster, UK. https://www.lancaster.ac.uk/fass/resources/sociology-online-papers/papers/law-enacting-naturecultures.pdf.

Les parcs nationaux et la protection de la nature. Brussels: Institute des Parcs Nationaux du Congo Belge, 1937.

Lewis, Keri. "Negotiating for Nature: Conservation Diplomacy and the Convention on Nature Protection and Wildlife Preservation in the Western Hemisphere, 1929–1976." PhD diss., University of New Hampshire, 2001.

"Ley de reforma agraria y colonizacion." Decreto no. 1480, *Registro oficial,* 23 July 1964.

Li, Tania Murray. *The Will to Improve: Governmentality, Development, and the Practice of Politics.* Durham, NC: Duke University Press, 2007.

Linnaeus, Carl. *Systema Naturae* (1758).

"Lonesome George." Re-Tortoise Pinta: Field Experiences from a Galapagos Restoration Project. 25 June 2012. http://retortoisepinta.blogspot.com/.

Lorimer, Jamie. "Nonhuman Charisma." *Environment and Planning D: Society and Space* 25 (2007): 911–932.

———. *Wildlife in the Anthropocene: Conservation After Nature.* Minneapolis: University of Minnesota Press, 2015.

Lorimer, Jamie, and Clemens Driessen. "Bovine Biopolitics and the Promise of Monsters in the Rewilding of Heck Cattle." *Geoforum* 30 (2011): 1–11.

Lowe, Celia. *Wild Profusion: Biodiversity Conservation in an Indonesian Archipelago.* Princeton, NJ: Princeton University Press, 2006.

Lowe, Percy. "The Finches of the Galapagos in Relation to Darwin's Conception of Species." *Ibis* 78, no. 2 (April 1936): 310–321.

Lu, Flora, Gabriela Valdivia, and Wendy Wolford. "Social Dimensions of 'Nature at Risk' in the Galápagos Islands, Ecuador." *Conservation and Society* 11, no. 1 (2013): 83–95.

Luna Tobar, Alfredo. *Historía Política Internacional de las Islas Galápagos.* Quito: Abya-Yala, 1997.

Lundh, Jacob. *Galápagos: A Brief History.* http://galapagos.to/TEXTS/LUNDH1-1.php.

Lyell, Charles. *Principles of geology, being an attempt to explain the former changes of the Earth's surface, by reference to causes now in operation.* Volume 1. London: John Murray, 1830. Available at Darwin Online. The Complete Work of Charles Darwin Online, edited by John van Wyhe. http://darwin-online.org.uk.

Lyons, Barry J. *Remembering the Hacienda: Religion, Authority, and Social Change in Highland Ecuador.* Austin: University of Texas Press, 2006.

MacArthur, R. H., and E. O. Wilson. *The Theory of Island Biogeography.* Princeton, NJ: Princeton University Press, 1967.

MacFarland, Craig. "Giant Tortoises: Goliaths of Galapagos." *National Geographic* 142, no. 5, November 1972, 632–649.

MacFarland, Craig, José Villa, and Basilio Toro. "The Galápagos Giant Tortoises (*Geochelone elephantopus*) Part I: Status of the Surviving Populations." *Biological Conservation* 6 (1974): 118–133.

Maganiello, Dominic. *Joyce's Politics.* London: Routledge & Kegan Paul, 1980.

Mann, Alexander. *Yachting on the Pacific: Together with Notes on Travel in Peru, and an Account of the Peoples and Products of Ecuador.* London: Duckworth, 1909.

Márquez, Cruz, Linda Cayot, and Solanda Rea. *La Crianza de Tortugas Gigantes en Cautiverio: Un Manual Operativo.* Puerto Ayora, Santa Cruz, Galápagos: Fundación Charles Darwin, 1999.

Márquez, Cruz, Jacinto Gordillo, and Arnaldo Tupiza. "The Fire of 1994 and Herpetofauna of Southern Isabela." *Noticias de Galápagos* 54 (1995): 8–10.

Márquez, Cruz, David A. Wiedenfeld, Sandra Landázuri, and Juan Chávez. "Human-Caused and Natural Mortality of Giant Tortoises in the Galápagos Islands During 1995–2004." *Oryx* 41, no. 3 (2007): 337–342.

Marris, Emma. *The Rambunctious Garden: Saving Nature in a Post-Wild World*. New York: Bloomsbury, 2013.

Martínez, Nicolás Guillermo. *Impresiones de un Viaje a Galápagos*. Quito: Talleres Gráficos Nacionales, 1934.

Massey, Doreen. *For Space*. London: SAGE, 2005.

McCarthy, James, and Scott Prudham. "Neoliberal Nature and the Nature of Neoliberalism." *Geoforum* 35 (2004): 275–283.

McClintock, Anne. *Imperial Leather: Race, Gender, and Sexuality in the Colonial Contest*. New York: Routledge, 1995.

McCook, Stuart. *States of Nature: Science, Agriculture, and Environment in the Spanish Caribbean, 1760–1940*. Austin: University of Texas Press, 2002.

McOuat, Gordon R. "Species, Rules and Meaning: The Politics of Language and the Ends of Definitions in 19th Century Natural History." *Studies in History and Philosophy of Science* 27, no. 4 (1996): 473–519.

McPhee, John. *Basin and Range*. New York: Farrar, Straus and Giroux, 1981.

Melville, Herman. *The Encantadas, or the Enchanted Isles*. London: Hesperus, 2002 (originally published 1854).

Merchant, Carolyn. *Reinventing Eden: The Fate of Nature in Western Culture*. New York: Routledge, 2003.

Merlen, Godfrey. *Restoring the Tortoise Dynasty: The Decline and Recovery of the Galápagos Giant Tortoise*. Quito: Charles Darwin Foundation, 1999.

Milinkovitch, Michel C., Ricardo Kanitz, Ralph Tiedemann, Washington Tapia, Fausto Llerena, Adalgisa Caccone, James P. Gibbs, and Jeffrey R. Powell. "Recovery of a Nearly Extinct Galápagos Tortoise Despite Minimal Genetic Variation." *Evolutionary Applications* 6, no. 2 (February 2013): 377–383.

Milinkovitch, Michel C., D. Monteyne, J. P. Gibbs, T. H. Fritts, W. Tapia, H. L. Snell, R. Tiedemann, A. Caccone, and J. R. Powell. "Genetic Analysis of a Successful Repatriation Programme: Giant Galápagos Tortoises." *Proceedings: Biological Sciences* 271, no. 1537 (2004): 341–345.

Miller, Shawn William. *An Environmental History of Latin America*. Cambridge: Cambridge University Press, 2007.

Mintz, Sidney. *Sweetness and Power: The Place of Sugar in Modern History*. New York: Viking, 1985.

Mitchell, Timothy. "Can the Mosquito Speak?" In *Rule of Experts: Egypt, Techno-Politics, Modernity*, 19–53. Berkeley: University of California Press, 2002.

———. *Rule of Experts: Egypt, Techno-Politics, Modernity*. Berkeley: University of California Press, 2002.

———. "The Work of Economics: How a Discipline Makes Its World." *European Journal of Sociology/Archives Européennes de Sociologie/Europäisches Archiv für Soziologie* 46, no. 2 (2005): 297–320.

Mitman, Gregg. "Pachyderm Personalities: The Media of Science, Politics, and Conservation." In *Thinking with Animals: New Perspectives on Anthropomorphism*, edited by Lorraine Daston and Gregg Mitman, 175–195. New York: Columbia University Press, 2005.

———. *Reel Nature: America's Romance with Wildlife on Film*. Cambridge, MA: Harvard University Press, 1999.

Mol, Annemarie. *The Body Multiple: Ontology in Medical Practice*. Durham, NC: Duke University Press, 2003.

Monbiot, George. *Feral: Rewilding the Land, the Sea, and Human Life*. Chicago: University of Chicago Press, 2016.

Moore, Robert T. "The Protection and Conservation of the Zoological Life of the Galapagos Archipelago." *Science* 82, no. 2135 (November 1935): 519–521.

Moraña, Mabel, Enrique Dussel, and Carlos A. Jáuregui. "Colonialism and Its Replicants." In *Coloniality at Large: Latin America and the Postcolonial Debate*, edited by Mabel Moraña, Enrique Dussel, and Carlos Jáuregui, 1–22. Durham, NC: Duke University Press, 2008.

Morrell, Benjamin. *A Narrative of Four Voyages to the South Sea, North and South Pacific Ocean, Chinese Sea, Ethiopic and Southern Atlantic Ocean, Indian and Antarctic Ocean, from the Year 1822 to 1831*. New York: J. & J. Harper, 1832.

Mountfort, G. "The Problems of Tourism to Island Reserves." *Noticias de Galápagos* 15/16 (November 1970): 11–13.

Murphy, Robert Cushman. *Oceanic Birds of South America: A Study of Species of the Related Coasts and Seas, Including the American Quadrant of Antarctica, Based upon the Brewster-Sanford Collection in the American Museum of Natural History*. New York: Macmillan, 1936.

Nadasdy, Paul. "Transcending the Debate over the Ecologically Noble Indian: Indigenous Peoples and Environmentalism." *Ethnohistory* 52 (2005): 291–331.

Nelson, S. H. "Resilience and the Neoliberal Counter-Revolution: From Ecologies of Control to Production of the Common." *Resilience* 2, no. 1 (2014): 1–17.

Neumann, Roderick P. *Imposing Wilderness: Struggles over Livelihood and Nature Preservation in Africa*. Berkeley: University of California Press, 1998.

Nicholls, Henry. "How Sea Level Influenced Evolution in the Galápagos." *Guardian*, 24 April 2014. https://www.theguardian.com/science/animal-magic/2014/apr/24/sea-level-evolution-galapagos.

———. "Invasive Species: The 18-km² Rat Trap." *Nature* 497, no. 7449 (May 2013): 306–308.

———. "The Legacy of Lonesome George: Tortoise's Death Spurs Conservation Efforts." *Nature* 487 (July 2012): 279–280.

————. *Lonesome George: The Life and Loves of the World's Most Famous Tortoise*. Basingstoke: Pan Macmillan, 2006.

Nordlohne, Marcel. "The Seven-Year Search for Nicholas Oliver Lawson." John Woram, Las Encantadas: Human and Cartographic History of the Galápagos Islands. https://www.galapagos.to/TEXTS/NORDLOHNE.HTM.

Ospina, Pablo. *Galápagos, naturaleza y sociedad: Actores sociales y conflictos ambientales.* Quito: Universidad Andina Simón Bolívar and Corporación Editora Nacional, 2006.

————. "El Hada del Agua: Etica ambiental y actores sociales en Galápagos." Quito: Instituto de Estudios Ecuatorianos, 2004.

————. "Migraciones, Actores e Identidades en Galápagos." Informe final del concurso: Culturas e identidades en América Latina y el Caribe. Programa Regional de Becas. CLACSO, 2001. http://bibliotecavirtual.clacso.org.ar/ar/libros/becas/2000/ospina.pdf.

Ospina, Pablo, and Cecilia Facloni, eds. *Galápagos: Migraciones, Economía, Cultura, Conflictos, y Acuerdos.* Programma de Naciones Unidas para el Desarollo. Quito: Corporación Editora Nacional, 2007.

Otterman, Lillian. *Clinker Islands: A Complete History of the Galapagos Archipelago.* Bradenton, FL: McGuinn & McGuire, 1993.

Paley, William. *Natural Theology: or, Evidences of the Existence and Attributes of the Deity, Collected from the Appearances of Nature.* London: Printed for R. Faulder, 1802.

Palmer, Steven Paul. "Confinement, Policing, and the Emergence of Social Policy in Costa Rica, 1880–1935." In *The Birth of the Penitentiary in Latin America,* edited by Ricardo Donato Salvatore and Carlos Aguirre, 224–253. Austin: University of Texas Press, 1996.

Parks, E. Taylor, and J. Fred Rippy. "The Galapagos Islands: A Neglected Phase of American Strategy Diplomacy." *Pacific Historical Review* 9, no. 1 (March 1940): 37–45.

Parreñas, Juno Salazar. *Decolonizing Extinction: The Work of Care in Orangutan Rehabilitation.* Durham, NC: Duke University Press, 2018.

Perlman, David. "Galapagos Quest Is On." *San Francisco Chronicle,* 11 January 1964, 6.

————. "The Good Life on the Galapagos." *San Francisco Chronicle,* 16 February 1964, 1, 14.

Perry, Roger. *Island Days.* London: Stacey International, 2004.

Pinchot, Gifford. *To the South Seas; the Cruise of the Schooner Mary Pinchot to the Galapagos, the Marquesas, and the Tuamotu Islands, and Tahiti.* New York: Blue Ribbon Books, 1930.

Pineo, Ronn. *Ecuador and the United States: Useful Strangers.* Athens: University of Georgia Press, 2007.

————. *Social and Economic Reform in Ecuador: Life and Work in Guayaquil.* Gainesville: University Press of Florida, 1996.

Pitelka, Frank A. "Rollo Beck: Old-School Collector, Member of an Endangered Species." *American Birds* 40, no. 3 (1986): 385–387.

"Plan de Manejo Participativo." Galápagos National Park. Santa Cruz, Galápagos: PNG, 2005.

"Plan Maestro para la Protección y Uso del Parque Nacional Galápagos." Programa de las Naciones Unidas para el Desarrollo (PNUD), Organización de las Naciones Unidas para la Agricultura y la Alimentacion (FAO), Organización de las Naciones Unidas para la Educacion, la Ciencia, y la Cultura (UNESCO), PNUD/FAO ECU/71/022, Santiago, Chile: FAO Oficina Regional para America Latina, 1974.

Poliquin, Rachel. *Breathless Zoo: Taxidermy and the Cultures of Longing.* University Park: Pennsylvania State University Press, 2012.

Porter, David. *Journal of a Cruise Made to the Pacific.* New York: Wiley and Halsted, 1822.

Poulakakis, Nikos, Danielle L. Edwards, et al. "Description of a New Galápagos Giant Tortoise Species (*Chelonoidis;* Testudines: Testudinidae) from Cerro Fatal on Santa Cruz Island." *PLoS ONE* 10, no. 10 (2015): e0138779. https://doi.org/10.1371/journal.pone.0138779.

Poulakakis, Nikos, Scott Glaberman, Michael Russello, Luciano B. Beheregaray, Claudio Ciofi, Jeffrey R. Powell, and Adalgisa Caccone. "Historical DNA Analysis Reveals Living Descendants of an Extinct Species of Galápagos Tortoise." *PNAS* 105, no. 40 (October 2008): 15464–15469.

Poulakakis, Nikos, Michael Russello, Dennis Geist, and Adalgisa Caccone. "Unraveling the Peculiarities of Island Life: Vicariance, Dispersal and the Diversification of the Extinct and Extant Giant Galápagos Tortoises." *Molecular Ecology* 21 (2012): 160–173.

Pratt, Mary Louise. *Imperial Eyes: Travel Writing and Transculturation.* London: Routledge, 1992.

Pritchard, Peter. "Further Thoughts on 'Lonesome George.'" *Noticias de Galápagos* 39 (1984): 20–23.

———. "The Galápagos Tortoises." *Chelonian Research Monographs* 1 (1996): 17–83.

Pugh, Jonathan. "Island Movements: Thinking with the Archipelago." *Island Studies Journal* 8, no. 1 (2013): 9–24.

Quammen, David. *Song of the Dodo: Island Biogeography in an Age of Extinctions.* New York: Scribner, 1996.

Quijano, Aníbal. "Coloniality of Power, Eurocentrism, and Social Classification." In *Coloniality at Large,* 181–224. Durham, NC: Duke University Press, 2008.

Quiroga, Diego. "Crafting Nature: The Galápagos and the Making and Unmaking of a 'Natural Laboratory.'" *Journal of Political Ecology* 16 (2009): 123–140.

Raby, Megan. *American Tropics: The Caribbean Roots of Biodiversity Science.* Chapel Hill: University of North Carolina Press, 2017.

Radcliffe, Sarah A. "Development for a Postneoliberal Era? *Sumak Kawsay,* Living Well and the Limits to Decolonisation in Ecuador." *Geoforum* 43 (2012): 240–249.

Raffles, Hugh. *In Amazonia: A Natural History*. Princeton, NJ: Princeton University Press, 2002.

———. "'Local Theory': Nature and the Making of an Amazonian Place." *Cultural Anthropology* 14, no. 3 (1999): 323–360.

Redford, K. "The Ecologically Noble Savage." *Orion* 9 (1991): 24–29.

Ritter, Friedrich. "Adam and Eve in the Galapagos." *Atlantic Monthly*, October 1931, 409–418.

———. "Satan Walks in the Garden." *Atlantic Monthly*, November 1931, 565–575.

Ritvo, Harriet. "Animal Planet." *Environmental History* 9 (2004): 204–220.

———. *The Platypus and the Mermaid and Other Figments of the Classifying Imagination*. Cambridge, MA: Harvard University Press, 1997.

Rivera, S., S. J. Divers, S. E. Knafo, P. Martinez, L. J. Cayot, W. Tapia Aguilera, and J. Flanagan. "Sterilisation of Hybrid Galápagos Tortoises (*Geochelone nigra*) for Island Restoration. Part 2: Phallectomy of Males Under Intrathecal Anaesthesia with Lidocaine." *Veterinary Record* 168, no. 78 (2011). doi: 10.1136/vr.c6361.

Robbins, Paul. *Political Ecology*. 2nd ed. Malden, MA: Wiley-Blackwell, 2012.

Robbins, Paul, and Sarah Moore. "Ecological Anxiety Disorder: Diagnosing the Politics of the Anthropocene." *Cultural Geographies* 20, no. 1 (2013): 3–19.

Roberts, Lois Crawford de. *El Ecuador en la Época Cacaotera*. Quito: Editorial Universitaria, 1980.

Robertson, Morgan. "The Nature That Capital Can See: Science, State, and Market in the Commodification of Ecosystem Services." *Environment and Planning D: Society and Space* 24, no. 3 (2006): 367–387.

Robins, Nicholas. *Mercury, Mining and Empire: The Human and Ecological Cost of Colonial Silver Mining in the Andes*. Bloomington: Indiana University Press, 2011.

Rogers, Woodes. *A Cruising Voyage Round the World*. London: Cassell, 1928 (originally published 1712).

Romer, Simon. "In the Galápagos Islands: A Battle Between a Man and a Goat." *New York Times*, 1 May 2007.

Rose, Deborah Bird. "Judas Work: Four Modes of Sorrow." *Environmental Philosophy* 5, no. 2 (2008): 51–66.

Rose, Ruth. "The Last Raid." In *Galápagos: World's End*, by William Beebe, 190–203. New York: G. P. Putnam's Sons, 1924.

Ross, Corey. *Ecology and Power in the Age of Empire*. Oxford: Oxford University Press, 2017.

Rothschild, Miriam. *Dear Lord Rothschild: Birds, Butterflies, and History*. Rehovot, Israel: Balaban International Science Services, 1983.

Rothschild, Walter. "Description of a New Species of Gigantic Tortoise from Indefatigable Island." *Novitates Zoologicae* 10 (1903): 119.

———. "On Giant Land Tortoises." *Novitates Zoologicae* 1 (September 1894): 676–677.

Rothschild, Walter, and Ernst Hartert. "A Review of the Ornithology of the Galapagos Islands, with Notes on the Webster-Harris Expedition." *Novitates Zoologicae* 6, no. 2 (1899): 85–142.

Rudwick, Martin J. S. *Earth's Deep History: How It Was Discovered and Why It Matters.* Chicago: University of Chicago Press, 2014.

Russell, Edmund. *Evolutionary History: Uniting History and Biology to Understand Life on Earth.* Cambridge: Cambridge University Press, 2011.

Rutherford, Stephanie. *Governing the Wild: Ecotours of Power.* Minneapolis: University of Minnesota Press, 2011.

———. "Green Governmentality: Insights and Opportunities in the Study of Nature's Rule." *Progress in Human Geography* 31, no. 3 (2007): 291–307.

Salvatore, Ricardo Donato, and Carlos Aguirre. *The Birth of the Penitentiary in Latin America: Essays on Criminology, Prison Reform, and Social Control, 1830–1940.* Austin: University of Texas Press, 1996.

———. "The Birth of the Penitentiary in Latin America: Toward an Interpretative Social History of Prisons." In *The Birth of the Penitentiary in Latin America: Essays on Criminology, Prison Reform, and Social Control, 1830–1940*, 1–43. Austin: University of Texas Press, 1996.

Santos, Tristana. "Galápagos al límite." *Vistazo*, 2 July 2015, 28–31.

Sarmiento de Gamboa, Pedro. *History of the Incas.* Edited by Clements Markham. Surrey, England: Ashgate and the Hakluyt Society, 2010 (originally published 1907).

Schmidt, Stanley. "Huxley Packs House." *Daily Californian* 183, no. 72 (9 January 1964).

Scott, David. "Colonial Governmentality." *Social Text* 43 (1995): 195–220.

Scott, James. *Weapons of the Weak: Everyday Forms of Peasant Resistance.* New Haven: Yale University Press, 1987.

"Sea Lions Massacred in Galapagos." BBC News, last updated 29 January 2008. http://news.bbc.co.uk/2/hi/americas/7214860.stm.

Secord, James A. "The Discovery of a Vocation: Darwin's Early Geology." *British Journal for the History of Science* 24, no. 2 (June 1991): 133–157.

Sevilla, Elisa. "Darwinians, Anti-Darwinians and the Galápagos, 1835–1935." In *Darwin, Darwinism and the Galápagos: The Legacy of Darwin and Its New Applications*, edited by Diego Quiroga and Ana Sevilla, 41–63. Basel: Springer International, 2017.

Sevilla Perez, Ana. "The Galápagos Islands and the Ecuadorian State: Early Encounters." In *Darwin, Darwinism and Conservation in the Galápagos Islands: The Legacy of Darwin and Its New Applications*, edited by Diego Quiroga and Ana Sevilla, 23–39. Basel: Springer International, 2017.

———. "La incompleta nacionalización de la amazonía ecuatoriana en el siglo XIX vista desde el mapa de Theodor Wolf (1892)." *Apuntes. Revista de estudios sobre patrimonio cultural* 26, no. 1 (2013): 102–113.

Shapiro, Beth. *How to Make a Mammoth: The Science of De-Extinction.* Princeton, NJ: Princeton University Press, 2015.

Shaw, David, ed. "A Way with Animals." *History and Theory* 52, theme issue, December 2013.

Shoemaker, Nancy. "Whalemeat in American History." *Environmental History* 10, no. 2 (April 2005): 269–294.

Simberloff, D. "Flagships, Umbrellas and Keystones: Is Single-Species Management Passé in the Landscape Era?" *Biological Conservation* 83 (1998): 247–257.

Simkin, Tom. "Geology of Galapagos." *Biological Journal of the Linnean Society* 21 (1984): 61–75.

Slater, Candace. "Amazonia as Edenic Narrative." In *Uncommon Ground: Rethinking the Human Place in Nature,* edited by William Cronon, 114–131. New York: W. W. Norton, 1996.

Sluga, Glenda. "UNESCO and the (One) World of Julian Huxley." *Journal of World History* 21, no. 3 (2010): 393–418.

Smith, Neil. *Uneven Development: Nature, Capital, and the Production of Space.* New York: Verso, 1984.

Soluri, John. *Banana Cultures: Agriculture, Consumption, and Environmental Change in Honduras and the United States.* Austin: University of Texas Press, 2005.

———. "On Edge: Fur Seals and Hunters Along the Patagonian Littoral, 1860–1930." In *Centering Animals in Latin American History,* edited by Deborah Few and Zeb Tortorici, 243–269. Durham, NC: Duke University Press, 2013.

Soulé, Michale. "What Is Conservation Biology?" *BioScience* 35, no. 11 (December 1985): 727–734.

Star, Susan Leigh, and James R. Griesemer. "Institutional Ecology, 'Translations' and Boundary Objects: Amateurs and Professionals in Berkeley's Museum of Vertebrate Zoology, 1907–39." *Social Studies of Science* 19 (1989): 387–420.

Stengers, Isabelle. *Power and Invention: Situating Science.* Minneapolis: University of Minnesota Press, 1997.

Stepan, Nancy Leys. *Picturing Tropical Nature.* Ithaca, NY: Cornell University Press, 2001.

Stewart, Alban. "Further Observations on the Origins of the Galapagos Islands." *Plant World* 18, no. 7 (July 1915): 192–200.

Strauch, Dore. *Satan Came to Eden: A Survivor's Account of the "Galapagos Affair."* North Charleston, SC: Troise Publishing, 2014.

Striffler, Steve. *In the Shadows of State and Capital: The United Fruit Company, Popular Struggle, and Agrarian Restructuring in Ecuador, 1900–1995.* Durham, NC: Duke University Press, 2002.

Sulloway, Frank. "Darwin and His Finches: The Evolution of a Legend." *Journal of the History of Biology* 15, no. 1 (Spring 1982): 1–53.

———. "Darwin and the Galapagos." *Biological Journal of the Linnaean Society* 21 (1984): 29–59.

————. "Darwin's Conversion: The Beagle Voyage and Its Aftermath." *Journal of the History of Biology* 15, no. 3 (Autumn 1982): 325–396.

————. "Tantalizing Tortoises and the Darwin-Galápagos Legend." *Journal of the History of Biology* 42 (2009): 3–31.

Sundberg, Juanita. "Conservation Encounters: Transculturation in the 'Contact Zones' of Empire." *Cultural Geographies* 13 (2006): 239–265.

Takacs, David. *The Idea of Biodiversity: Philosophies of Paradise.* Baltimore: Johns Hopkins University Press, 1996.

Tapia, Washington. "A Return to Española Island to Understand Its Ecosystem Dynamics." *Galapagos Conservancy,* 21 July 2016. https://www.galapagos.org/blog/return-to-Espanola-2016/.

Tapia, Washington, José Antonio González Novoa, Pablo Ospina, Diego Quiroga, Günther Reck, and Carlos Montes del Olmo. "Entendiendo Galápagos como un sistema socioecológico complejo: Implicaciones para la investigación científica en el archipiélago." Quito: Parque Nacional Galápagos, Universidad Andina Simón Bolívar, Universidad Autónoma de Madrid, Universidad San Francisco de Quito, 2009, 127–140.

Tapia, Washington, Jeffreys Málaga, and James P. Gibbs. "Giant Tortoises Hatch on Galapagos Island." *Nature* 517, no. 271 (January 2015). https://www.nature.com/articles/517271a.

Tapia, Washington, Pablo Ospina, et al. "Toward a Shared Vision of Galápagos: The Archipelago as a Socioecological System." Galápagos Report 2007–2008. Puerto Ayora, Galápagos, Ecuador, 2008.

Taylor, J. Edward, Jared Hardner, and Micki Steward. "Ecotourism and Economic Growth in the Galápagos: An Island Economy-wide Analysis." Working paper no. 06-001, Department of Agricultural and Resource Economics, University of California, Davis, August 2006. https://core.ac.uk/download/pdf/7011091.pdf?repositoryId=153.

Thompson, Charis. "When Elephants Stand for Competing Philosophies of Nature: Amboseli National Park, Kenya." In *Complexities: Social Studies of Knowledge Practices,* edited by John Law and Annemarie Mol, 166–190. Durham, NC: Duke University Press, 2002.

Thompson, Lanny. "Heuristic Geographies: Territories and Areas, Islands and Archipelagoes." In *Archipelagic American Studies,* edited by Brian Russell Roberts and Michelle Ann Stephens, 57–73. Durham, NC: Duke University Press, 2017.

Townsend, Charles Haskins. "Fur Seals and the Seal Fisheries." *Bulletin of the Bureau of Fisheries* 28 (February 1910): 315–322.

————. "The Galapagos Islands Revisited." *Bulletin of the New York Zoological Society* 31, no. 5 (September–October 1928): 165–169.

————. *The Galapagos Tortoises in Their Relation to the Whaling Industry.* New York: New York Zoological Society, 1925.

————. "Giant Tortoise Finds That Man Is Friendly." *New York Times*, 20 May 1928.

————. "Impending Extinction of the Galapagos Tortoises." *Zoological Society Bulletin* 27 (March 1924): 55–56.

————. "New Information on the Galapagos Tortoises." *Zoological Society Bulletin* 27 (July 1924): 89–90.

Tsing, Anna. *Friction: An Ethnography of Global Connection*. Princeton, NJ: Princeton University Press, 2004.

————. *The Mushroom at the End of the World: On the Possibility of Life in Capitalist Ruins*. Princeton, NJ: Princeton University Press, 2015.

Tye, Alan. "La flora endémica de Galápagos: Aumentan las especies amenazadas." In *Informe Galápagos 2006–2007*, 101–107. Puerto Ayora, Galápagos, Ecuador: FCD, PNG, INGALA, 2007.

Tye, Alan, Howard L. Snell, Stewart B. Peck, and Henning Adsersen. "Outstanding Terrestrial Features of the Galápagos Archipelago." In *A Biodiversity Vision for the Galápagos Islands*, edited by R. Bensted-Smith, 12–13. Puerto Ayora, Galápagos: Charles Darwin Foundation and World Wildlife Fund, 2002.

Tyrell, Ian. *Crisis of a Wasteful Nation*. Chicago: University of Chicago Press, 2015.

"UC Island Safari Complete." *Berkeley Daily Gazette*, 16 March 1964, 3.

Urry, John. *The Tourist Gaze*. 2nd ed. London: SAGE, 2002.

Valdivia, Gabriela, Wendy Wolford, and Flora Lu. "Border Crossings: New Geographies of Protection and Production in the Galápagos Islands." *Annals of the Association of American Geographers* 104, no. 3 (2014): 686–701.

Valencia, Alexandra. "Ecuador to Open Amazon's Yasuni Basin to Oil Drilling." Reuters, 15 August 2013. https://in.reuters.com/article/us-ecuador-oil/ecuador-to-open-amazons-yasuni-basin-to-oil-drilling-idINBRE97E15220130816.

Van Denburgh, John. "Expedition of the California Academy of Sciences to the Galapagos Islands, 1905–1096. X. The Gigantic Land Tortoises of the Galapagos Archipelago." *Proceedings of the California Academy of Sciences*, 4th series, vol. 2, pt. 1 (September 1914): 203–374.

————. "Preliminary Description of Four New Races of Gigantic Land Tortoises from the Galapagos Islands." *Proceedings of the California Academy of Sciences*, 4th series, vol. 1 (1907): 1–6.

van Dooren, Thom. "Breeding Cranes: The Violent-Care of Captive Life." In *Flight Ways: Life and Loss at the Edge of Extinction*, 87–124. New York: Columbia University Press, 2014.

————. *Flight Ways: Life and Loss at the Edge of Extinction*. New York: Columbia University Press, 2014.

————. "Invasive Species in Penguin Worlds: An Ethical Taxonomy of Killing for Conservation." *Conservation and Society* 9, no. 4 (2011): 286–298.

van Dooren, Thom, Eben Kirksey, and Ursula Münster. "Multispecies Studies: Cultivating Arts of Attentiveness." *Environmental Humanities* 8, no. 1 (2016): 1–23.

Vetter, Jeremy. "Labs in the Field? Rocky Mountain Biological Stations in the Early Twentieth Century." *Journal of the History of Biology* 45, no. 4 (2012): 587–611.

Vince, Gaia. "Embracing Invasives." *Science* 331, no. 6023 (2011): 1383–1384.

Viveiros de Castro, Eduardo. "Perspectival Anthropology and the Method of Controlled Equivocation." *Tipití: Journal of the Society for the Anthropology of Lowland South America* 2, no. 1 (2004): 3–22.

Von Hagen, Victor Wolfgang. *Ecuador and the Galápagos Islands*. Norman: University of Oklahoma Press, 1949.

———. *Ecuador the Unknown: Two and a Half Years' Travels in the Republic of Ecuador and Galápagos Islands*. London: Jarrolds, 1939.

Waggoner, Ben. "Louis Agassiz (1807–1873)." University of California Museum of Paleontology. Last modified 2 March 2011. http://www.ucmp.berkeley.edu/history/agassiz.html.

Wakild, Emily. "A Panorama of Parks: Deep Nature, Depopulations, and the Cadence of Conserving Nature." In *A Living Past: Environmental Histories of Modern Latin America*, edited by John Soluri, Claudia Leal, and José Augusto Pádua, 246–265. New York: Berghahn Books, 2018.

———. "Protecting Patagonia: Science, Conservation and the Pre-History of the Nature State on a South American Frontier, 1903–1934." In *The Nature State: Rethinking the History of Conservation*, edited by Wilko Graf von Hardenberg, Matthew Kelley, Claudia Leal, and Emily Wakild, 37–54. Oxon: Routledge Earthscan, 2017.

———. *Revolutionary Parks: Conservation, Social Justice, and Mexico's National Parks, 1910–1940*, Tucson: University of Arizona Press, 2011.

Walker, Brian, Lance Gunderson, Ann Kinzig, Carl Folke, Steve Carpenter, and Lisen Schultz. "A Handful of Heuristics and Some Propositions for Understanding Resilience in Social-Ecological Systems." *Ecology and Society* 11, no. 1 (2006). http://www.ecologyandsociety.org/vol11/iss1/art13/.

Wallace, Alfred Russel. *Island Life: Or, the phenomena and causes of insular faunas and floras: Including a Revision and Attempted Solution of the Problem of Geological Climates*. London: MacMillan, 1880.

Walpole, M., and N. Leader-Williams. "Tourism and Flagship Species in Conservation." *Biodiversity and Conservation* 11 (2002): 543–547.

Walsh, Catherine. "(Post)Coloniality in Ecuador: The Indigenous Movement's Practices and the Politics of (Re)signification and Decolonization." In *Coloniality at Large: Latin America and the Postcolonial Debate*, edited by Mabel Moraña, Enrique Dussel, and Carlos Jáuregui, 506–518. Durham, NC: Duke University Press, 2008.

———. "'Staging Encounters': The Educational Decline of U.S. Puerto Ricans in (Post)-Colonial Perspective." *Harvard Educational Review* 68, no. 2 (1998): 218–243.

Walsh-Dilley, Marygold, Wendy Wolford, and James McCarthy. "Rights for Resilience: Bringing Power, Rights, and Agency into the Resilience Framework." Oxfam America, 17 June 2013. http://www.atkinson.cornell.edu/Assets/ACSF/docs/collaborations/oxfam/R4R%20Conceptual%20Framework.pdf.

Watkins, Graham, and Felipe Cruz. "Galápagos at Risk: A Socioeconomic Analysis of the Situation in the Archipelago." Puerto Ayora, Galápagos, Ecuador: Charles Darwin Foundation, 2007.

Weiner, Jonathan. *The Beak of the Finch: A Story of Evolution in Our Time.* New York: Vintage Books, 1995.

West, Paige. *Conservation Is Our Government Now: The Politics of Ecology in Papua New Guinea.* Durham, NC: Duke University Press, 2006.

West, Paige, and James Carrier. "Ecotourism and Authenticity: Getting Away from It All?" *Current Anthropology* 45 (2004): 483–498.

West, Paige, James Igoe, and Dan Brockington. "Parks and Peoples: The Social Impact of Protected Areas." *Annual Review of Anthropology* 35 (2006): 251–277.

Westwood, Ben. *Moon Ecuador and the Galápagos Islands.* Berkeley, CA: Avalon Travel, 2014.

Whatmore, Sarah. *Hybrid Geographies: Natures Cultures Spaces.* Oxford: SAGE, 2002.

Williams, Raymond. *The Country and the City.* London: Hogarth, 1985.

Wilshusen, Peter R., Steven R. Brechin, Crystal L. Fortwangler, and Patrick C. West. "Reinventing a Square Wheel: Critique of a Resurgent 'Protection Paradigm' in International Biodiversity Conservation." *Society and Natural Resources* 15 (2002): 17–40.

Wittmer, Margret. *Floreana: A Woman's Pilgrimage to the Galapagos.* New York: Beaufort Books, 1989.

Wolch, Jennifer, and Jody Emel. *Animal Geographies: Place, Politics, and Identity in the Nature-Culture Borderlands.* New York: Verso, 1998.

Wolf, Eric R., and Sidney W. Mintz. "Haciendas and Plantations in Middle America and the Antilles." *Social and Economic Studies* 6, no. 3 (1957): 380–412.

Wolf, Teodoro. *Memoria de las Islas Galapagos.* Quito: Imprenta de Gobierno, 1887.

Wolford, Wendy. "The Difference Ethnography Can Make: Understanding Social Mobilization and Development in the Brazilian Northeast." *Qualitative Sociology* 29, no. 3 (September 2006): 335–352.

Wolford, Wendy, Flora Lu, and Gabriela Valdivia. "Environmental Crisis and the Production of Alternatives: Conservation Practice(s) in the Galapagos Islands." In *Science and Conservation in the Galápagos Islands: Frameworks and Perspectives,* edited by Stephen Walsh and Carlos Mena, 87–104. New York: Springer, 2013.

Woram, John. "Who Killed the Iguanas?" In *Charles Darwin Slept Here,* 299–314. Rockville Centre, NY: Rockville Press, 2005.

World Commission on Environment and Development. *Our Common Future.* Oxford: Oxford University Press, 1987.

Yackulic, Charles B., Stephen Blake, and Guillaume Bastille-Rousseau. "Benefits of the Destinations, Not Costs of the Journeys, Shape Partial Migration Patterns." *Journal of Animal Ecology* 86 (2017): 972–982.

Zimmerer, Karl. *Globalization and New Geographies of Conservation*. Chicago: University of Chicago Press, 2006.

INDEX

Page numbers in *italics* refer to illustrations.